INTRODUCTION TO FEEDBACK CONTROL

Kirsten Morris
University of Waterloo

San Diego San Francisco New York Boston London Toronto Sydney Tokyo

Sponsoring Editor	Barbara Holland
Production Editor	Julio Esperas
Editorial Coordinator	Karen Frost
Marketing Manager	Marianne Rutter
Cover Design	Richard Hannus, Hannus Design Associates
Copyeditor	Kristin Landon
Composition	TechBooks
Printer	Maple-Vail Book Manufacturing Group

This book is printed on acid-free paper. ∞

ACADEMIC PRESS
A Harcourt Science and Technology Company
525 B Street, Suite 1900, San Diego, California 92101-4495, USA
http://www.academicpress.com

Academic Press
Harcourt Place, 32 Jamestown Road, London NW1 7BY, UK
http://www.academicpress.com

Harcourt/Academic Press
A Harcourt Science and Technology Company
200 Wheeler Road, Burlington, Massachusetts 01803, USA
http://www.harcourt-ap.com

Library of Congress Catalog Card Number: 00-108488
International Standard Book Number: 0-12-507660-6

PRINTED IN THE UNITED STATES OF AMERICA
00 01 02 03 04 05 MB 9 8 7 6 5 4 3 2 1

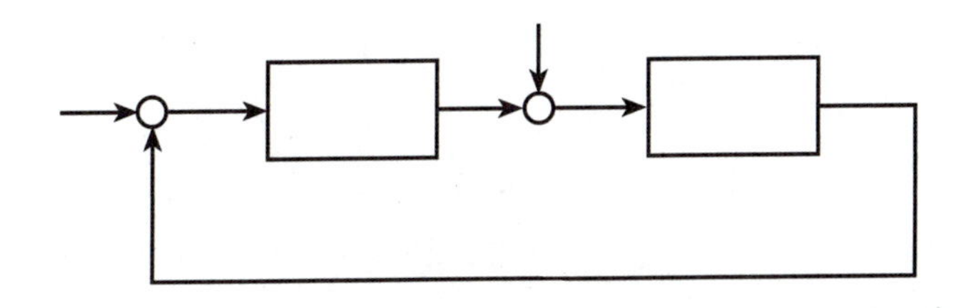

CONTENTS

III

STABILITY

IV

BASIC LOOPSHAPING

V

BASIC STATE FEEDBACK AND ESTIMATION

VI

CONTROLLER PARAMETRIZATION

VII

GENERALIZED PLANTS

VIII

ESTIMATOR-BASED $\mathcal{H}_\infty$ CONTROLLER DESIGN

IX

MODEL MATCHING

APPENDIX A

NORMED LINEAR SPACES

APPENDIX B

ALGEBRA

APPENDIX C

SYSTEM MANIPULATIONS

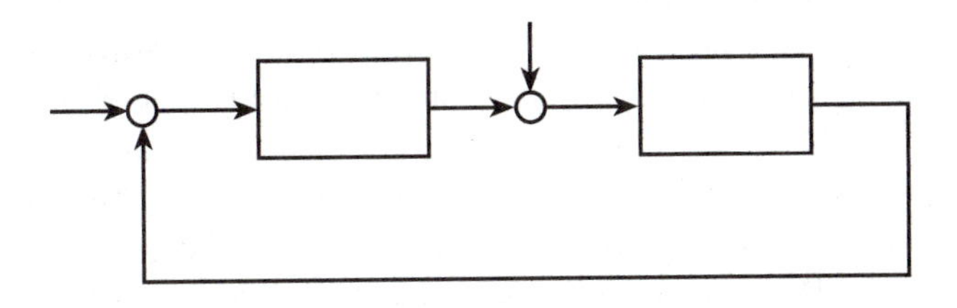

PREFACE

Feedback control is present in every aspect of modern life, from the simple thermostats in our house that maintain a specified temperature, to the complex devices that maintain the position of communications satellites. Feedback occurs in both natural and manufactured situations. It is a concept independent of technology or the system being controlled. A number of biological systems involve very complex feedback. For instance, regulation of blood sugar and balance both involve feedback control. Feedback control is a crucial part of almost all engineering systems. One of the earliest recorded examples of feedback control in engineering is a float valve used to regulate water level. This method was used in Greece more than 2000 years ago and is still used today to control water level in boilers and reservoirs. Another historically important example is the governor on James Watt's steam engine. This steam governor was key to the successful operation of the engine. An increase in engine speed causes the rotational force of two flyweights to increase. This increase in force moves a shaft that alters a valve position, reducing the flow of steam and lowering the engine speed. In this manner, a desired engine speed is maintained. In this century, Bode's work on feedback in operational amplifiers led to the ability to obtain accurate output from electronic devices.

Along with improvement in control technology this century came advances in theory. There is now a very large body of knowledge dealing with stability and control of systems. Introductory textbooks for mathematics students generally deal with optimal control, particularly linear quadratic regulators. However, most controllers implemented in practical situations are based on input/output design. This is the approach taught to engineering undergraduate students. The theory behind this approach is generally not covered in introductory engineering books.

A second difference between traditional mathematics and engineering control courses is that mathematicians are taught using a differential equation description of the system being controlled. This is known as the "time domain" since all analysis is done with time as the independent variable. Another method of analysis uses system transfer functions. The transfer function can be obtained by taking Laplace transforms of the differential equations modeling the system. Transfer functions can also be obtained from frequency responses, and for this reason this analysis is said to be in the "frequency domain." This is the approach used in introductory engineering courses. The two approaches generally merge at the graduate level. This book is an effort to bridge the gap between the two approaches.

This book leads from elementary systems theory to an introduction to advanced multivariable $\mathcal{H}_\infty$ control. $\mathcal{H}_\infty$ controller design is one of the most popular methods of modern controller design. A familarity with frequency domain analysis and classical input/output controller design such as loop shaping is very useful in understanding this design method. This is because $\mathcal{H}_\infty$ controller design developed as an attempt to formalize classical methods so that they could be extended to multi-input/multi-output systems.

A common thread in this book is the trade-off between performance and maintaining stability in the presence of system uncertainty. Both time domain and frequency domain methods of system description are used so that students become familar with both viewpoints. Only linear systems are considered and all explanations and most examples are single-input/single-output for ease of exposition. However, all results apply to multivariable systems and details of the generalization are given. All material is rigorously presented, with proofs of all major results. Applications and examples are used to illustrate the material. This is useful for communicating the main points and also for demonstrating relevance to physical problems. The reader is assumed to have taken introductory courses in linear algebra, linear ordinary differential equations, and complex analysis.

This book is accessible to senior mathematics undergraduate students and to theoretically inclined engineering students. It has been used for a number of years as the text for a fourth-year mathematics course and for a graduate course composed of both engineering and mathematics graduate students. Chapters 1–4 plus possibly Chapters 5 or 6 form a solid one-term undergraduate course. Material marked with an asterix (*) may be omitted with no loss of continuity. At the graduate level, Chapters 1–4 can be covered very quickly. Most engineering graduate students have seen the material without the theory in undergraduate courses. Chapters 1–4 and 7–8 can be used to teach an introduction to $\mathcal{H}_\infty$ controller design at the first-year graduate level. An alternative syllabi is to concentrate on the theoretical aspects of recent frequency-domain results and to cover Chapters 6 and 9 after Chapters 1–4. Other syllabi are, of course, possible.

This book started as lecture notes for a course in feedback control offered to engineering and mathematics students at the University of Waterloo. Thanks are due to Helen Warren for typing early drafts of this book from my almost illegible lecture notes. Janet Grad and Geir Dullerud taught from the notes that became this book and provided me with useful suggestions for improvement. The feedback (!) of the many students in engineering and applied mathematics at Waterloo who used the notes was invaluable in improving its readibility. I also want to thank the following reviewers for their careful reading of the manuscript and for helpful suggestions for improvement: Robert Paz (New Mexico State University), Wolfgang Kliemann (Iowa State University), Joe H. Chow (Rensselaer Polytechnic University), Pablo A. Iglesias (Johns Hopkins University), Jon H. Davis (Queen's University, Canada), Jurdevic Velimir (University of Toronto), Xiaochang Wang (Texas Tech University), and Alan Edelman (MIT). The final stages of publication of this book went smoothly thanks to the professional editorial and production staff at Academic Press. Most importantly, I thank my husband Dave. I wouldn't have finished this book without his support and encouragement.

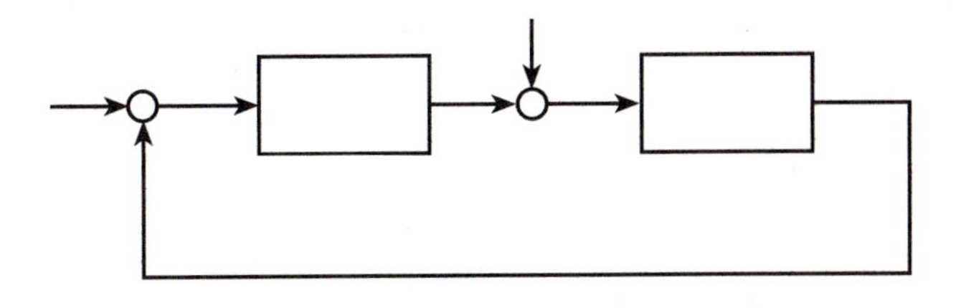

I

WHAT IS FEEDBACK CONTROL?

One of the earliest known examples of feedback in engineering is fluid level control via a float valve. As fluid level in a vessel drops, a float attached to a lever falls. The other end of the lever is raised, opening a valve that allows fluid to flow into the vessel. The first known system was developed by the Greek mechanician Heron about 2000 years ago. The basic principle is still used in boiler drums and flush toilets.

Feedback occurs in a wide variety of situations. A number of biological systems involve feedback. One example is the regulation of blood sugar level via insulin (Fig. 1.3). Driving (Fig. 1.1) and economic supply and demand (Fig. 1.2) are two feedback control systems involving human behavior.

All these systems are examples of feedback control systems. Measurements of the system being controlled are used to correct the outputs: the system is "self-correcting." An open-loop control system is one where the controlled inputs are chosen without regard to the actual system outputs. One example of an open-loop system is a driver (Fig. 1.1) closing his or

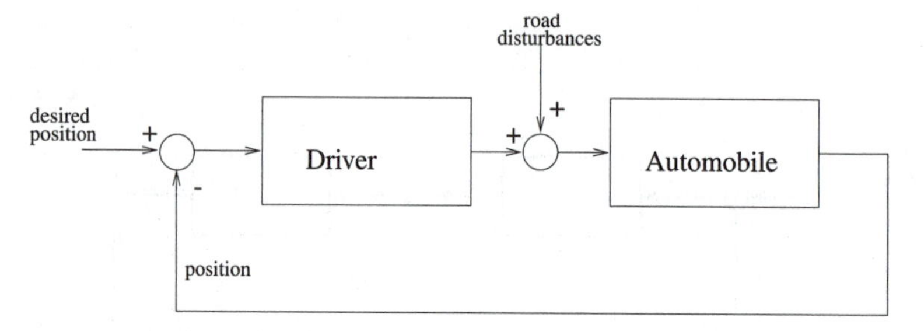

FIGURE 1.1 Steering of an automobile.

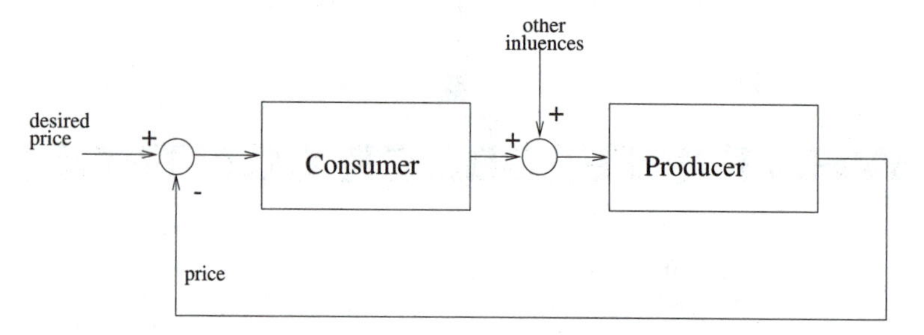

FIGURE 1.2 Economic feedback model of supply and demand. Consumer demand increases prices. This decreases demand, which leads to a decrease in price and so on.

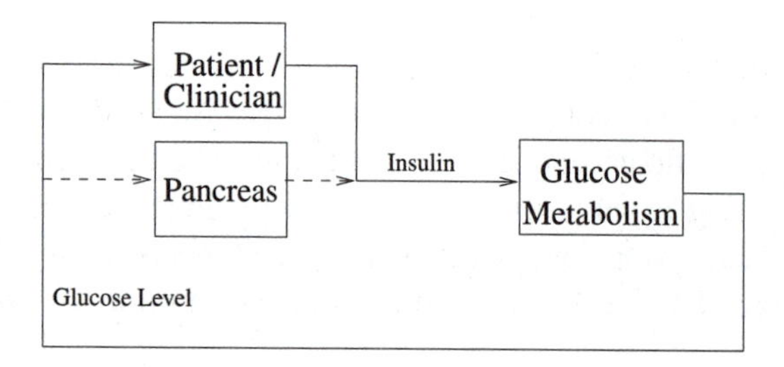

FIGURE 1.3 Glucose control in diabetic patients. The dashed line shows feedback in the normal situation. When impairment of the control loop opens the loop, the external controller is used (the solid line).

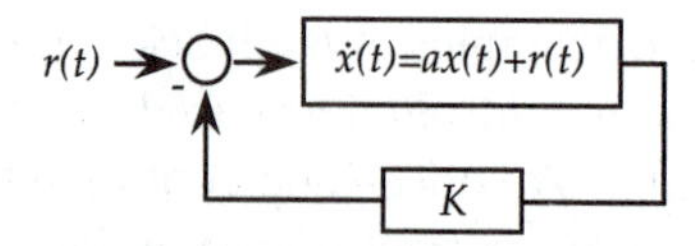

FIGURE 1.4 Example of simple feedback system.

her eyes once a direction is chosen. Another example is shown in Fig. 1.3. Diabetes arises when insulin is no longer released in an amount sufficient to control blood sugar at an appropriate level: The loop is "opened." When information on a model is available, open-loop control can be useful. Since no mathematical model is perfect, and unforeseen disturbances also generally occur, feedback is also required to correct errors.

Another advantage of feedback is that it can be used to change an unstable system to a stable system. (Formal definitions of stability are given in a later chapter.) For example, consider a simple system modeled by

$$\dot{x}(t) = ax(t) + r(t)$$

where $a > 0$. The trajectories of this system will grow with time. However, if the system is placed in the closed loop shown in Fig. 1.4, any value of $K > a$ will lead to trajectories where $\lim_{t\to\infty} x(t) = 0$.

All feedback systems have certain elements in common. The system that is being controlled is usually referred to as the *plant*, while the system that effects the control is called the *controller*. Measured quantities from the plant (outputs) are used as inputs to the controller. That is, the plant outputs are "fedback" to the controller.

Performance of a control system is often specified in terms of the size of certain signals of interest. For instance, in Fig. 1.1, the aim of the control system is to make the error as small as possible. In the chapters that follow we will determine what we mean by "small," and also develop formal definitions of stability.

NOTES AND REFERENCES

Control of diabetes is discussed in [5]. Table 1.1 is based on information in [37]. Other examples of feedback control can be found in Chapter 1 of [40].

TABLE 1.1 Automotive computer control systems

1977	First engine control computer
1983	Air suspension system
1984	Antilock braking
1985	Computer-controlled airbags
1988	Full electronic transmission control
1990	Traction control
?	Compliance with new U.S. emission laws
?	Active control of acoustic noise
?	"Smart" cars

EXERCISES

1. For the control systems shown in Figs. 1.1–1.3:

   (a) State performance specifications.
   (b) Place the control system into the standard feedback loop (Fig. 1.5). That is, state what the signals r and e are, and also define the systems P and H.

2. Describe a feedback control system not mentioned in this chapter. Repeat question 1 for this control system.

3. Consider the following simple model of insulin/sugar interaction:

$$\frac{dS}{dt} = -a_1 SY + a_2 U(S_e - S)(S_e - S)^2 + a_3 f(t)$$

$$\frac{dY}{dt} = b_1 U(S - S_e)(S_e - S)^2 - b_2 Y + b_3 A(t)$$

   where S and Y are the sugar and insulin concentrations, respectively, f is sugar introduced into the bloodstream, and A is artificial introduction of insulin. The parameters $a_i, b_i, i = 1..3$ are nonnegative. The function $U(g)$ is the Heaviside step function.

   (a) What parameter(s) is small or zero in a case where insulin production is too low?
   (b) What is the main effect on the dynamics when low natural production of insulin is compensated by introducing insulin at 24-hour intervals? How could this effect be reduced?

4. A home thermostat is a simple feedback control device. Draw a block diagram of a home heating system.

5. Give several examples, with block diagrams, of feedback systems involving human behavior.

6. Draw a block diagram of an automobile cruise control system.

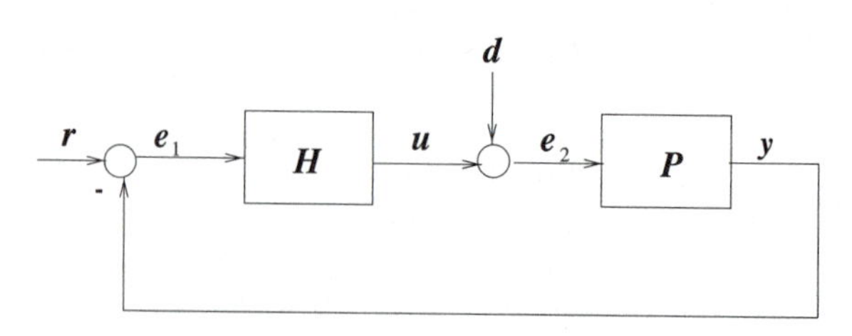

FIGURE 1.5 Feedback system. P is the plant (the system to be controlled) and H is the controller.

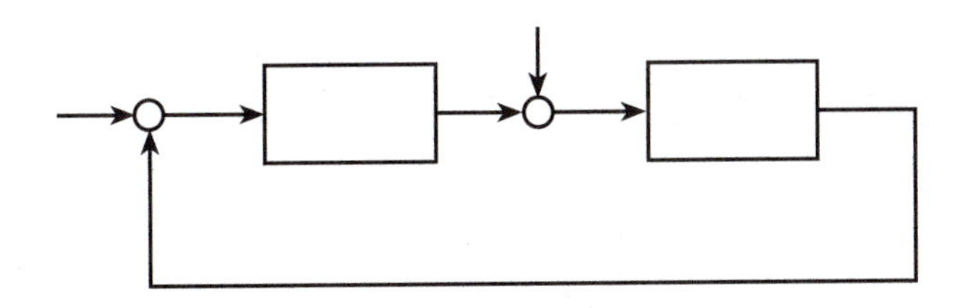

II SYSTEMS THEORY

2.1 STATE-SPACE DESCRIPTIONS

Any system modeled by a set of linear ordinary differential equations can be written in the following form:

$$\begin{aligned} \dot{x}(t) &= Ax(t) + Bu(t), \quad x(0) = x_o, \\ y(t) &= Cx(t) + Eu(t). \end{aligned} \tag{2.1}$$

where $A \in R^{n \times n}$, $B \in R^{n \times m}$, $C \in R^{p \times n}$, and $E \in R^{p \times m}$. Roughly, A describes the internal dynamics; B the effect of the controlled input on the state; and C, E describe the sensors. The function u is the *input*, y is the *output*, and x is called the *state*. Note that $x(t) \in R^n$ so in general it will be a vector. The number of state variables n is called the *order* of the system. The number of columns m in B is the number of inputs and the number of rows p in C is the number of outputs. We will primarily be concerned with single-input/single-output (SISO) systems where $m = p = 1$.

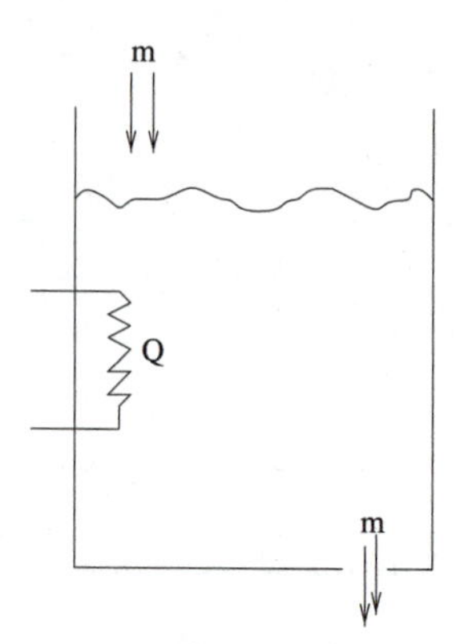

FIGURE 2.1 Water tank.

A system modeled by the set of equations (2.1) will be indicated by (A, B, C, E), or

$$\left[\begin{array}{c|c} A & B \\ \hline C & E \end{array}\right].$$

If we are only interested in the state we often write (A, B).

Example 2.1 (*Water Heating Tank*): Assume that the water in a water heating tank is perfectly mixed and that the ambient temperature is constant. Also assume the flow of water out of the tank, m, equals the flow into the tank so that the mass of water in the tank is constant (Fig. 2.1). Let T indicate the temperature above ambient and Q the rate of heat transfer applied. Let C_p indicate the heat capacity per unit mass and M the mass of water in the tank. An energy balance on the tank yields

$$C_p M \frac{dT}{dt} = Q(t) - C_p m T(t) - kT(t) \tag{2.2}$$

where k is a constant relating heat loss across the walls of the tank to the tank temperature. Suppose that the temperature in the tank is measured. Then a state-space formulation is

$$\begin{aligned} \frac{dT}{dt} &= \left(\frac{-C_p m - k}{C_p M}\right) T(t) + \left(\frac{1}{C_p M}\right) Q(t) \\ y(t) &= T(t). \end{aligned}$$

This is a simple model with one input (Q), one output (T), and a single state (T).

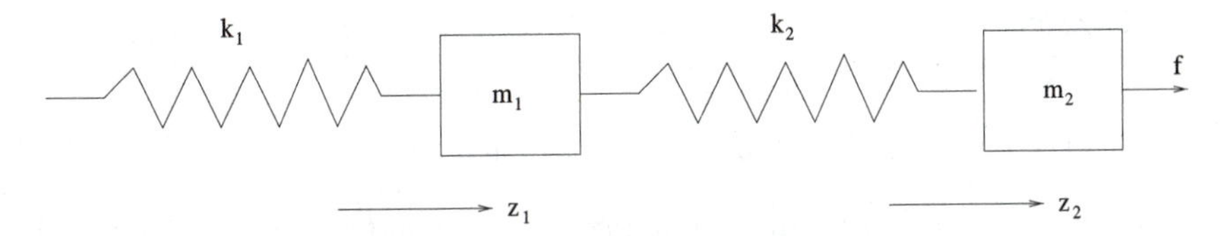

FIGURE 2.2 Double spring–mass system.

Example 2.2 (*Double Spring–Mass*): A spring–mass system is shown in Fig. 2.2. Letting z_1 indicate the deviation from equilibrium position of the first mass and z_2 the deviation from equilibrium of the second,

$$m_1\ddot{z}_1(t) = -k_1 z_1(t) + k_2(z_2(t) - z_1(t)) - d_1\dot{z}_1(t) + d_2(\dot{z}_2(t) - \dot{z}_1(t))$$
$$m_2\ddot{z}_2(t) = -k_2(z_2(t) - z_1(t)) - d_2(\dot{z}_2(t) - \dot{z}_1(t)) + f(t)$$

where $f(t)$ is an external force applied to the mass m_2. For physical reasons, k_1, k_2, m_1, and m_2 are all positive constants, while $0 < d_1 < 2\sqrt{k_1 m_1}$, and $0 < d_2 < 2\sqrt{k_2 m_2}$. To write in standard first-order state-space form, define the state vector x of length 4 as $x_1 = z_1, x_2 = \dot{z}_1, x_3 = z_2, x_4 = \dot{z}_2$. Rewriting the foregoing equations, we obtain

$$\dot{x}(t) = \begin{bmatrix} 0 & 1 & 0 & 0 \\ -\frac{k_1+k_2}{m_1} & -\frac{d_1+d_2}{m_1} & \frac{k_2}{m_1} & \frac{d_2}{m_1} \\ 0 & 0 & 0 & 1 \\ \frac{k_2}{m_2} & \frac{d_2}{m_2} & -\frac{k_2}{m_2} & -\frac{d_2}{m_2} \end{bmatrix} x(t) + \begin{bmatrix} 0 \\ 0 \\ 0 \\ \frac{1}{m_2} \end{bmatrix} f(t). \tag{2.3}$$

If we are measuring the position of the second mass,

$$y(t) = [0 \quad 0 \quad 1 \quad 0]\, x(t).$$

Alternatively, we could define $x_1 = z_1$, $x_2 = z_2 - z_1$, $x_3 = \dot{z}_1$, $x_4 = \dot{z}_2 - \dot{z}_1$. With this definition of the state,

$$\dot{x}(t) = \begin{bmatrix} 0 & 0 & 1 & 0 \\ 0 & 0 & 0 & 1 \\ -\frac{k_1}{m_1} & \frac{k_2}{m_1} & -\frac{d_1}{m_1} & \frac{d_2}{m_1} \\ \frac{k_1}{m_1} & -k_2\left[\frac{1}{m_1} + \frac{1}{m_2}\right] & \frac{d_1}{m_1} & -d_2\left[\frac{1}{m_1} + \frac{1}{m_2}\right] \end{bmatrix} x(t) + \begin{bmatrix} 0 \\ 0 \\ 0 \\ \frac{1}{m_2} \end{bmatrix} f(t),$$

$$y(t) = [1 \quad 1 \quad 0 \quad 0]\, x(t). \tag{2.4}$$

EXAMPLE 2.3 (*General nth Order Differential Equation*): Letting z^n indicate the nth derivative of z, consider

$$z^n(t) + b_{n-1}z^{n-1}(t) + b_{n-2}z^{n-2}(t) \cdots + b_o z(t) = u(t). \tag{2.5}$$

Observation is made of a linear combination of z and the derivatives of z:

$$y(t) = a_o z(t) + a_1 z^1(t) + \cdots a_{n-1} z^{n-1}(t). \tag{2.6}$$

Define $x_1 = z$, $x_2 = z^1$, and so on to $x_n = z^{n-1}$. We can rewrite (2.5, 2.6) in a standard state-space form (2.1) with

$$A = \begin{bmatrix} 0 & 1 & 0 & 0 & \cdots \\ 0 & 0 & 1 & 0 & \cdots \\ 0 & 0 & 0 & 1 & \cdots \\ \vdots & \vdots & \vdots & \vdots & \vdots \\ 0 & 0 & \cdots & 0 & 1 \\ -b_0 & -b_1 & \cdots & -b_{n-2} & -b_{n-1} \end{bmatrix},$$

$$B = \begin{bmatrix} 0 \\ 0 \\ \vdots \\ 0 \\ 1 \end{bmatrix},$$

$$C = [a_0 \quad a_1 \quad \cdots \quad a_{n-2} \quad a_{n-1}],$$

and $E = 0$.

We now consider the solution to the differential equations describing the system:

$$\dot{x}(t) = Ax(t) + Bu(t), \qquad x(0) = x_0. \tag{2.7}$$

Consider first the uncontrolled case where $B = 0$:

$$\dot{x}(t) = Ax(t), \qquad x(0) = x_0.$$

If the order of a system is 1, then we have

$$\dot{x}(t) = ax(t), \qquad x(0) = x_0$$

for a scalar a. The solution to this simple linear differential equation is

$$x(t) = \exp(at)x_0.$$

The exponential function has the important properties that

1. $\exp(0) = 1$
2. $\frac{d\exp(at)}{dt} = a\exp(at)$.

In order to solve the more general equation (2.7) we need a function of the matrix A with the same properties. This proposed function "$\exp(At)$" will be a $n \times n$ matrix for each time t and it should have the properties:

1. $\exp(0) = I$
2. $\frac{d\exp(At)}{dt} = A\frac{d\exp(At)}{dt}$.

Since for a real number a,

$$\exp(at) = \sum_0^\infty \frac{a^i t^i}{i!},$$

we define the matrix exponential

$$\exp(At) = \sum_0^\infty \frac{A^i t^i}{i!}. \tag{2.8}$$

Note that

$$\begin{aligned}\exp(A0) &= A^0 \\ &= I\end{aligned}$$

as required. Also, it can be shown that term-by-term differentiation of this series is justified and so

$$\begin{aligned}\frac{d}{dt}\exp(At) &= \sum_0^\infty \frac{d}{dt}\frac{A^i t^i}{i!} \\ &= \sum_1^\infty \frac{A^i t^{i-1}}{(i-1)!} \\ &= \sum_0^\infty A\frac{A^i t^i}{(i)!} \\ &= A\exp(At).\end{aligned}$$

It follows that $x(t) = \exp(At)x_o$ is the solution to

$$\dot{x}(t) = Ax(t), \qquad x(0) = x_o.$$

We now consider the controlled case. If the order of the system is 1,

$$\dot{x}(t) = ax(t) + bu(t), \qquad x(0) = x_o$$

for scalars a, b then the solution is

$$x(t) = \exp(at)x_0 + \int_0^t \exp(a(t-\tau))bu(\tau)d\tau.$$

This suggests that the solution to the general case (2.7) is

$$x(t) = \exp(At)x_0 + \int_0^t \exp(A(t-\tau))Bu(\tau)d\tau, \tag{2.9}$$

It can easily be verified that (2.9) is the solution to (2.7). The solution to (2.1) is thus

$$y(t) = C\exp(At)x_0 + \int_0^t C\exp(A(t-\tau))Bu(\tau)d\tau + Eu(t).$$

Let $\delta(t)$ indicate the impulse defined by

$$\int_0^t \delta(t)u(t)dt = u(0)$$

for all functions u well-defined at 0. Define

$$g(t) = C\exp(At)B.$$

If the input is an impulse $\delta(t)$ and $E = 0$ then for zero initial condition

$$\begin{aligned} y(t) &= \int_0^t g(t-\tau)\delta(\tau)d\tau \\ &= C\exp(At)B. \end{aligned}$$

For this reason, g is called the system *impulse response*. In the more general case where $E \neq 0$,

$$g(t) = C\exp(At)B + E\delta(t)$$

is still called the impulse response. The solution to (2.1) is

$$y(t) = C\exp(At)x(0) + \int_0^t g(t-\tau)u(\tau)d\tau. \tag{2.10}$$

2.2 CONTROLLABILITY AND OBSERVABILITY

Intuitively, a control system should be designed so that the input u "controls" all the states; and also so that all the states can be "observed" from the output y. The concepts of *controllability* and *observability* formalize these ideas.

DEFINITION 2.4: A system (2.7) (or (A, B)) is *controllable* if for any initial condition x_0, final condition x_f, and time T, there exists a piecewise continuous input $u(\cdot)$ so that $x(T) = x_f$.

THEOREM 2.5: The following statements are equivalent:

1. (A, B) is controllable.
2. The matrix

$$P(t) = \int_0^t e^{At} B B^* e^{A^* t} dt$$

is positive definite for all $t > 0$.

3. The Kalman controllability matrix

$$\mathcal{C} = [B \quad AB \quad \ldots \quad A^{n-1}B]$$

has rank n (where A is $n \times n$).

□ *Proof:* 1 ⇔ 2 Suppose that $P(T) > 0$ for any $t > 0$ and let $T > 0$, $x_o, x_f \in R^n$ be arbitrary. It can be verified that the control

$$u(t) = -B^* \exp(A^*(T - t)) P(T)^{-1} (\exp(AT)x_o - x_f)$$

takes the system with initial state x_o to x_f at time T. Now suppose that for some $T > 0$, $P(T)$ is not positive definite, that is, it is singular. Since the integrand is positive semidefinite, this means that there exists $q \in R^n$, $q \neq 0$, such that

$$q^* \exp(At)B = 0, \text{ for all } \quad 0 \leq t \leq T.$$

Now choose initial state $x_o = \exp(-AT)q$ and let u be any control. At time T we have state

$$x(T) = \exp(AT)\exp(-AT)q + \int_0^T \exp A(T - r)Bu(r)dr.$$

Multiplying on the left by q^*,

$$q^* x(T) = q^* q.$$

Thus, it is impossible to control the system to $x(T) = 0$, and so if $P(T)$ is not positive definite, the system is not controllable.

2 ⇔ 3 Using the Cayley–Hamilton Theorem (Appendix B), for each matrix A there exists functions $\alpha_i(t)$ so that the matrix exponential (2.8) can be written

$$\exp(At) = \sum_{i=0}^{n-1} \alpha_i(t) A^i. \tag{2.11}$$

Now, suppose that $\mathcal{C}$ has rank less than n. That is, there is $q \neq 0$ such that

$$q^* \mathcal{C} = 0,$$

or,

$$q^* A^i B = 0, \qquad i = 0 \ldots n-1.$$

From (2.11) this implies that

$$q^* \exp(At)B \tag{2.12}$$

for all t and so

$$q^* P(t) = 0$$

for all t, and $P(t)$ is not positive definite.

If for some $T > 0$, $P(T)$ is not positive definite, there is $q \neq 0$ such that

$$q^* \exp(At)B = 0, \text{ for all } 0 \leq t \leq T.$$

Define $f(t) = q^* \exp(At)B$. All the derivatives of f must also equal zero on this interval:

$$q^* A^i \exp(At)B = 0, i = 0, 1, ..$$

Evaluating at $t = 0$,

$$q^* A^i B = 0, i = 0, 1, ..$$

Or,

$$q^* \mathcal{C} = 0.$$ ■

Item (3) in the preceding theorem provides a simple test for controllability. The matrix $\mathcal{C}$ is easily constructed, and the rank of $\mathcal{C}$ calculated. This is known as the *Kalman* test. Another matrix rank test is given next. We first need a lemma.

Lemma 2.6: Consider the system (A, B, C, E) of order n. If the controllability matrix has rank $n_1 < n$, then there exists a state transformation $z = Px$ (where P is a nonsingular matrix) that transforms the system into

$$\begin{bmatrix} \dot{z}_c \\ \dot{z}_u \end{bmatrix} = \begin{bmatrix} \bar{A}_c & \bar{A}_{12} \\ 0 & \bar{A}_u \end{bmatrix} \begin{bmatrix} z_c \\ z_u \end{bmatrix} + \begin{bmatrix} \bar{B}_c \\ 0 \end{bmatrix} u \tag{2.13}$$

$$y = [\bar{C}_c \quad \bar{C}_u] \begin{bmatrix} z_c \\ z_u \end{bmatrix} + Eu.$$

□ *Proof:* By assumption,

$$\mathcal{C} = [B \quad AB \quad \ldots \quad A^{n-1}B]$$

has rank n_1. Let $v_1, \ldots, v_{n_1}$ be any n_1 linearly independent columns. Choose $n - n_1$ additional vectors $v_{n_1+1}, \ldots, v_n$ and define

$$V = \begin{bmatrix} \vdots & \vdots & \vdots & \vdots & \cdots & \vdots \\ v_1 & \vdots & v_{n_1} & v_{n_1+1} & \cdots & v_n \\ \vdots & \vdots & \vdots & \cdots & \vdots & \vdots \end{bmatrix}.$$

The additional vectors need to be chosen so that V is nonsingular, but they are otherwise arbitrary. Define the state transformation matrix $P = V^{-1}$. Recall from linear algebra that the columns of $P^{-1} = V$ are the basis vectors of the new state space.

We now show that P is the required transformation. Each v_i, $i = 1, \ldots, n_1$ is a column of $\mathcal{C}$. For such v_i, the vector Av_i is a linear combination of the columns of $\mathcal{C}$ and hence it is a linear combination of the v_i, $i = 1, \ldots, n_1$. Hence, $\bar{A} = PAP^{-1}$ has the form in (2.13). Also, since the columns of B are in $\mathcal{C}$, B is a linear combination of the v_i, $i = 1, \ldots, n_1$. It follows that the columns of $\bar{B} = PB$ depend only on v_i, $i = 1, \ldots, n_1$. Since $B = V\bar{B}$, $\bar{B}$ has the form in (2.13). Define $\bar{C} = CP^{-1}$. The controllability matrix $\bar{\mathcal{C}}$ for $(\bar{A}, \bar{B}, \bar{C}, E)$ is

$$\begin{aligned} \bar{\mathcal{C}} &= \left[\begin{bmatrix} \bar{B}_c \\ 0 \end{bmatrix} \vdots \begin{bmatrix} \bar{A}_c \bar{B}_c \\ 0 \end{bmatrix} \cdots \begin{bmatrix} \bar{A}_c^{n_1-1} \bar{B}_c \\ 0 \end{bmatrix} \cdots \begin{bmatrix} \bar{A}_c^{n-1} \bar{B}_c \\ 0 \end{bmatrix} \right] \\ &= \left[\bar{\mathcal{C}}_c \vdots \begin{bmatrix} \bar{A}_c^{n_1} \bar{B}_c \\ 0 \end{bmatrix} \vdots \cdots \vdots \begin{bmatrix} \bar{A}_c^{n-1} \bar{B}_c \\ 0 \end{bmatrix} \right]. \end{aligned}$$

The columns of

$$\begin{aligned} \begin{bmatrix} \bar{A}_c^i \bar{B}_c \\ 0 \end{bmatrix} &= \bar{A}^i \bar{B} \\ &= PA^i B \end{aligned}$$

for $i = n_1, \ldots, n$ and these columns are linearly dependent on $v^i = A^i B$; $i = 0, \ldots, n_1 - 1$; that is, on the columns of $\bar{\mathcal{C}}_c$. Since

$$\operatorname{rank} \bar{\mathcal{C}} = \operatorname{rank} \mathcal{C} = n_1,$$

$$\operatorname{rank} \bar{\mathcal{C}}_c = n_1$$

and the subsystem $(\bar{A}_c, \bar{B}_c, \bar{C}_c, E)$ is controllable. ■

Theorem 2.7 (*Popov–Belevitch–Hautus (PBH)*): (A, B) is controllable if and only if the matrix

$$[(sI - A) \quad B]$$

has rank n for all complex numbers s.

□ *Proof:* Suppose first that $[(\lambda I - A) \quad B]$ does not have full row rank. That is, there is $q \neq 0$ such that

$$q^*[(\lambda I - A) \quad B] = 0.$$

Then

$$\begin{aligned} q^*\mathcal{C} &= q^*[B \quad AB \quad A^2B \quad \dots \quad A^{n-1}B] \\ &= q^*[B \quad \lambda B \quad \lambda^2 B \quad \dots \quad \lambda^{n-1}B] \\ &= 0. \end{aligned}$$

Thus by Theorem 2.5, (A, B) is not controllable.

Now suppose that (A, B) is not controllable. From Theorem 2.6 there is a transformation P with

$$\bar{A} = PAP^{-1} = \begin{bmatrix} \bar{A}_c & \bar{A}_{12} \\ 0 & \bar{A}_u \end{bmatrix}$$

$$\bar{B} = PB = \begin{bmatrix} \bar{B}_c \\ 0 \end{bmatrix}.$$

Let λ be an eigenvalue of $\bar{A}_u$ and choose a left eigenvector v_u so that $v_u^* \bar{A}_u = v_u^* \lambda$. Define the $n \times 1$ vector

$$v = \begin{bmatrix} 0 \\ v_u \end{bmatrix}.$$

Then

$$\begin{aligned} v^*[(\lambda I - \bar{A})\bar{B}] &= [0 \quad v_u]^* \begin{bmatrix} \lambda I - \bar{A}_c & -\bar{A}_{12} & \bar{B}_c \\ 0 & \lambda I - \bar{A}_u & 0 \end{bmatrix} \\ &= 0. \end{aligned}$$

Thus,

$$0 = v^* P [(\lambda I - A)P^{-1} \quad B].$$

Since P is nonsingular, $v^* P \neq 0$. Also, because P^{-1} is nonsingular,

$$(v^* P)(\lambda I - A)P^{-1} = 0$$

implies that

$$(v^* P)(\lambda I - A) = 0.$$

Hence,

$$v^* P[(\lambda I - A) \quad B] = 0.$$

Thus, if (A, B) is not controllable, the matrix

$$[(sI - A) \quad B]$$

loses rank at $s = \lambda$. ■

Example 2.8: Recall the spring–mass–dashpot system in Example 2.2. Using the first state-space realization (2.3) calculate the controllability matrix,

$$\mathcal{C} = [B \quad AB \quad A^2B \quad A^3B].$$

We have

$$\det \mathcal{C} = \frac{1}{m_2^4 m_1^3}\left[d_2^2 k_1 - d_2 k_2 d_1 + k_2^2 m_1\right].$$

This determinant is quadratic in d_2 and it has a real root if and only if $d_1^2 \geq 4k_1 m_1$. This is physically impossible and so we conclude that $\mathcal{C}$ has full rank. The system is controllable. Note that in this example we are able to control 4 states with a single control.

Example 2.9: Consider the system of two parallel spring–dashpots shown in Fig. 2.3 and modeled by

$$m\ddot{x}_1(t) = -k_1 x_1(t) + u(t)$$
$$m\ddot{x}_2(t) = -k_2 x_2(t) + u(t).$$

In state-space form,

$$\begin{bmatrix} \dot{x}_1(t) \\ \ddot{x}_1(t) \\ \dot{x}_2(t) \\ \ddot{x}_1(t) \end{bmatrix} = \begin{bmatrix} 0 & 1 & 0 & 0 \\ -\frac{k_1}{m} & 0 & 0 & 0 \\ 0 & 0 & 0 & 1 \\ 0 & 0 & -\frac{k_2}{m} & 0 \end{bmatrix} \begin{bmatrix} x_1(t) \\ \dot{x}_1(t) \\ x_2(t) \\ \dot{x}_2(t) \end{bmatrix} + \begin{bmatrix} 0 \\ \frac{1}{m} \\ 0 \\ \frac{1}{m} \end{bmatrix} u(t).$$

If the initial displacements are nonzero and $u \equiv 0$, the platforms will vibrate forever. Can the control u be chosen to control the system to some $x_f \in R^4$ such as $[0, 0, 0, 0]^*$ in a finite amount of time?

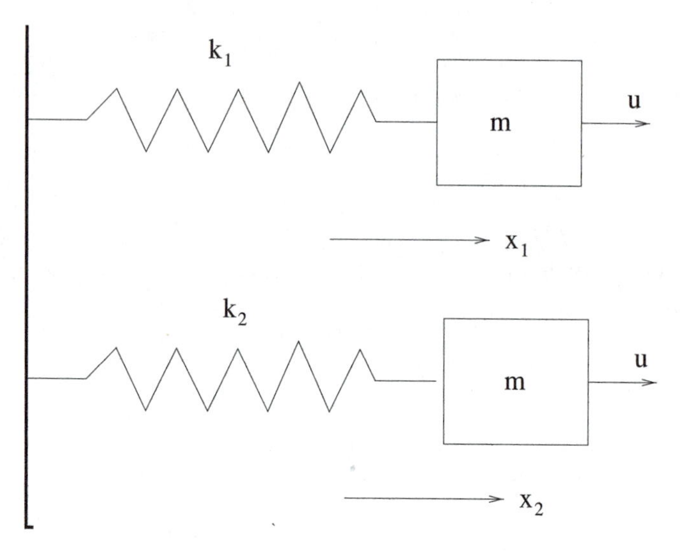

FIGURE 2.3 Parallel spring–mass system.

We are asking whether the system is controllable. The controllability matrix is

$$\mathcal{C} = \begin{bmatrix} 0 & \frac{1}{m} & 0 & -\frac{k_1}{m^2} \\ \frac{1}{m} & 0 & -\frac{k_1}{m^2} & 0 \\ 0 & \frac{1}{m} & 0 & -\frac{k_2}{m^2} \\ \frac{1}{m} & 0 & -\frac{k_2}{m^2} & 0 \end{bmatrix}.$$

Here, $\det \mathcal{C} = (-k_2 + k_1)^2/m^6$ and so

$$\text{rank } \mathcal{C} = 4$$

if and only if $k_1 \neq k_2$. The system is controllable if and only if the spring constants are not equal.

A system (2.1) is *observable* if there exists $T > 0$ such that any initial state x_o can be determined from knowledge of the input $u(t)$ and output $y(t)$ over the interval $[0, T]$. However, for any input u and initial condition x_o, we have

$$y(t) - \int_0^t C \exp(A(t-r))Bu(r)dr - Eu(t) = C \exp(At)x_o.$$

Defining

$$z(t) = y(t) - \int_0^t C\exp(A(t-r))Bu(r)dr - Eu(t),$$

$$z(t) = C\exp(At)x_o. \tag{2.14}$$

Thus, a system is observable if and only if knowledge of the output $z(\cdot)$ with zero input on an interval $[0, T]$ allows the initial condition x_o to be determined. This motivates the following formal definition of observability.

Definition 2.10: A system (A, C)

$$\dot{x}(t) = Ax(t), \qquad x(0) = x_0,$$
$$y(t) = Cx(t),$$

is *observable* if any initial condition x_0 can be determined by knowledge of $y(t)$ over any interval $[0, T]$ where $T > 0$.

Theorem 2.11: The following statements are equivalent:

1. (A, C) is observable.

2. The matrix

$$Q(t) = \int_0^t e^{A^*t}C^*Ce^{At}dt$$

is positive definite for all $t > 0$.

3. The Kalman observability matrix

$$\mathcal{O} = \begin{bmatrix} C \\ CA \\ \vdots \\ CA^{n-1} \end{bmatrix}$$

has rank n (where A is $n \times n$).

4. The matrix

$$\begin{bmatrix} sI - A \\ C \end{bmatrix}$$

has rank n for all complex numbers s.

5. (A^*, C^*) is controllable.

□ *Proof:* We will show that (1) and (3) are equivalent.
Taking derivatives of $z(t) = C\exp(At)x_o$, and setting $t = 0$ we obtain

$$\begin{bmatrix} z(0) \\ \dot{z}(0) \\ \vdots \\ z^{n-1}(0) \end{bmatrix} = \begin{bmatrix} C \\ CA \\ \vdots \\ CA^{n-1} \end{bmatrix} x_o. \tag{2.15}$$

If condition (3) holds, then (2.15) has a unique solution x_o for any z, and the system is observable. Now suppose (3) does not hold. That is, there is $p \neq 0$ such that $\mathcal{O}p = 0$. If the initial state $x(0) = p$ then, using (2.11), it follows that $z(t) = C\exp(At)p = 0$ for all t and the initial state cannot be determined.

From condition (3) above and condition (3) in Theorom 2.5 we have that (A, C) is observable if and only if (A^*, C^*) is controllable. Items (2) and (4) then follow. ■

The following lemma can be proven by duality (Thm. 2.11, condition 5) or by a proof identical to that of Lemma 2.6.

Lemma 2.12: Consider the system (A, B, C, E) of order n. If the observability matrix has rank $n_1 < n$, then there exists a state transformation $z = Px$ (where P is a nonsingular matrix) which transforms the system into

$$\begin{bmatrix} \dot{z}_o \\ \dot{z}_u \end{bmatrix} = \begin{bmatrix} A_o & A_{12} \\ 0 & A_u \end{bmatrix} \begin{bmatrix} z_o \\ z_u \end{bmatrix} + \begin{bmatrix} B_o \\ B_u \end{bmatrix} u \tag{2.16}$$

$$y = \begin{bmatrix} C_o & 0 \end{bmatrix} \begin{bmatrix} z_o \\ z_u \end{bmatrix} + Eu.$$

Example 2.13: Referring back to the double spring–mass system (Fig. 2.2) we will check whether the all states are observable from measurement of the second mass. Using the first state-space description of the system, the observability matrix is

$$\mathcal{O} = \begin{bmatrix} C \\ CA \\ CA^2 \\ CA^3 \end{bmatrix}$$

$$= \begin{bmatrix} 0 & 0 & 1 & 0 \\ 0 & 0 & 0 & 1 \\ \frac{k_2}{m_2} & \frac{d_2}{m_2} & -\frac{k_2}{m_2} & -\frac{d_2}{m_2} \\ -\left(\frac{d_2(k_1+k_2)}{m_1 m_2} + \frac{d_2 k_2}{m_2^2}\right) & \left(\frac{k_2}{m_2} - \frac{d_2(d_1+d_2)}{m_1 m_2} - \frac{d_2^2}{m_2^2}\right) & \left(\frac{d_2 k_2}{m_1 m_2} + \frac{d_2 k_2}{m_2^2}\right) & \left(\frac{d_2^2}{m_1 m_2} - \frac{k_2}{m_2} + {\frac{d_2^2}{m_2}}^2\right) \end{bmatrix}.$$

In this example,

$$\det \mathcal{O} = \frac{d_2^2 k_1 - k_2 d_1 d_2 + k_2^2 m_1}{m_1 m_2^2}.$$

The determinant is quadratic in d_2 and it has a real root if and only if

$$d_1^2 \geq 4 k_1 m_1 .$$

This is physically impossible and we conclude that the system is observable.

Example 2.14: A satellite with mass m in orbit about the earth is controlled by three small rocket engines. Let (r, θ, ϕ) indicate the position of the satellite in spherical coordinates with the center of the earth as origin. The equations of motion are

$$\begin{aligned}
\ddot{r} &= r\dot{\theta}^2 \cos^2 \theta + r\dot{\phi}^2 - k/r^2 + u_r/m \\
\ddot{\theta} &= -2\dot{r}\dot{\theta}/r + 2\dot{\theta}\dot{\phi} \sin \phi / \cos \phi + u_\theta/(mr) \cos \phi \\
\ddot{\phi} &= -\dot{\theta}^2 \cos \phi \sin \phi - 2\dot{r}\dot{\phi}/r + u_\phi/(mr).
\end{aligned}$$

The control forces are u_r, u_θ, and u_ϕ. Define the state

$$z = [r, \theta, \phi, \dot{r}, \dot{\theta}, \dot{\phi}].$$

The linearized state equations are

$$\dot{z} = \begin{bmatrix} 0 & 0 & 0 & 1 & 0 & 0 \\ 0 & 0 & 0 & 0 & 1 & 0 \\ 0 & 0 & 0 & 0 & 0 & 1 \\ 3\omega_0^2 & 0 & 0 & 0 & 2\omega_0 r_0 & 0 \\ 0 & 0 & 0 & -2\omega_0/r_0 & 0 & 0 \\ 0 & 0 & -\omega_0^2 & 0 & 0 & 0 \end{bmatrix} z + \begin{bmatrix} 0 & 0 & 0 \\ 0 & 0 & 0 \\ 0 & 0 & 0 \\ \frac{1}{m} & 0 & 0 \\ 0 & \frac{1}{mr_0} & 0 \\ 0 & 0 & \frac{1}{mr_0} \end{bmatrix} u$$

where $u = (u_r, u_\theta, u_\phi)$. The position is measured:

$$y = \begin{bmatrix} 1 & 0 & 0 & 0 & 0 & 0 \\ 0 & 1 & 0 & 0 & 0 & 0 \\ 0 & 0 & 1 & 0 & 0 & 0 \end{bmatrix} z.$$

Notice that $r, \theta, \dot{r}, \dot{\theta}$ are unaffected by $\phi, \dot{\phi}$ and vice versa. Since u_ϕ only enters through the last equation for $\dot{\phi}$, this suggests that the system with the single control u_ϕ is uncontrollable. The controllability matrix is, with

$$B = \begin{bmatrix} 0 \\ 0 \\ 0 \\ 0 \\ 0 \\ \frac{1}{mr_0} \end{bmatrix},$$

$$\mathcal{C} = [B \quad AB \quad \ldots \quad A^5 B]$$

$$= \frac{1}{mr_0} \begin{bmatrix} 0 & 0 & 0 & 0 & 0 & 0 \\ 0 & 0 & 0 & 0 & 0 & 0 \\ 0 & 1 & 0 & -\omega_0^2 & 0 & \omega_0^4 \\ 0 & 0 & 0 & 0 & 0 & 0 \\ 0 & 0 & 0 & 0 & 0 & 0 \\ 1 & 0 & -\omega_0^2 & 0 & \omega_0^4 & 0 \end{bmatrix}.$$

Clearly, rank $\mathcal{C} = 2$. This indicates that there are two controllable states. If we use all three inputs,

$$B = \begin{bmatrix} 0 & 0 & 0 \\ 0 & 0 & 0 \\ \frac{1}{m} & 0 & 0 \\ 0 & \frac{1}{mr_0} & 0 \\ 0 & 0 & 0 \\ 0 & 0 & \frac{1}{mr_0} \end{bmatrix}, \qquad AB = \begin{bmatrix} \frac{1}{m} & 0 & 0 \\ 0 & \frac{1}{mr_0} & 0 \\ 0 & 0 & \frac{1}{mr_0} \\ 0 & 2\frac{\omega_0}{m} & 0 \\ -2\frac{\omega_0}{r_0 m} & 0 & 0 \\ 0 & 0 & 0 \end{bmatrix}.$$

We obtain immediately that the Kalman controllability matrix has rank 6. The system is controllable.

Also, since we are measuring all components of position,

$$C = \begin{bmatrix} 1 & 0 & 0 & 0 & 0 & 0 \\ 0 & 1 & 0 & 0 & 0 & 0 \\ 0 & 0 & 1 & 0 & 0 & 0 \end{bmatrix},$$

$$CA = \begin{bmatrix} 0 & 0 & 0 & 1 & 0 & 0 \\ 0 & 0 & 0 & 0 & 1 & 0 \\ 0 & 0 & 0 & 0 & 0 & 1 \end{bmatrix}.$$

The Kalman observability matrix has rank 6. The system is observable.

2.3 TRANSFER FUNCTIONS

The Laplace transform of a function $x(t)$ is defined as

$$\hat{x}(s) = \int_0^\infty x(t)e^{-st}dt$$

for values of s for which this improper integral converges. Basic results in linear differential equations show that the solution $x(t)$ to the linear differential equation (2.1) satisfies, for some real M and α,

$$\|x(t)\| \leq Me^{\alpha t}.$$

Here $\| \quad \|$ indicates any vector norm. This result assumes only that the input u also satisfies an exponential growth bound. Thus, the solution to (2.1) has a Laplace transform $\hat{x}(s)$ defined for $|s| > \alpha$. Taking Laplace transforms in (2.1) with $x(0) = 0$ we obtain

$$s\hat{x}(s) = A\hat{x}(s) + B\hat{u}(s)$$
$$\hat{y}(s) = C\hat{x}(s) + E\hat{u}(s).$$

Thus, defining

$$G(s) = C(sI - A)^{-1}B + E,$$
$$\frac{\hat{y}(s)}{\hat{u}(s)} = G(s).$$

The function $G(s)$ describes the ratio between the output $\hat{y}(s)$ and the input $\hat{u}(s)$. This function is called the system *transfer function.*

The inverse transform of G is the impulse response

$$g(t) = Ce^{At}B + E\delta(t)$$

where δ indicates the impulse. Using Eq. (2.10), we obtain that if $x(0) = 0$,

$$y(t) = \int_0^t g(t - \tau)u(\tau)d\tau.$$

By Cramer's Rule (Appendix B),

$$G(s) = \frac{C\mathrm{adj}(sI - A)B}{\det(sI - A)} + E \tag{2.17}$$

and so G is a ratio of polynomials with real coefficients. The degree of the numerator polynomial is no greater than that of the denominator polynomial.

Definition 2.15: The *poles* of a rational function $G(s)$ are the points where G is not analytic. Alternatively, they are the points p where $\lim_{s\to p} G(s) = \infty$.

The function G is analytic, except at its poles. The poles are a subset of the zeros of $\det(sI - A)$. That is, every pole of G is an eigenvalue of A. (The converse may not be true, because of cancellations with zeros of the numerator. This is discussed later.) Also, the poles of G are either real, or occur in complex-conjugate pairs.

Definition 2.16: The *relative degree* of a transfer function G is defined as follows:

$$\text{Relative degree of } G = \deg(\text{denominator}\,(G)) - \deg(\text{numerator}\,(G)). \tag{2.18}$$

Definition 2.17: A system with transfer function for which the relative degree is nonnegative is said to be *proper*. If the relative degree is positive the system is *strictly proper*.

Note that (2.17) implies that all transfer functions of a system (2.1) are proper. They are strictly proper if and only if $E = 0$.

Example 2.18 (*Example 2.1 Cont.*): Taking Laplace transforms of the differential equation (2.2) for the tank, with zero initial conditions, we obtain

$$C_p M s\hat{T}(s) = (-C_p m - k)\hat{T}(s) + \hat{Q}(s)$$

where $\hat{T}$, $\hat{Q}$ indicate the Laplace transforms of temperature T and heat transfer Q, respectively. Rearranging, we obtain

$$\frac{\hat{T}}{\hat{Q}} = \frac{\frac{1}{C_p M}}{s + \tau}$$

where

$$\tau = \frac{C_p m + k}{C_p M}.$$

Notice that the eigenvalue of the "A" matrix associated with the ordinary differential equation (2.2) becomes a pole in the transfer function.

Example 2.19 (*Example 2.2 Cont.*): Taking Laplace transforms of the system of equations (2.3) for the double spring–mass, with zero initial conditions for all states, we obtain

$$s\begin{bmatrix}\hat{x}_1(s)\\ \hat{x}_2(s)\\ \hat{x}_3(s)\\ \hat{x}_4(s)\end{bmatrix} = A\hat{x}(s) + \begin{bmatrix}0\\0\\0\\ \frac{1}{m_2}\end{bmatrix}\hat{f}(s)$$

$$\hat{y}(s) = \hat{x}_3(s).$$

Rearranging,

$$\frac{\hat{y}(s)}{\hat{f}(s)} = [0 \quad 0 \quad 1 \quad 0](sI - A)^{-1}\begin{bmatrix}0\\0\\0\\ \frac{1}{m_2}\end{bmatrix}$$

$$= \frac{m_1 s^2 + (d_1 + d_2)s + k_1 + k_2}{m_1 m_2 s^4 + (d_1 m_2 + m_1 d_2 + d_2 m_2)s^3 + a_2 s^2 + a_1 s + k_1 k_2}.$$

where $a_2 = (m_1 k_2 + k_1 m_2 + m_2 k_2 + d_1 d_2 + d_2^2)$, $a_1 = (d_2 k_1 + d_1 k_2 + d_2 k_2)$.

Example 2.20 (*Example 2.3 Cont.*): Taking Laplace transforms of the general nth-order differential equation (2.5), with zero initial conditions,

$$(s^n + b_{n-1}s^{n-1} + \cdots b_o)\hat{z}(s) = \hat{u}(s).$$

Or, defining

$$d(s) = s^n + b_{n-1}s^{n-1} + \cdots b_o,$$

$$\hat{z}(s) = \frac{1}{d(s)}\hat{u}(s). \tag{2.19}$$

Taking Laplace transforms of (2.6),

$$\hat{y}(s) = (a_o + a_1 s + \cdots a_{n-1}s^{n-1})\hat{z}(s). \tag{2.20}$$

Putting (2.19) and (2.20) together, we obtain

$$\frac{\hat{y}(s)}{\hat{u}(s)} = \frac{(a_o + a_1 s + \cdots a_{n-1}s^{n-1})}{d(s)}.$$

There are a number of reasons why transfer functions are useful. One is that a differential equation model where the input u appears as a convolution becomes a rational function multiplied by the input $\hat{u}$. Multiplication is a simpler operation than convolution. Also, the transfer function is analytic, except at a finite number of poles. As we shall see, analytic function theory will be very useful in analyzing control systems.

Example 2.21 (*Idle Speed Control* [45]): A controller for a fuel-injected engine should adjust a valve setting so that a constant engine speed is maintained, despite changes in load, such as turning on the air conditioning. The engine speed is measured. The components of the engine model are shown in Fig. 2.4.

The valve dynamics are neglected:

$$V = 1.$$

The manifold dynamics are modeled by a single first-order o.d.e.,

$$\dot{x}_m(t) = -a_m x_m(t) + k_m u_m(t)$$
$$y_m(t) = x_m(t),$$

where $a_m > 0$ and u_m is the actual valve position. Therefore the transfer function of the manifold is

$$M = \frac{k_m}{s + a_m}.$$

The induction to power delay is T seconds. The transfer function e^{-Ts} of the delay is usually neglected or approximated by a rational function. A common choice is

$$IP = \frac{1}{1 + Ts}.$$

The engine dynamics are modeled by

$$E = \frac{k_E}{s + a_E}.$$

Experiments were performed to estimate the various parameters: $a_m = 1.1$, $a_E = 1.3$, $T = .1$, $k_m = 1.1$, and $k_E = 1.3$. (The gains k_m and k_E are normalized.)

Define the overall transfer function from demanded valve position to engine speed

$$\begin{aligned} P &= E \cdot IP \cdot M \cdot V \\ &= \frac{k_m k_E}{(s + a_m)(s + a_E)(1 + Ts)}. \end{aligned}$$

Let H indicate the controller, as yet undetermined. Figure 2.5 shows the control system written in the standard feedback form.

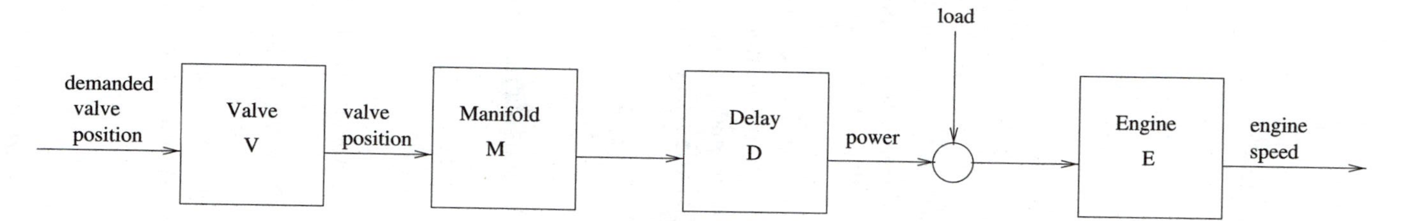

FIGURE 2.4 Idle speed control system.

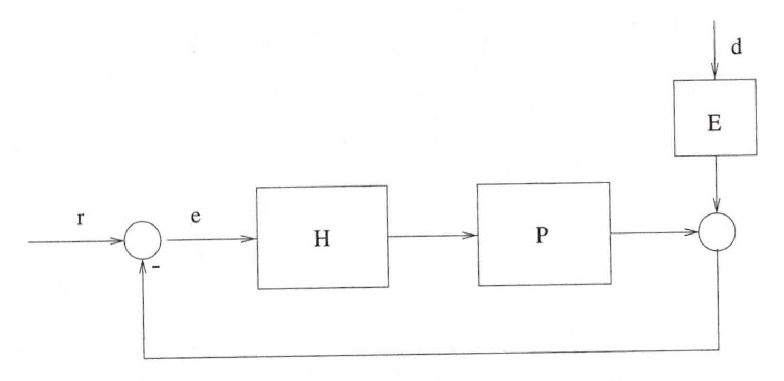

FIGURE 2.5 Block diagram of engine speed model.

The transfer functions M, IP, and E are all strictly proper: Each has relative degree 1. The valve function V is non-strictly proper: It has relative degree 0. The overall system transfer function P has relative degree 3.

The controller design problem is to find H so that the closed loop is stable and the error e is small, for a demanded valve position r and arbitrary disturbance d. Also, since the model is a simplification of a complex system, the controller should also perform "reasonably" well on systems that are "close" to the nominal model derived earlier. These points are discussed in the chapters to follow.

2.4 REALIZATIONS

Let G be a given transfer function. A set of matrices (A, B, C, E) with the property

$$G(s) = C(sI - A)^{-1}B + E$$

is called a *realization* of G. Since the Laplace transform of any function is unique, a particular realization (A, B, C, E) (or system of equations (2.1)) uniquely defines the corresponding transfer function.

However, every transfer function has many realizations. For instance, two realizations of the same spring–mass control system are given in Example 2.2. Although the matrices A and C are different, the equations describe the same system.

Definition 2.22: If (A_1, B_1, C_1, E_1) and (A_2, B_2, C_2, E_2) are realizations of systems, and both realizations have the same transfer function, we say these realizations are *zero-state equivalent*.

Any two realizations related by an invertible state transformation are zero-state equivalent. The proof of the following theorem is a straightforward calculation and is omitted.

THEOREM 2.23: Let (A, B, C, E) be a given realization of a transfer function G. The time-domain description of the system is (2.1). Let $P \in R^{n\times n}$ be any invertible matrix. Defining $z = Px$, and substituting into (2.1) we obtain another system of equations:

$$\dot{z}(t) = (PAP^{-1})z(t) + (PB)u(t), \quad z(0) = 0,$$
$$y(t) = (CP^{-1})z(t) + Eu(t).$$

The system realization $(PAP^{-1}, PB, CP^{-1}, E)$ also has transfer function G.

Clearly, every transfer function has an infinite number of realizations. The transformation that decomposes a system into controllable and uncontrollable states (Lemma 2.6) is an invertible state transformation, as is the transformation that separates out the observable and unobservable states (Lemma 2.12).

THEOREM 2.24: Two realizations (A_1, B_1, C_1, E_1) and (A_2, B_2, C_2, E_2) are zero-state equivalent if and only if $E_1 = E_2$ and

$$C_1 A_1^i B_1 = C_2 A_2^i B_2$$

for all integers $i \geq 0$.

□ *Proof:* It is clear that we must have $E_1 = E_2$ and, in order for the transfer functions to be identical,

$$C_1 \exp(A_1 t) B_1 = C_2 \exp(A_2 t) B_2.$$

Using the series expansion of the matrix exponential,

$$\sum_{i=0}^{\infty} \frac{C_1 A_1^i B_1}{i!} t^i = \sum_{i=0}^{\infty} \frac{C_2 A_2^i B_2}{i!} t^i.$$

Since this must hold for all t,

$$C_1 A_1^i B_1 = C_2 A_2^i B_2$$

for all i. ■

THEOREM 2.25: The controllable realization $(\bar{A}_c, \bar{B}_c, \bar{C}_c, E)$ in Theorem 2.6 is zero-state equivalent to the original realization (A, B, C, E).

□ *Proof:* The system $(\bar{A}, \bar{B}, \bar{C}, E)$ in the proof of Theorem 2.6 is obtained from the original realization by an invertible transformation. Thus, these two realizations are zero-state equivalent and have the same transfer function.

We will now show that $(\bar{A}_c, \bar{B}_c, \bar{C}_c, E)$ and $(\bar{A}, \bar{B}, \bar{C}, E)$ have the same transfer function to complete the proof. Now,

$$\begin{bmatrix} sI - \bar{A}_c & \bar{A}_{12} \\ 0 & sI - \bar{A}_u \end{bmatrix} \begin{bmatrix} (sI - \bar{A}_c)^{-1} & -(sI - \bar{A}_c)^{-1}\bar{A}_{12}(sI - \bar{A}_u)^{-1} \\ 0 & (sI - \bar{A}_u)^{-1} \end{bmatrix} = I$$

The transfer function matrix of $(\bar{A}, \bar{B}, \bar{C}, E)$ is

$$[\bar{C}_c \quad \bar{C}_u] \begin{bmatrix} sI - \bar{A}_c & \bar{A}_{12} \\ 0 & sI - \bar{A}_u \end{bmatrix}^{-1} \begin{bmatrix} \bar{B}_c \\ 0 \end{bmatrix} + E$$

$$= [\bar{C}_c \quad \bar{C}_u] \begin{bmatrix} (sI - \bar{A}_c)^{-1} & (sI - \bar{A}_c)^{-1}\bar{A}_{12}(sI - \bar{A}_u)^{-1} \\ 0 & (sI - \bar{A}_u)^{-1} \end{bmatrix} \begin{bmatrix} \bar{B}_c \\ 0 \end{bmatrix} + E$$

$$= \bar{C}_c(sI - \bar{A}_c)^{-1}\bar{B}_c + E.$$

This is the transfer function of $(\bar{A}_c, \bar{B}_c, \bar{C}_c, E)$. ■

Thus, the controllable n_1-dimensional subsystem (A_c, B_c, C_c, E) has the same transfer function as the original system. A similar result holds for the observable realization in Lemma 2.12.

Theorem 2.26: The observable realization (A_o, B_o, C_o, E) in Lemma 2.12 is zero-state equivalent to the original realization (A, B, C, E).

If the poles of a transfer function G are known or can be easily found, it is straightforward to obtain the partial fraction expansion of G. The inverse Laplace transform of G then yields a realization. For instance, if there are no repeated poles in a transfer function G, it can be written as

$$G(s) = \sum_{i=1}^{n} a_i/(s - p_i) + E$$

where p_i are the poles and a_i is the residue at p_i. Then simply set $A = \operatorname{diag}(p_i)$, $[B]_i = a_i$, $[C]_i = 1$.

Example 2.27: Consider

$$G(s) = \frac{10}{(s+2)(s+3)}$$
$$= 10/(s+2) - 10/(s+3).$$

The following is a realization of G:

$$A = \begin{bmatrix} -2 & 0 \\ 0 & -3 \end{bmatrix}, \; B = \begin{bmatrix} 1 \\ 1 \end{bmatrix}, \; C = [10 \quad -10], \; E = 0.$$

Using the state transformation $Px = z$ where

$$P = \begin{bmatrix} 2 & 4 \\ 1 & 1 \end{bmatrix}$$

we obtain another realization for G:

$$\tilde{A} = \begin{bmatrix} -4 & 4 \\ -1/2 & -1 \end{bmatrix}, \; \tilde{B} = \begin{bmatrix} 6 \\ 2 \end{bmatrix}, \; \tilde{C} = [-10 \quad 30], \; \tilde{E} = 0.$$

Example 2.28: Consider the transfer function

$$P = \frac{1.43}{(s+1.1)(s+1.3)(1+.1s)}.$$

Here the denominator is already factored and so it is convenient to obtain a modal realization. The partial fraction expansion has the form

$$P = \frac{a_1}{s+1.1} + \frac{a_2}{s+1.3} + \frac{a_3}{s+10}.$$

The coefficients a_i are obtained by calculating the residue associated with each pole:

$$P = \frac{8.0337}{s+1.1} - \frac{8.2184}{s+1.3} + \frac{.18465}{s+10}.$$

Thus, a realization of this transfer function is

$$A = \begin{bmatrix} -1.1 & 0 & 0 \\ 0 & -1.3 & 0 \\ 0 & 0 & -10 \end{bmatrix}, \; B = \begin{bmatrix} \sqrt{8.0337} \\ \sqrt{8.2184} \\ \sqrt{.18465} \end{bmatrix},$$

$$C = \begin{bmatrix} \sqrt{8.0337} & -\sqrt{8.2184} & \sqrt{.18465} \end{bmatrix}, \; E = 0.$$

Another type of realization of a transfer function can be obtained without factoring the denominator. Isolate the feedthrough term E and write the transfer function in the form

$$E + \frac{a_{n-1}s^{n-1} + a_{n-2}s^{n-2} \cdots + a_1 s + a_0}{s^n + b_{n-1}s^{n-1} + b_{n-2}s^{n-2} \cdots + b_1 s + b_0}.$$

Note that the coefficient of s^n in the denominator is 1.

Let G be the transfer function of a strictly proper SISO system. Write $G = n/d$ where

$$\begin{aligned} n &= a_0 + a_1 s + \cdots a_{n-1} s^{n-1} \\ d &= b_0 + b_1 s + \cdots b_{n-1} s^{n-1} + s^n. \end{aligned}$$

Indicating the input and output as usual by u and y, respectively, and introducing an intermediate function v, we have

$$\begin{aligned} dv &= u \\ y &= nv. \end{aligned} \tag{2.21}$$

Define the Laplace transforms of the states by

$$\begin{aligned} \hat{x}_1(s) &= v(s) \\ \hat{x}_2(s) &= sv = s\hat{x}_1 \\ &\vdots \\ \hat{x}_n(s) &= s^{n-1}v = s\hat{x}_{n-1}. \end{aligned}$$

Substituting into (2.21) we have

$$s^n \hat{x}_1(s) + b_{n-1} s^{n-1} \hat{x}_1(s) + \cdots b_0 \hat{x}_1(s) = u(s).$$

Inverting the Laplace transform (with zero initial conditions) we obtain

$$\dot{x}_n + b_{n-1} x_n + b_{n-2} x_{n-1} + \cdots b_0 x_1 = u(t).$$

Thus,

$$\dot{\mathbf{x}}(t) = \begin{bmatrix} 0 & 1 & 0 & 0 & \cdots \\ 0 & 0 & 1 & 0 & \cdots \\ 0 & 0 & 0 & 1 & \cdots \\ \vdots & \vdots & \vdots & \vdots & \vdots \\ 0 & 0 & \cdots & 0 & 1 \\ -b_0 & -b_1 & \cdots & -b_{n-2} & -b_{n-1} \end{bmatrix} \mathbf{x}(t) + \begin{bmatrix} 0 \\ 0 \\ 0 \\ \vdots \\ 1 \end{bmatrix} u(t). \tag{2.22}$$

Also,

$$y(t) = [a_0 \ \ a_1 \ldots a_{n-1}]\, \mathbf{x}(t).$$

This kind of realization is always controllable and is known as the *controllable canonical* realization. It has satisfactory numerical properties for problems where the dimension of x is less than 50, but problems in computation may result with larger dimensions.

Definition 2.29: A realization (A, B, C, E) of a transfer matrix G is *minimal* if no other realization of G has lower order, i.e., fewer state variables.

The following theorem applies to SISO systems. (Two polynomials $a(s)$ and $b(s)$ are *coprime* if they have no common zeros.)

Theorem 2.30: A realization (A, B, C, E) of a transfer function G is minimal if and only if $b(s) = \det(sI - A)$ and $a(s) = C\text{adj}(sI - A)B$ are coprime polynomials.

□ *Proof:* Let n denote the order of the realization. Suppose first that a and b are not coprime. Then canceling common factors we obtain a reduced transfer function with $\deg \tilde{b}(s) < \deg b(s) = n$. Using this reduced transfer function, we obtain a realization with order less than n. Thus, (A, B, C, E) is not minimal.

Now suppose that (A, B, C, E) is not minimal. Then there is another realization $(\tilde{A}, \tilde{B}, \tilde{C}, E)$ with order $\tilde{n} < n$ and corresponding transfer function $\tilde{a}(s)/\tilde{b}(s)$ where $\deg \tilde{b}(s) \leq \tilde{n} < n$. Since the transfer function of a system is unique,

$$\frac{a(s)}{b(s)} = \frac{\tilde{a}(s)}{\tilde{b}(s)}.$$

This implies that a and b must have common factors and are not coprime. ■

The following lemma can be verified using Cramer's Rule.

Lemma 2.31: Consider a SISO system with realization (A, B, C, E). Then

$$C\ \text{adj}(sI - A)B + E \det(sI - A) = \det \begin{bmatrix} sI - A & -B \\ -C & -E \end{bmatrix}.$$

Theorem 2.32: A realization (A, B, C, E) of order n is minimal if and only if (A, B) is controllable and (A, C) is observable.

□ *Proof:* Suppose that the system is not observable. Then there is $p \in R^n$, $\lambda \in C$ such that

$$\begin{bmatrix} \lambda I - A \\ C \end{bmatrix} p = \begin{bmatrix} 0 \\ 0 \end{bmatrix}.$$

Thus,

$$\begin{bmatrix} \lambda I - A & -B \\ -C & -E \end{bmatrix} \begin{bmatrix} p \\ 0 \end{bmatrix} = \begin{bmatrix} 0 \\ 0 \end{bmatrix}.$$

Thus, if the system is not observable then some λ is both a zero of the numerator $a(s)$ and denominator $b(s)$ polynomials defined in the previous theorem. It follows that the realization is not minimal.

A similar argument shows that if the realization is not controllable it is not minimal.

Now suppose that a realization (A, B, C, E) with order n is not minimal. Then there is an equivalent realization (A_1, B_1, C_1, E_1) with order $n_1 < n$ that is minimal. Let $\mathcal{C}$ and $\mathcal{O}$ be the controllability and observability matrices, respectively, for the nonminimal realization, and similarly define $\mathcal{C}_1$ and $\mathcal{O}_1$ for the minimal realization. Define the matrix

$$H = \mathcal{C}\mathcal{O} = \begin{bmatrix} CB & CAB & CA^2B & \dots CA^{n-1}B \\ CAB & CA^2B & \dots & CA^nB \\ \vdots & \vdots & \vdots & \\ CA^{n-1}B & CA^nB & \dots & CA^{2n-2}B \end{bmatrix}.$$

Since the two realizations have the same transfer function, Theorem 2.24 implies that

$$CA^iB = C_1A_1^iB_1$$

for all $i = 0, 1, 2, \dots$. Thus, H is independent of the realization and

$$\mathcal{C}\mathcal{O} = \mathcal{C}_1\mathcal{O}_1.$$

Since

$$\text{rank } \mathcal{C}_1\mathcal{O}_1 \leq n_1,$$

$$\min(\text{rank } \mathcal{C}, \text{rank } \mathcal{O}) \leq n_1 < n.$$

Using Theorems 2.5 and 2.11 we conclude that the system is uncontrollable and/or unobservable. ■

The range of $\mathcal{C}$ is the space of all controllable states, or equivalently, all states reachable from 0. The nullspace of $\mathcal{O}$ is the the subspace of unobservable states. Then it follows that

$$\text{rank}(\mathcal{O}\mathcal{C}) = \dim \text{Range}(\mathcal{O}\mathcal{C})$$

is the dimension of all states that are both controllable and observable.

The proof of the following is a straightforward construction and is left as an exercise.

Theorem 2.33: Two minimal equivalent realizations with states x_1, x_2 are related via an invertible state transformation T where $x_1 = Tx_2$. Letting $\mathcal{C}_1$ and $\mathcal{O}_1$ be the controllability and observability matrices, respectively, for the first realization, and similarly defining $\mathcal{C}_2$ and $\mathcal{O}_2$ for the second realization,

$$T = (\mathcal{O}_2^*\mathcal{O}_2)^{-1}\mathcal{O}_2^*\mathcal{O}_1,$$
$$T^{-1} = \mathcal{C}_1\mathcal{C}_2^*(\mathcal{C}_2\mathcal{C}_2^*)^{-1}.$$

2.5 ZEROS AND POLES

The definitions of the zeros and poles of a scalar transfer function $G(s)$ are exactly the definitions from complex analysis. However, there are a number of ways to generalize these definitions to multi-input/multi-output systems where G is a transfer matrix. The definitions are less straightforward. This section will only be concerned with SISO systems (where B is $n \times 1$ and C is $1 \times n$).

THEOREM 2.34: Let (A, B, C, E) be a system with transfer function G. Every pole of G is an eigenvalue of A.

□ *Proof:* From (2.17), it is clear that a transfer function $G(s)$ can only have a pole at $s = p$ if $\det(pI - A) = 0$. ■

DEFINITION 2.35: The *zeros* of the transfer function G are the points z where $G(z) = 0$. They are known as the *transmission zeros* or the *zeros* of the system.

Just as the poles of G correspond to the eigenvalues of A in the state-space description, the zeros of G correspond to the system *invariant zeros*.

DEFINITION 2.36: The complex number λ is an *invariant zero* of the system (A, B, C, E) of order n if

$$\operatorname{rank}\begin{bmatrix} \lambda I - A & -B \\ -C & -E \end{bmatrix} < n + 1. \tag{2.23}$$

THEOREM 2.37: Let (A, B, C, E) be a system with transfer function G. Every transmission zero is a system invariant zero.

□ *Proof:* Let z be a zero of G. Then from (2.17),

$$C\operatorname{adj}(sI - A)B + E\det(sI - A) = 0$$

Using Lemma 2.31 we obtain that

$$\det\begin{bmatrix} zI - A & -B \\ -C & -E \end{bmatrix} = 0.$$

Thus, z is an invariant zero. ■

Recall that if λ is an eigenvalue of a matrix A, then by definition, there is a vector $x_o \neq 0$ so that $Ax_o = \lambda x_o$. Using the definition of the matrix exponential we obtain that

$$\exp(At)x_o = \exp(\lambda t)x_o.$$

Thus, there is a constant $c_1 = Cx_o$ so that the output of the system (A, B, C, E) with zero input and initial condition x_o is $c_1 \exp(\lambda t)$. Zeros play an analogous role in the time response.

LEMMA 2.38: Let (A, B, C, E) be a system with transfer function G. If the input $u(t) = e^{\lambda t}$ where λ is not an eigenvalue of A and the initial condition $x(0) = -(\lambda I - A)^{-1}B$ then the output $y(t) = G(\lambda)e^{\lambda t}$.

□ *Proof:* Using Laplace transforms,

$$\begin{aligned}\hat{y}(s) &= C(sI - A)^{-1}x(0) + C(sI - A)^{-1}B\hat{u}(s) + E\hat{u}(s)\\ &= C(sI - A)^{-1}x(0) + C(sI - A)^{-1}B(s - \lambda)^{-1} + E(s - \lambda)^{-1}\\ &= C(sI - A)^{-1}x(0) + C[(sI - A)^{-1} - (\lambda I - A)^{-1}]B(s - \lambda)^{-1}\\ &\quad + G(\lambda)(s - \lambda)^{-1}.\end{aligned}$$

Using the identity

$$(sI - A)^{-1} - (\lambda I - A)^{-1} = (\lambda - s)(sI - A)^{-1}(\lambda I - A)^{-1}$$

and subsituting in the initial condition we obtain

$$\begin{aligned}\hat{y}(s) &= -C(sI - A)^{-1}(\lambda I - A)^{-1}B + C(sI - A)^{-1}(\lambda I - A)^{-1}B\\ &\quad + G(\lambda)(s - \lambda)^{-1}\\ &= G(\lambda)(s - \lambda)^{-1}.\end{aligned}$$

Inverting the Laplace transform $(s - \lambda)^{-1}$, we obtain $y(t) = G(\lambda)e^{\lambda t}$. ■

THEOREM 2.39: If λ is a transmission zero then for $u = u_o e^{\lambda t}$ and some initial condition x_o, where u_o, x_o are not both zero, the output $y(t) = 0$ for all $t > 0$.

□ *Proof:* If λ is a transmission zero, it cannot be an eigenvalue. Assume then that λ is not an eigenvalue, and choose initial condition $x_0 = -(\lambda I - A)^{-1}B$, $u_o = e^{\lambda t}$. From the previous lemma, $y(t) = 0$ for all $t > 0$. ■

Thus, given a realization (A, B, C, E) with transfer function G, every pole of G is an eigenvalue of A and every zero of G is a system invariant zero. However, the converse is not always true.

For instance, let λ be an eigenvalue of A and x_λ the corresponding eigenvector. Suppose that $Cx_\lambda = 0$. For initial condition $x(0) = x_\lambda$, and zero input, the output is $y = Ce^{\lambda t}x_\lambda = 0$. Thus the output is zero. (This means the system is unobservable.) It follows that

$$\begin{bmatrix} \lambda I - A & -B \\ -C & -E \end{bmatrix} \begin{bmatrix} x_\lambda \\ 0 \end{bmatrix} = \begin{bmatrix} 0 \\ 0 \end{bmatrix}$$

and so λ is also an invariant zero. In this case λ will cancel in the numerator and denominator of the transfer function, and it is neither a pole nor a zero.

A similar situation occurs if $x_\lambda B = 0$ for some left eigenvector $x_\lambda^* \lambda = x_\lambda^* A$ of A. (In this case the system is uncontrollable.) We have

$$[x_\lambda^* \quad 0] \begin{bmatrix} \lambda I - A & -B \\ -C & -E \end{bmatrix} = \begin{bmatrix} 0 \\ 0 \end{bmatrix}$$

and so λ is also an invariant zero.

If λ is an invariant zero and also an eigenvalue, then it will appear in both the numerator and the denominator of

$$\frac{C \operatorname{adj}(sI - A)B + E \det(sI - A)}{\det(sI - A)}.$$

The two factors will cancel and λ will be neither a pole or a zero of the transfer function. The next theorem shows that this does not happen if the realization is minimal.

THEOREM 2.40: Let (A, B, C, E) be a minimal realization of a transfer function G.

1. A number λ is a pole of G if and only if it is an eigenvalue of A.
2. A number λ is a zero of G if and only if it is an invariant zero of the realization.

□ *Proof:* Let n be the order of the realization. (1) We showed earlier that every pole is an eigenvalue. If some eigenvalue is not a pole, then the degree of the denominator of $G(s)$ is less than the order of the realization. That is, the realization is not minimal.

(2) We showed earlier that every transmission zero is an invariant zero. We now show that every invariant zero is a transmission zero. Let λ be some invariant zero of the realization. Then

$$\det \begin{bmatrix} \lambda I - A & -B \\ -C & -E \end{bmatrix} = 0$$

and so the numerator polynomial of $G(s)$ is zero. If λ is also an eigenvalue, then the denominator polynomial is also zero. We can cancel the factor

$(s - \lambda)$ and obtain an identical transfer function with degree of the denominator less than n. This contradicts the assumption that the realization is minimal. Suppose then that λ is not an eigenvalue and write

$$\begin{bmatrix} A - \lambda I & B \\ C & E \end{bmatrix} = \begin{bmatrix} A - \lambda I & 0 \\ C & I \end{bmatrix} \begin{bmatrix} (A - \lambda I)^{-1} & 0 \\ 0 & G(\lambda) \end{bmatrix} \begin{bmatrix} A - \lambda I & B \\ 0 & I \end{bmatrix}.$$

Since the realization is minimal, the first and third matrices on the right-hand-side are nonsingular. This implies that the middle matrix must lose rank and $G(\lambda) = 0$. ■

2.6 TIME RESPONSE

Many interesting aspects of the transient behavior of a system can be observed from the *step response*. The step response is the output of the system when the input is

$$u(t) = \begin{cases} 1 & t \geq 0 \\ 0 & t < 0 \end{cases}.$$

(If there is more than one input, only one input is a step and the other inputs are 0.)

For a first-order system with transfer function

$$\frac{p}{s + p},$$

the response to a unit step is $1 - e^{-pt}$.

If $p > 0$, the 98% settling time (T_s) is defined as the time taken to reach 98% of the steady-state value. This time can be calculated as $T_s = \frac{4}{p}$.

Consider now a second-order system with transfer function written

$$\frac{w_n^2}{s^2 + 2\xi w_n + w_n^2}.$$

The quantity $w_n > 0$ is called the natural frequency and ξ, the damping parameter, satisfies $0 \leq \xi \leq 1$.

The step response is

$$1 - \frac{1}{\beta} e^{-\xi w_n t} \sin(w_n \beta t + \theta)$$

where $\beta = \sqrt{1 - \xi^2}$ and $\theta = \arctan(\beta / \xi)$. The step response of several second-order systems with the same natural frequency and different

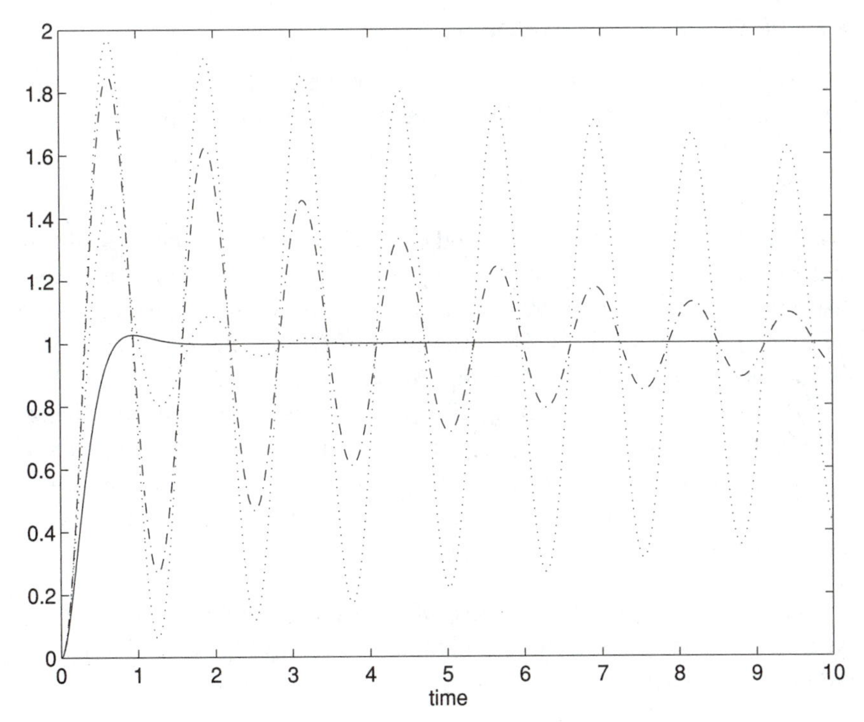

FIGURE 2.6 Step response of $G(s) = 25/(s^2 + 10\xi s + 25)$, $\xi = 0.01, 0.05, 0.25, 0.75$.

damping parameters is shown in Fig. 2.6. The 98% settling time is

$$T_s = \frac{4}{\xi w_n}.$$

Another characteristic of an underdamped system ($\xi < 1$) is the time to the first peak,

$$T_p = \frac{\pi}{w_n\sqrt{1-\xi^2}}.$$

The response of a second-order system is also characterized by the percent overshoot (*P.O.*)—the ratio of the peak response to the steady-state value. It can be shown that

$$P.O. = 100e^{-\xi\pi/\sqrt{1-\xi^2}}.$$

Higher-order systems are difficult to analyze in general. However, the response of these systems is often similar to that of a first-order or second-order system.

2.7 FREQUENCY RESPONSE

Let $G(s)$ be the transfer function of some system. Letting $j = \sqrt{-1}$, suppose a steady-state response of the system to the periodic input

$$e^{j\omega t} = \cos(\omega t) + j \sin(\omega t)$$

exists. It is a simple consequence of the definition of the Laplace transform that this steady-state response is $G(j\omega)$. (The proof of this is an exercise.) By varying the input over all frequencies ω, $G(j\omega)$ can be measured for many physical systems. The restriction of $G(s)$ to the imaginary axis, $G(j\omega)$, is therefore called the system *frequency response*.

We now show that a transfer function that is analytic in the closed right half-plane $\mathrm{Re}(s) \geq 0$ can be constructed uniquely from its frequency response. We first need *Cauchy's Integral Formula*. This result states that the value of a function analytic inside a closed curve is determined by its values on the curve.

THEOREM 2.41 (*Cauchy's Integral Formula*): Suppose that f is analytic on a simple closed curve C and in its interior. Let s_0 be a point inside, but not on, C. Then

$$f(s_0) = \frac{1}{2\pi j} \int_C \frac{f(s)}{s - s_0} ds, \tag{2.24}$$

where the curve C is traversed counterclockwise.

THEOREM 2.42: Let f be analytic and of bounded magnitude in $\mathrm{Re}(s) \geq 0$. Let $s_0 = \sigma_0 + j\omega_0$ be some point with $\sigma_0 > 0$. Then

$$f(s_0) = \frac{1}{\pi} \int_{-\infty}^{\infty} f(j\omega) \frac{\sigma_0}{\sigma_0^2 + (\omega - \omega_0)^2} d\omega. \tag{2.25}$$

□ *Proof:* For any point $s_0 = \sigma_0 + j\omega_0$ construct a contour C_r (see Fig. 2.7) with r large enough that s_0 is inside C_r.

Cauchy's integral formula gives

$$f(s_0) = \frac{1}{2\pi j} \int_{C_r} \frac{f(s)}{s - s_0} ds$$

and since $-\bar{s}_0$ is outside C_r

$$0 = \frac{1}{2\pi j} \int_{C_r} \frac{f(s)}{s + \bar{s}_0} ds.$$

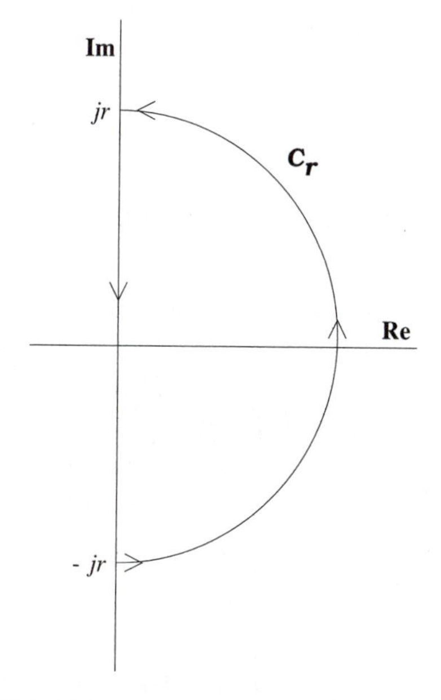

FIGURE 2.7 Contour for Theorem 2.42.

Subtracting the second integral from the first, we obtain

$$\begin{aligned} f(s_0) &= \frac{1}{2\pi j}\int_{C_r} \frac{f(s)(\bar{s}_0 + s_0)}{(s - s_0)(s + \bar{s}_0)} ds \\ &= \frac{1}{\pi j}\int_{C_r} \frac{f(s)\sigma_0}{(s - s_0)(s + \bar{s}_0)} ds \\ &= I_1 + I_2 \end{aligned}$$

where

$$\begin{aligned} I_1 &= \frac{1}{\pi j}\int_{-jr}^{jr} \frac{f(s)\sigma_0}{(s - s_0)(s + \bar{s}_0)} ds, \\ I_2 &= \frac{1}{\pi j}\int_{-\pi/2}^{\pi/2} f(re^{j\theta}) \frac{\sigma_0 \cdot rje^{j\theta}}{(re^{j\theta} - s_0)(re^{j\theta} + \bar{s}_0)} d\theta. \end{aligned}$$

The first integral can be rewritten as

$$\begin{aligned} I_1 &= \frac{-1}{\pi}\int_{-r}^{r} \frac{f(j\omega)\sigma_0}{(j\omega - s_0)(j\omega + \bar{s}_0)} d\omega \\ &= \frac{1}{\pi}\int_{-r}^{r} \frac{f(j\omega)\sigma_0}{\sigma_0^2 + (\omega - \omega_0)^2}\, d\omega. \end{aligned}$$

We now show that $\lim_{r\to\infty} I_2 = 0$. Define $M = \sup_{\mathrm{Re}(s)\geq 0} |f(s)|$. We have

$$|I_2| \leq \frac{M \cdot \sigma_0}{\pi r} \int_{-\pi/2}^{\pi/2} \frac{1}{|e^{j\theta} - s_0 r^{-1}||e^{j\theta} + \bar{s_0} r^{-1}|} d\theta.$$

For any $\epsilon > 0$, choose $r > |s_0|/\epsilon$. For such r,

$$|I_2| \leq \frac{M\sigma_0}{\pi r} \frac{\pi}{(1-\epsilon)^2}$$

$$\leq \frac{M\sigma_0 \cdot \epsilon}{|s_0|(1-\epsilon)^2}.$$

Since $\epsilon > 0$ is arbitrary,

$$\lim_{r\to\infty} I_2 = 0.$$

It follows that

$$f(s_0) = I_1$$

as was to be proven. ■

The value of a transfer function (or Laplace transform) at a point could in theory be calculated by computing the integral in (2.25). Unfortunately, this is an improper integral and calculation of a transfer function at more than a couple of points using this method is not feasible. Other techniques can be used to identify a system using only the experimentally measured frequency response. Simple transfer functions can be estimated from the graph of the frequency response. This will now be discussed.

At each frequency ω the frequency response $G(j\omega)$ is a complex number which we can separate into its magnitude r and phase ϕ:

$$G(j\omega) = r(\omega)e^{j\phi(\omega)}.$$

Typically, in control applications, the frequency response is displayed by graphing the magnitude and phase separately (see Fig. 2.8).

These are called *Bode plots*, after an early pioneer in controller design. The frequency is on a logarithmic (base 10) scale while the magnitude or gain in decibels (dB) is plotted on a linear scale:

$$\text{Gain} = 20 \log_{10} r.$$

Thus, the magnitude is also displayed on a logarithmic scale. The phase is plotted in degrees on a linear scale. Figure 2.8 shows the Bode plots of $1.1/(s+1.1)$.

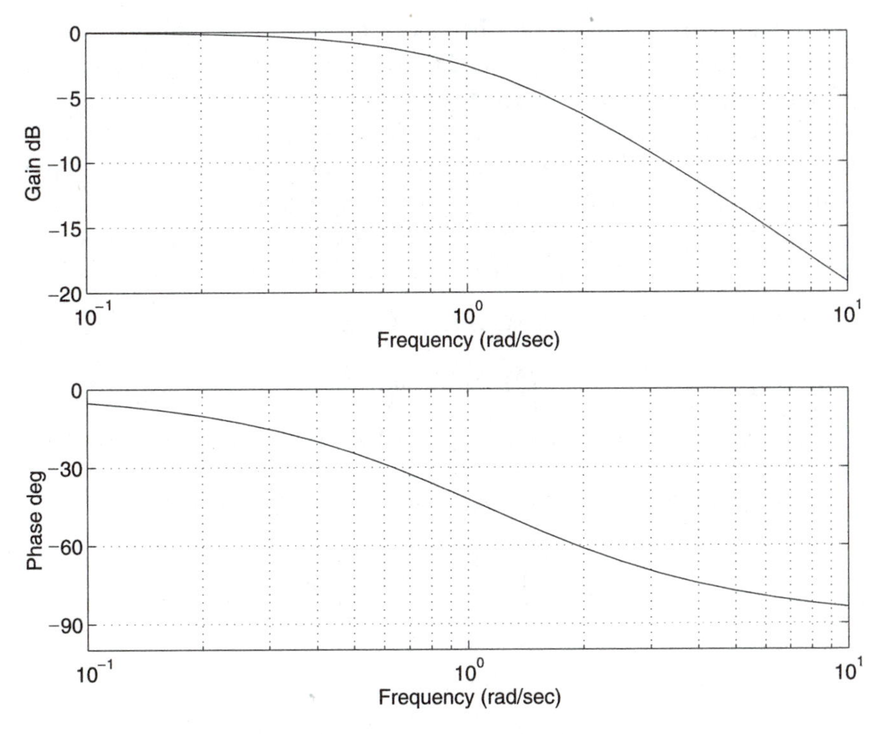

FIGURE 2.8 Bode plots of $1.1/(s+1.1)$.

Suppose we have two transfer functions, with

$$G_1(\omega) = M_1 e^{j\phi_1}, \qquad G_2(\omega) = M_2 e^{j\phi_2}$$

at some frequency ω. Then,

$$G_1(\omega)G_2(\omega) = M_1 M_2 e^{j(\phi_1+\phi_2)}.$$

Since

$$\log|G_1(\omega)G_2(\omega)| = \log|G_1| + \log|G_2|,$$
$$\text{mag}(G_1G_2)(\omega) = M_1 + M_2$$

and

$$\text{phase}(G_1G_2)(\omega) = \text{phase}G_1(\omega) + \text{phase}G_2(\omega).$$

The phase and the gain of the composite function can be found by adding the individual phases and gains (in decibels). The Bode plots of the product of two transfer functions can be obtained as the sum of the individual plots. Thus, the Bode plots of complicated transfer functions can be constructed from the plots of simple transfer functions.

Computer software can be used (for instance "bode" in MATLAB) to sketch the plots of transfer functions. We now show how to identify simple transfer functions from their Bode plots. We first describe the Bode plots of several simple transfer functions, and then illustrate how more complicated transfer functions can be identified from their Bode plots.

1. Bode diagram of a constant K (Fig. 2.9):

$$(K)_{dB} = 20 \log_{10} K.$$

The corresponding phase angle is a constant 0° (if $K > 1$) or 180° (if $K < 1$).

2. Bode diagram of integration $\frac{1}{s}$ (Fig. 2.10): The magnitude is

$$\left(\frac{1}{j\omega}\right)_{dB} = -20 \log_{10} \|\omega\|.$$

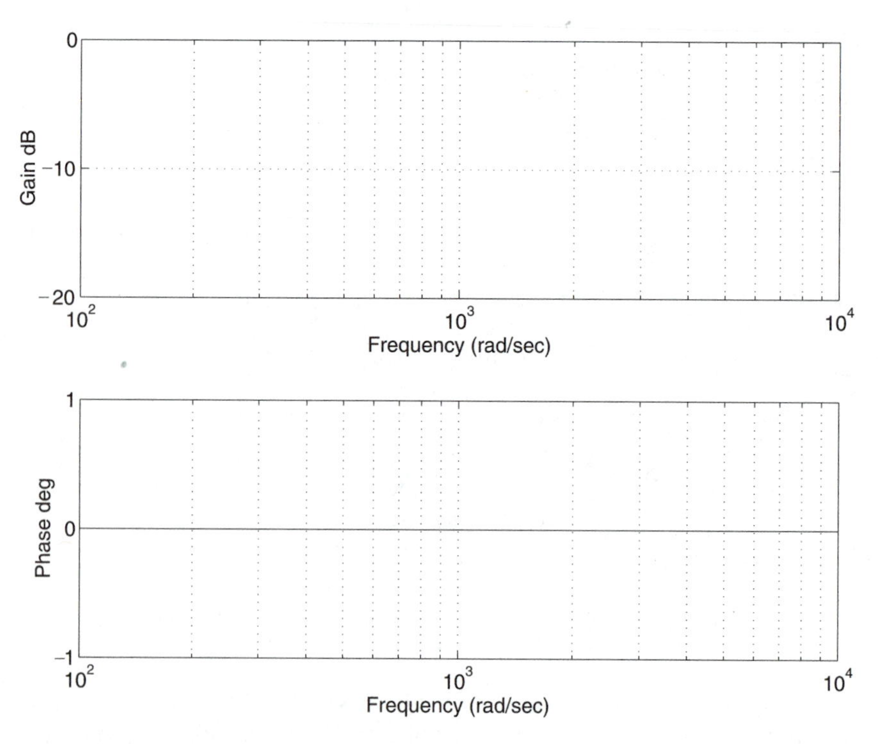

FIGURE 2.9 Bode diagram of a constant: $G(s) = \frac{1}{10}$.

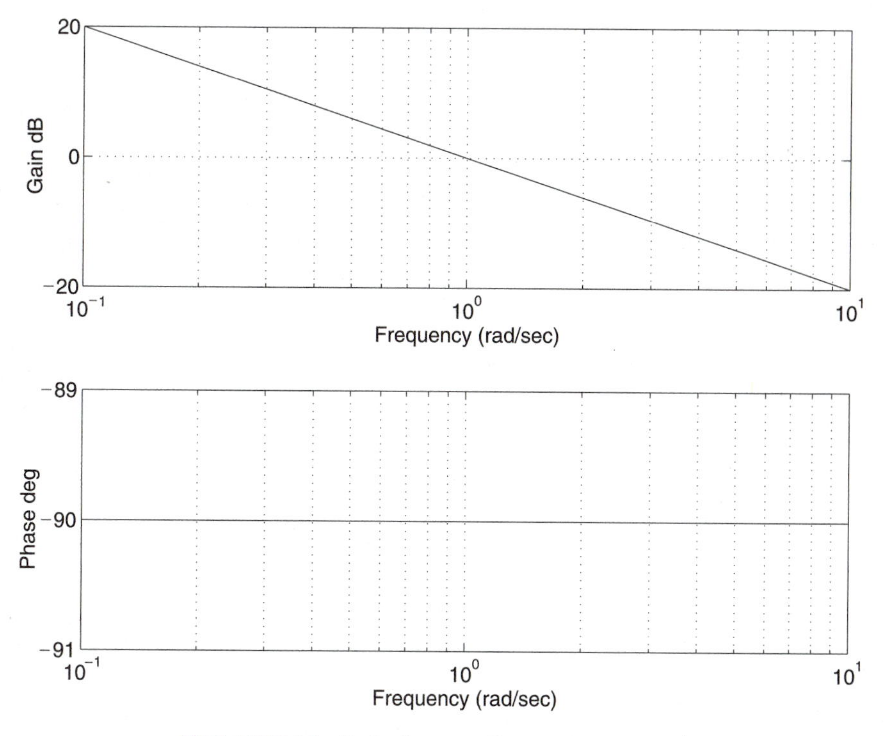

FIGURE 2.10 Bode diagram of integration $G(s) = \frac{1}{s}$.

Note that the amplitude curve is a linear function of frequency ω. Since $\frac{1}{j\omega}$ is always purely imaginary, the phase angle is a constant $-90°$.

3. Bode diagram of differentiation s (Fig. 2.11): The Bode diagram of s is similar to that of $\frac{1}{s}$, with the exception that the magnitude has a positive slope and the phase is 90°.

4. Bode diagram of a simple phase lag $\frac{1}{1+s/p}$ (Fig. 2.12): For small ω, the magnitude is nearly flat, while for ω much larger than the pole it has a slope of -20 dB/decade. The phase runs from 0 to $-90°$, passing through $-45°$ at the pole.

In general, the gain is $-20\log_{10}(1+\frac{\omega^2}{p^2})^{1/2}$ (dB) with a phase $-\tan^{-1}(\frac{\omega}{p})$. If $\frac{\omega}{p} \ll 1$, then the phase is close to zero and the phase is 0° with gain approximately 0 dB. Similarly, if $\frac{\omega}{p} \gg 1$, then the phase is close to $-90°$ and the gain is approximately $-20\log_{10}(\frac{\omega}{p})$ (dB). It can be shown that the difference between the two straight line asymptotes is at a maximum at $s = p$ and equals -3 dB.

The frequency at which the magnitude starts to drop rapidly is called the system *bandwidth*. This terminology is used also for higher-order systems. It relates to the dominant pole of the system.

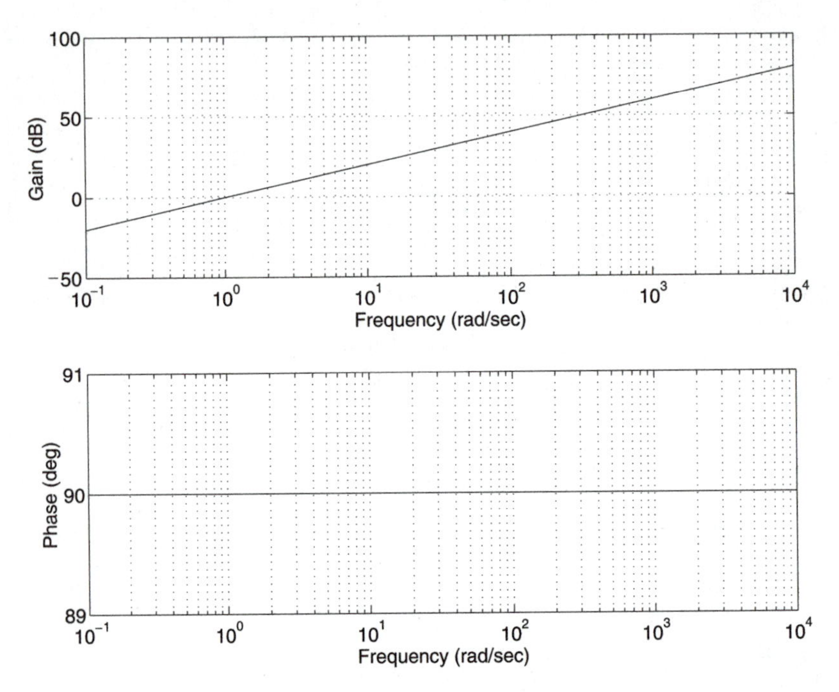

FIGURE 2.11 Bode diagram of differentiation $G(s) = s$.

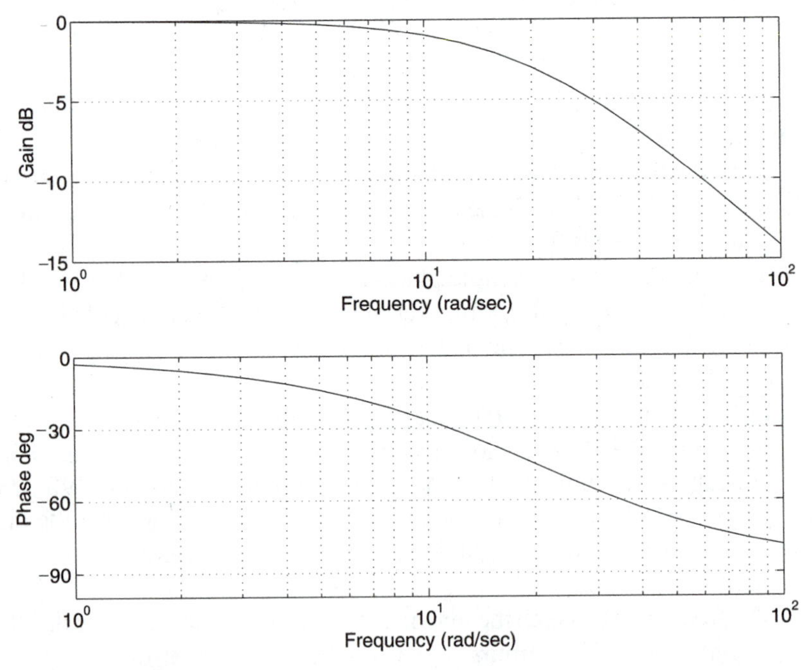

FIGURE 2.12 Bode diagram of simple phase lag: $G(s) = \frac{1}{1+s/20}$.

5. Bode diagram of a simple phase lead $1 + s/z$: The Bode diagram of a simple phase lead is similar to that of a simple phase lag. The difference is that the amplitude has a positive slope and the phase angle goes from 0° to 90° if $z > 0$ or 0° to −90° if $z < 0$. Such networks are illustrated in Figs. 2.13 and 2.14.

6. Bode diagram of a quadratic transfer function $w_n^2/(s^2 + 2\xi w_n s + w_n^2)$: This is the transfer function of a second-order system with natural frequency w_n and damping ξ. A time-domain realization of a system with this transfer function is

$$\ddot{y}(t) + 2\xi w_n \dot{y}(t) + \omega_n^2 y(t) = \omega_n^2 u(t).$$

(As in the examples above, a constant is chosen so that the system has magnitude 1 at $\omega = 0$.) The Bode diagram of a quadratic transfer function is shown in Fig. 2.16. Here $\omega_n = 5$ and $\xi = .05$. The Bode diagrams of several second-order transfer functions with the same natural frequency and different damping are shown in Fig. 2.15. (Compare the frequency responses in Fig. 2.15 with the step responses of the same systems in Fig. 2.6.)

Now the gain (in decibels) is

$$-20\log_{10}\left[\left(\frac{2\xi\omega}{\omega_n}\right)^2 + \left(1 - \frac{\omega^2}{\omega_n^2}\right)^2\right]^{1/2},$$

with phase angle

$$\tan^{-1}\frac{2\xi\omega_n\omega}{\omega_n^2 - \omega^2} = \tan^{-1}\frac{2\xi\frac{\omega}{\omega_n}}{1 - \left(\frac{\omega}{\omega_n}\right)^2}.$$

If $\frac{\omega}{\omega_n} \ll 1$, the phase is close to zero. The gain is approximately

$$20\log_{10}\sqrt{1} = 0\,\text{dB}.$$

For $\frac{\omega}{\omega_n} \gg 1$, the phase is close to 180° and the gain is approximately

$$-40\log_{10}\frac{\omega}{\omega_n}\,\text{dB}.$$

The slope of the straight line approximation for large ω is twice that of a simple phase lag, due to the presence of two poles. At $\omega = \omega_n$, the phase is 90° and the magnitude shows a peak. The magnitude of the peak increases with decreasing damping. The precise value of damping is usually found by trial and error.

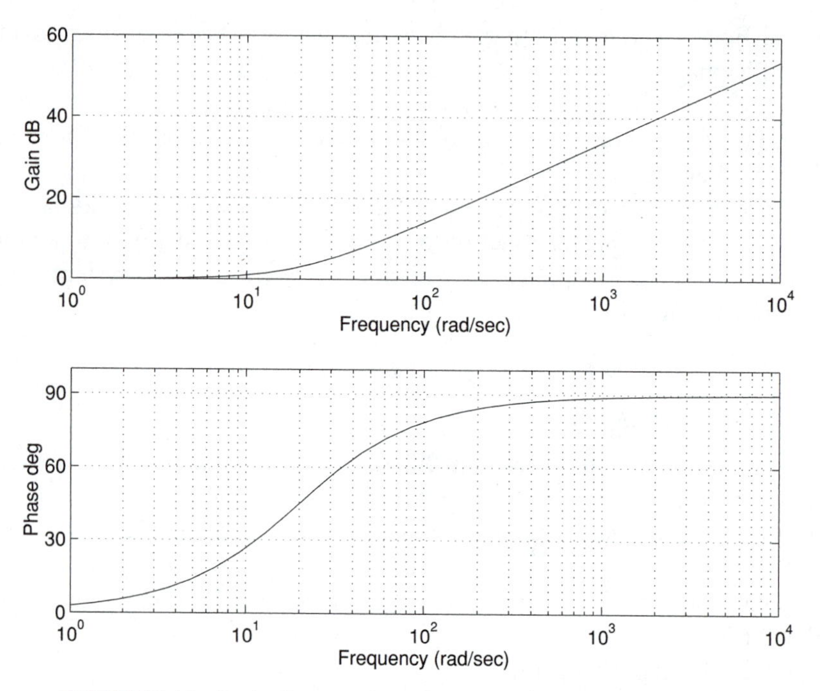

FIGURE 2.13 Bode diagram of simple phase lead (1): $G(s) = 1 + s/20$.

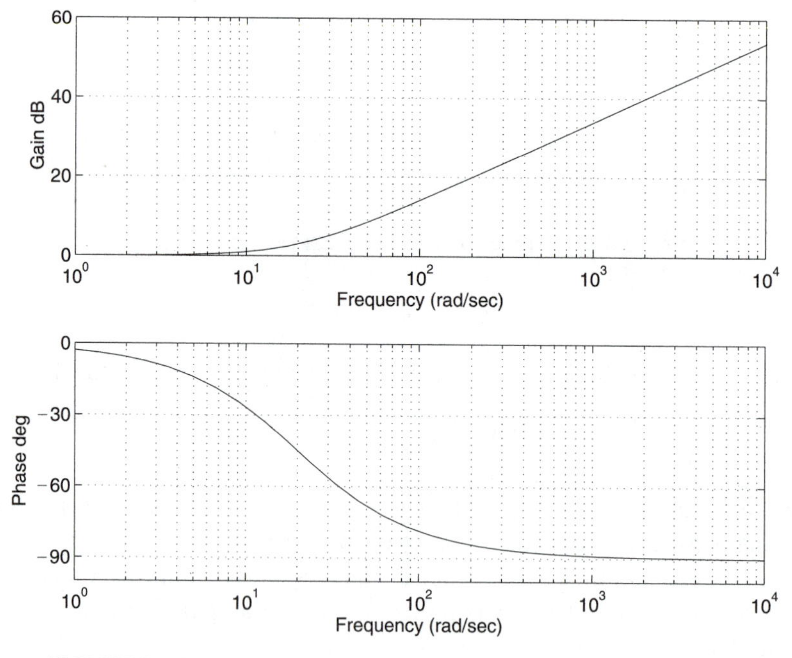

FIGURE 2.14 Bode diagram of simple phase lead (2): $G(s) = 1 - s/20$.

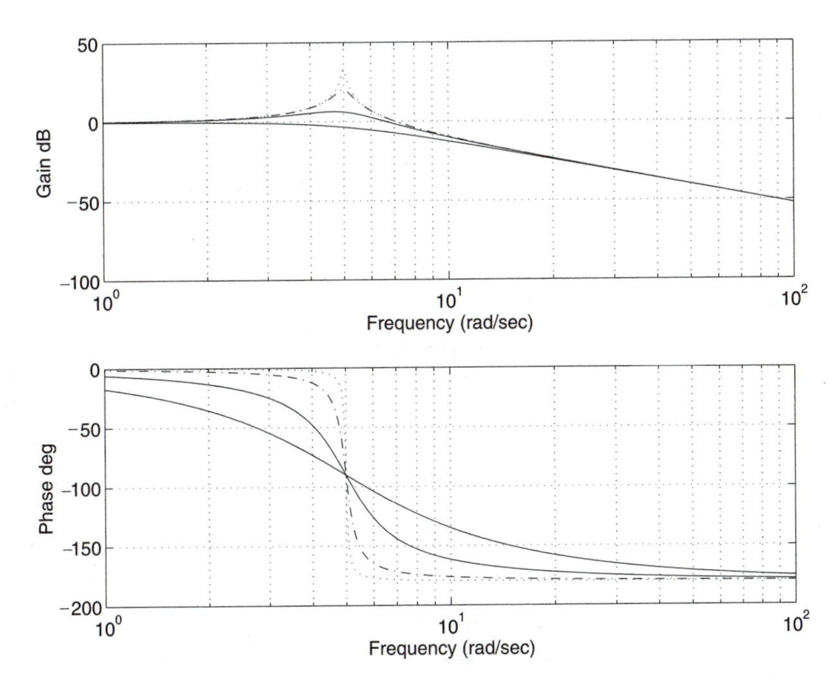

FIGURE 2.15 $G(s) = 25/(s^2 + 10\xi s + 25)$, $\xi = 0.01, 0.05, 0.25, 0.75$.

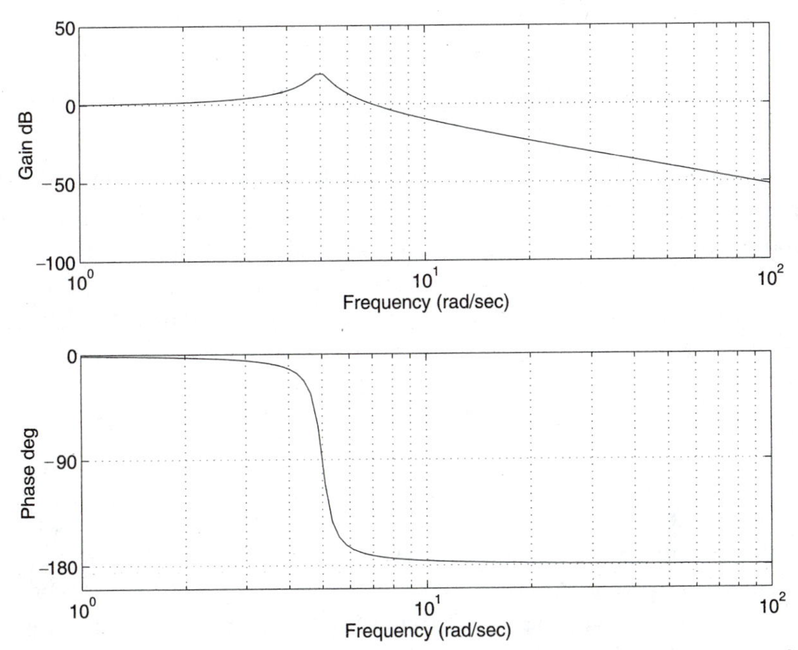

FIGURE 2.16 Bode diagram of a quadratic transfer function: $G(s) = 25/(s^2 + 0.5s + 25)$.

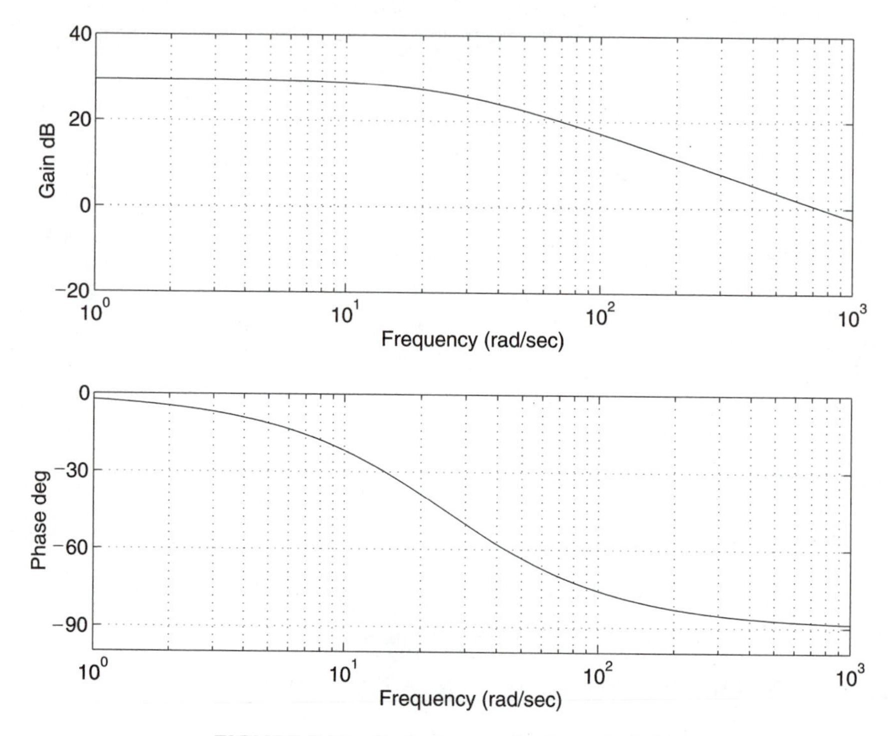

FIGURE 2.17 Bode diagram for Example 2.43.

We now use several examples to illustrate how a function can be identified from its Bode plot.

EXAMPLE 2.43 (*Fig. 2.17*): At small values of ω, the phase is 0°. At about $\omega = 25$ the slope changes and the phase is -45°. For large ω, the slope of the magnitude curve is -20 dB/decade and the phase is -90°.

This suggests that the system has a single pole at $s \approx -25$ and no zeros. For small frequencies, the magnitude is about 30 dB, and so (since $10^{30/20} \approx 30$)

$$G(s) = \frac{30}{1 + s/25}.$$

EXAMPLE 2.44 (*Fig. 2.18*): For small frequencies, the phase angle is -90° and the magnitude plot has a slope of -20 dB/decade. This implies that there is a pole at $s = 0$.

The slope increases at $\omega \approx 25$ and at this point the phase has increased by 45° to 135°. There is another pole at $s \approx -25$.

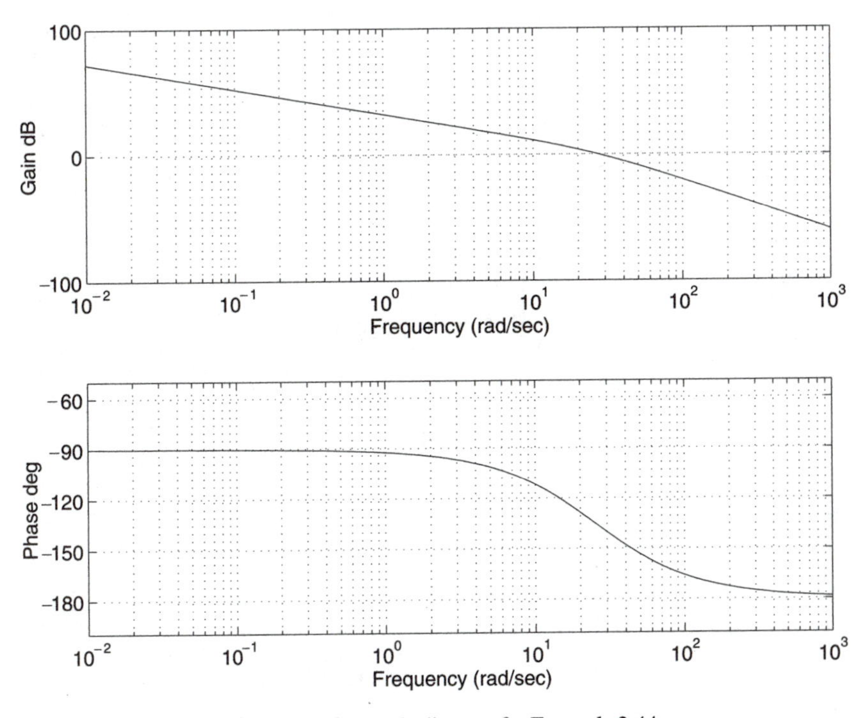

FIGURE 2.18 Bode diagram for Example 2.44.

For large ω, the phase is 135° and the slope of the magnitude curve is −40 dB/decade, so there are no more poles. The transfer function is

$$G(s) = \frac{k}{s(1 + s/25)}.$$

By matching the magnitude curve we obtain $k = 40$.

Example 2.45 (*Fig. 2.19*): The phase is 0° at $\omega = 0$ and increases to 180° at infinity. It passes through −90° at $\omega = 8$ with a steep slope. The magnitude plot is flat for small frequencies and has a slope of −40 dB/decade after $\omega = 8$. At $\omega = 8$ there is a sharp peak in the graph. This is the transfer function of a lightly damped second-order system with natural frequency 8:

$$G(s) = \frac{k}{s^2 + 2\xi 8s + 8^2}.$$

Since $G(jw) \approx 0.5$ for small ω, $k = 32$. Trial and error shows that the damping $\xi = 0.05$.

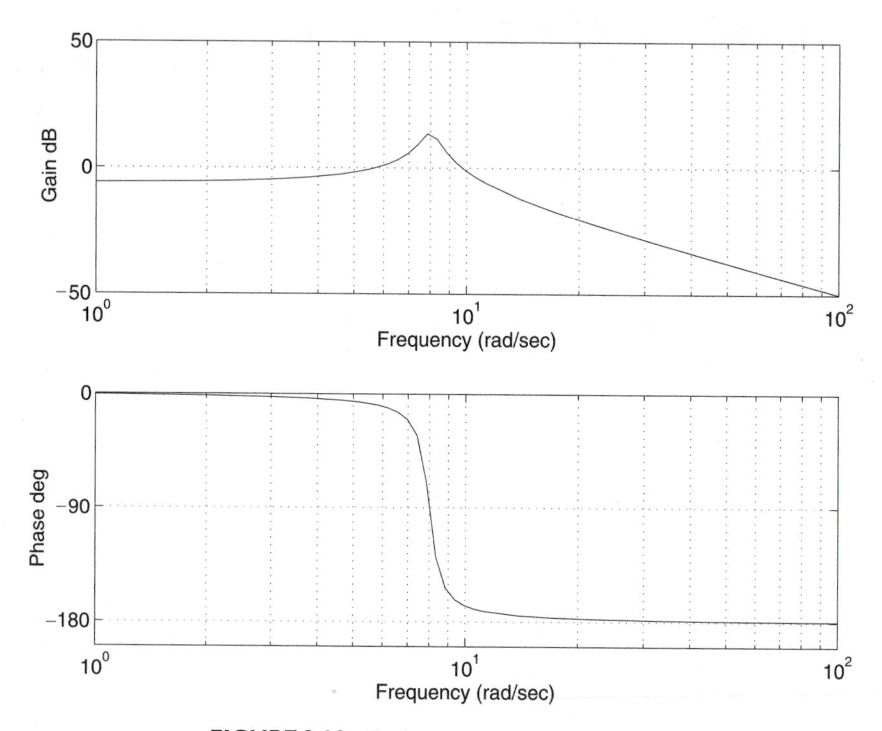

FIGURE 2.19 Bode diagram for Example 2.45.

EXAMPLE **2.46** (*Fig. 2.20*): The phase is 0° at low frequencies, then increases, passing through 45° at about $\omega = 1$. The slope of the magnitude plot also changes from being flat to a slope of 20 dB/decade at this point. This suggests a zero at $s = 1$ or $s = -1$. The phase angle increases here by 90°, so there is a zero at $s = -1$.

The slope of the magnitude plot flattens at $\omega = 25$ and the phase has decreased to 45° again. This suggests a pole at $s = -25$.

The phase angle is 0° and the magnitude plot flat at infinity, indicating that there are equal numbers of poles and zeros. Thus,

$$G(s) = 5\frac{(1+s)}{(1+s/25)},$$

where the value of $G(0)$ is found from the value of $|G(j\omega)|$ for small ω.

Compare the Bode plot of this function to that of

$$5\frac{(1-s)}{1+s/25},$$

shown in Fig. 2.21.

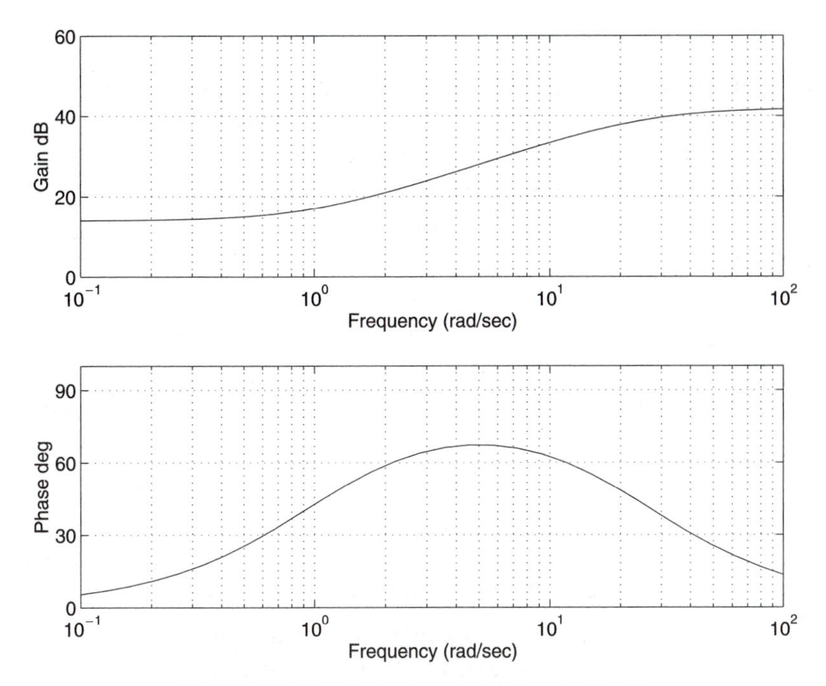

FIGURE 2.20 Bode diagram for Example 2.46.

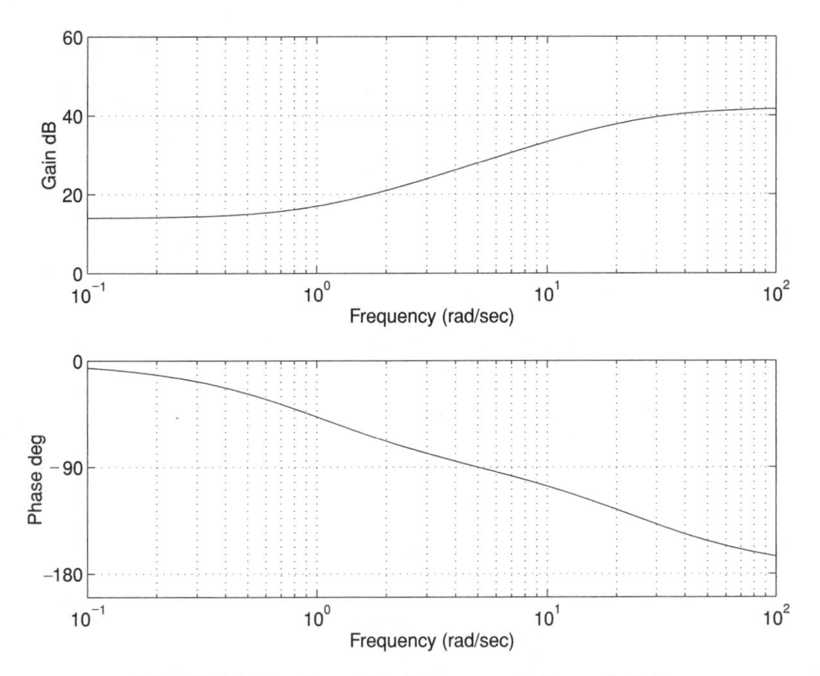

FIGURE 2.21 Second Bode diagram for Example 2.46.

NOTES AND REFERENCES

The state-space material, including realization theory, is covered in more detail in, for example, [8, 23]. Bode plots are discussed in detail in many undergraduate engineering control textbooks, such as [40, Chap. 6]. Generalizations of the concept of zeros to multi-input/multi-output systems can be found in, for instance, [29]. Example 2.14 and Exercise 9 are from [23].

EXERCISES

1. Let G_i, $i = 1..n$ be a set of transfer functions, each with relative degree r_i and define $G = \prod_{i=1}^{n} G_i$. Prove that the relative degree of G is $\sum_{i=1}^{n} r_i$.
2. Calculate the transfer function for the spring–mass system in Example 2.2 using the second realization (2.4). Verify that it is the same as that obtained using the first realization.
3. Write the following systems in state-space form (i.e., determine the matrices A, B, C, and E). Calculate the transfer function.

   (a)

$$
\begin{aligned}
\dot{x}_1(t) &= -2x_1(t) \\
\dot{x}_2(t) &= -11x_2(t) + u(t) \\
y(t) &= x_1(t) + x_2(t)
\end{aligned}
$$

   (b)

$$
\begin{aligned}
\dot{x}_1(t) &= -5x_1(t) + 1.5x_2(t) \\
\dot{x}_2(t) &= -15x_1(t) + 2.5x_2(t) + u(t) \\
y(t) &= 2x_1(t) - 3x_2(t)
\end{aligned}
$$

   (c)

$$
\begin{aligned}
\ddot{x}(t) &= 9x(t) + 2u(t) \\
y(t) &= x(t)
\end{aligned}
$$

   (d)

$$
\begin{aligned}
\ddot{x}(t) &= -9x(t) + 2u(t) \\
y(t) &= \ddot{x}(t)
\end{aligned}
$$

(e)

$$\ddot{x}(t) = 9x(t) + 2u(t)$$
$$y(t) = \ddot{x}(t)$$

(f)

$$\frac{d^4x}{dt^4} + 2\frac{d^3x}{dt^3} - \frac{dx}{dt} + 4x(t) = u(t)$$
$$y(t) = \dot{x}(t)$$

(g)

$$\dot{x}_1(t) = -x_1(t)$$
$$\dot{x}_2(t) = x_3(t)$$
$$\dot{x}_3(t) = 0.1x_1(t) - 0.4x_2(t) - 4x_3(t) + u(t)$$
$$y(t) = x_3(t)$$

(h)

$$\ddot{x}(t) + \dot{x}(t) + 5x(t) = 2v(t)$$
$$y(t) = x(t)$$

4. In the previous question, which systems are observable? Which are controllable? Which are minimal?

5. Define for any nonzero p,

$$A = \begin{bmatrix} p & 1 & 0 \\ 0 & p & 1 \\ 0 & 0 & p \end{bmatrix}.$$

Show that (A, B) is uncontrollable for any $B = [b_1 \quad b_2 \quad 0]^*$. Find a condition on the observation matrix C so that (A, C) is observable.

6. Let (A, B, C, D) be an uncontrollable and unobservable realization of a system. Use Lemmas 2.6 and 2.12 to obtain an algorithm for a realization in which the states are decomposed by whether they are controllable and/or observable. Then obtain a minimal realization.

7. Let $G(s)$ be the transfer function of a strictly proper system. Assume that all the poles of G have negative real parts. Consider the system with zero initial condition and a periodic input $e^{j\omega t}$.

Prove that for any $\epsilon > 0$, there is $T > 0$ so that for all $t > T$ the output $y(t)$ satisfies $|G(j\omega)e^{j\omega t} - y(t)| < \epsilon$. In other words, the steady-state response to $e^{j\omega t}$ is $G(j\omega)e^{j\omega t}$. Then show that the steady-state response to $\cos(\omega t)$ is $\mathrm{Re}(G(j\omega)e^{j\omega t})$ and that the steady-state response to $\sin(\omega t)$ is $\mathrm{Im}(G(j\omega)e^{j\omega t})$.

8. Find a realization of each of the following transfer functions.

(a) $\frac{1}{s+3} + \frac{2}{s+20}$

(b) $\frac{s}{s+1}$

(c) $\frac{32}{s^3+1.6s^2+16s}$

(d) $\frac{s^2+10s}{s^3+10s^2+2s+1}$

(e) $\frac{2+8s}{s^4+2s^2+2s+3}$

(f) $\frac{5s^2}{s^3+3s+2}$

(g)* $e^{-s}\frac{1}{s-10}$

9. The dynamics of a hot air balloon are modeled by

$$\dot{T}(t) = -k_1T(t) + k_2u(t),$$
$$\dot{v}(t) = -k_3(v(t) - w(t)) + \sigma T(t),$$
$$\dot{h}(t) = v(t),$$

where T is the temperature of the balloon (relative to equilibrium temperature), u is heat added, v is velocity, h is height, and w is wind speed. The parameters k_1, k_2, k_3, and σ are all positive constants.

(a) Choose state variables T,v, and h. Show that the state-space equations are

$$\frac{d}{dt}\begin{bmatrix} T(t) \\ v(t) \\ h(t) \end{bmatrix} = \begin{bmatrix} -k_1 & 0 & 0 \\ \sigma & -k_3 & 0 \\ 0 & 1 & 0 \end{bmatrix}\begin{bmatrix} T(t) \\ v(t) \\ h(t) \end{bmatrix} + \begin{bmatrix} k_2 \\ 0 \\ 0 \end{bmatrix}u(t) + \begin{bmatrix} 0 \\ k_3 \\ 0 \end{bmatrix}w(t).$$

(b) Can the temperature T and velocity v be observed by measurements of altitude h?

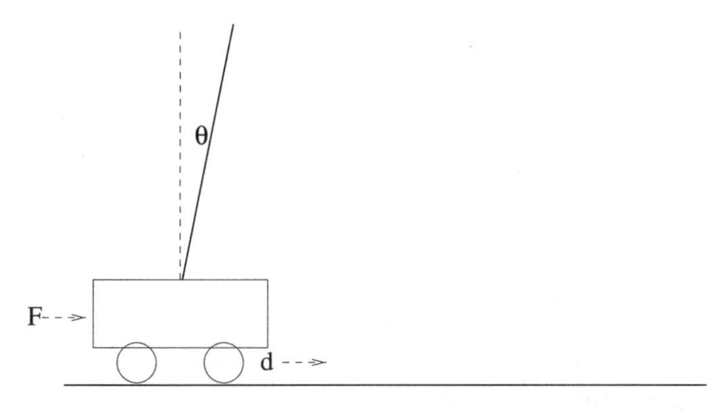

FIGURE 2.22 Inverted pendulum and cart.

(c) Can the variables v and h be observed by measure of T?

(d) Calculate the transfer function from u to h and from w to h. Is the system controllable by u? Controllable by w?

10. Consider an inverted pendulum balanced on a moving cart (Fig. 2.22). It is modeled by the two second-order differential equations:

$$(m_p + m_c)\ddot{d} + m_p\ddot{\theta} I_p \cos(\theta) - m_p\dot{\theta}^2 I_p \sin(\theta) = F(t)$$
$$m_p I_p \cos(\theta)\ddot{d} + m_p\ddot{\theta} I_p^2 - m_p g I_p \sin(\theta) = 0.$$

In the above equation, $F(t)$ is the input force to cart (N), $d(t)$ is the cart position (m), and $\theta(t)$ is the angle of the pendulum (rad). The angle is measured from the vertical up position, with clockwise angles positive. A force $u(t)$ may be applied to the cart to balance the pendulum about the vertical. The mass of the rod (m_p kg), mass of the cart (m_c kg), and the center of gravity of the rod (I_p m) are all positive constants.

(a) Show that after linearization of the angle θ about the vertical, the equations of motion may be written

$$x(t) = \begin{bmatrix} 0 & 0 & 1 & 0 \\ 0 & 0 & 0 & 1 \\ \frac{m_p g}{m_c I_p} & -\frac{m_p g}{m_c I_p} & 0 & 0 \\ -\frac{g}{I_p} & \frac{g}{I_p} & 0 & 0 \end{bmatrix} x(t) + \begin{bmatrix} 0 \\ 0 \\ \frac{1}{m_c} \\ 0 \end{bmatrix} F(t)$$

where

$$x(t) = \begin{bmatrix} d(t) \\ d(t) + I_p\theta(t) \\ \dot{d}(t) \\ \dot{d}(t) + I_p\dot{\theta}(t) \end{bmatrix}.$$

(b) Is the system controllable?

(c) Show that if only $\theta(t)$ is measured the system is unobservable. Explain on physical grounds.

(d) Show that with a different measurement, $d(t) + I_P\theta(t)$, the system is observable.

(e) Calculate the transfer functions in (c) and in (d). What are the poles? What are the zeros?

11. Prove Lemma 2.31.

12. Prove Theorem 2.33.

13. For the tank (Example 2.1) calculate the settling time in terms of the parameters. What happens to the settling time if the tank mass M is increased? If the flow rate m is increased?

14. Identify the transfer functions whose Bode plots are shown in Figs. 2.23–2.28.

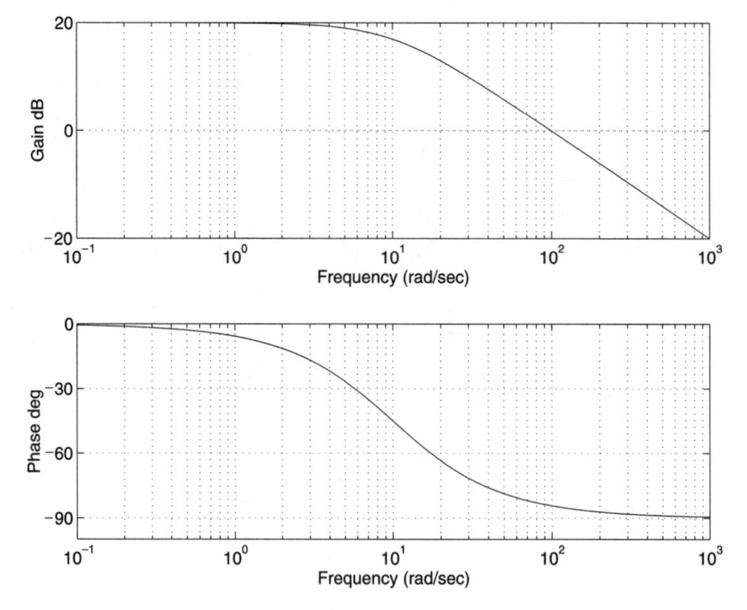

FIGURE 2.23 Bode plots for Exercise 14 (a).

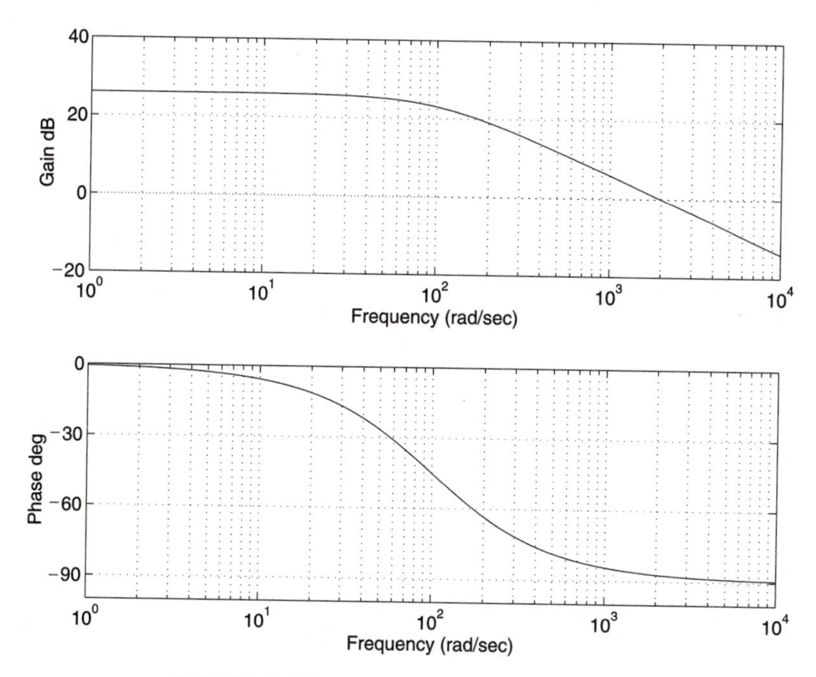

FIGURE 2.24 Bode plots for Exercise 14 (b).

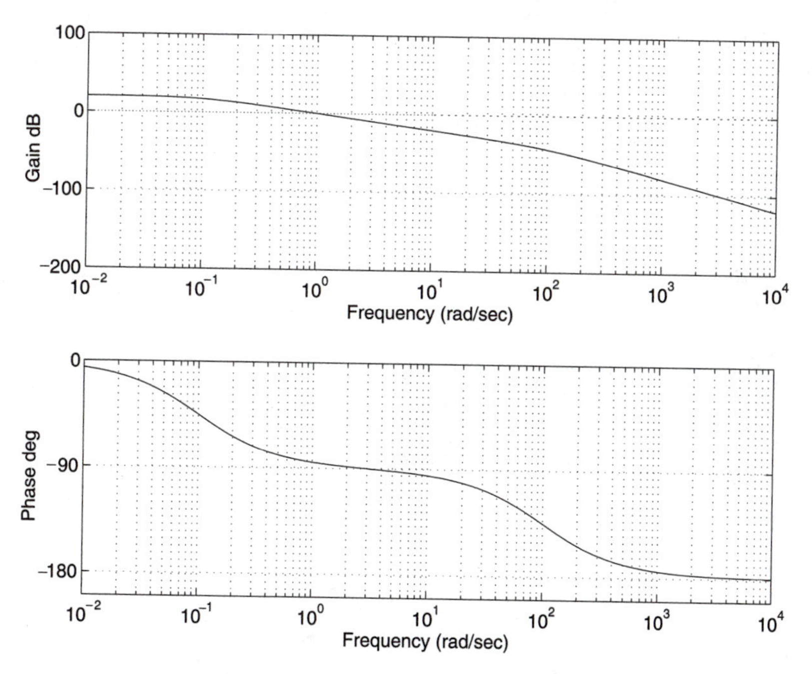

FIGURE 2.25 Bode plots for Exercise 14 (c).

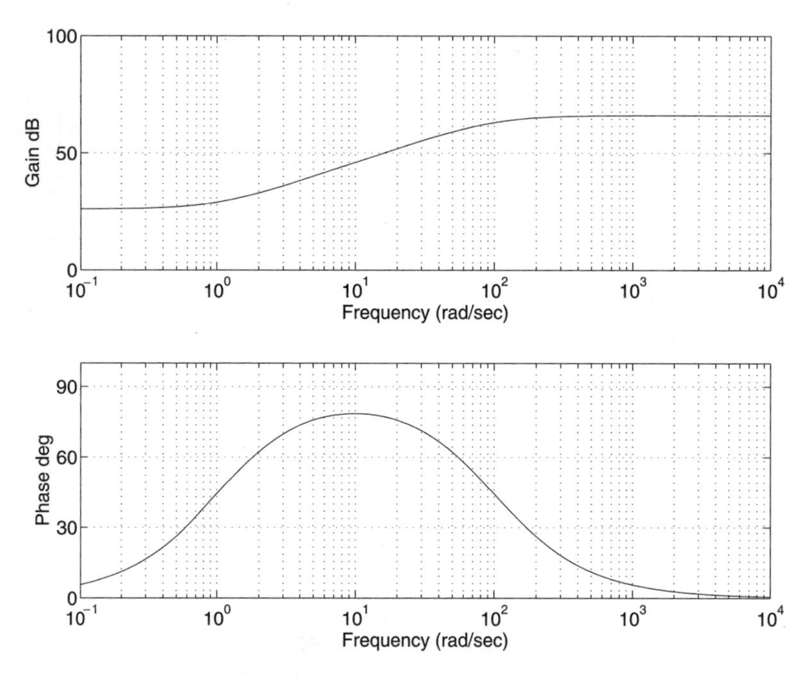

FIGURE 2.26 Bode plots for Exercise 14 (d).

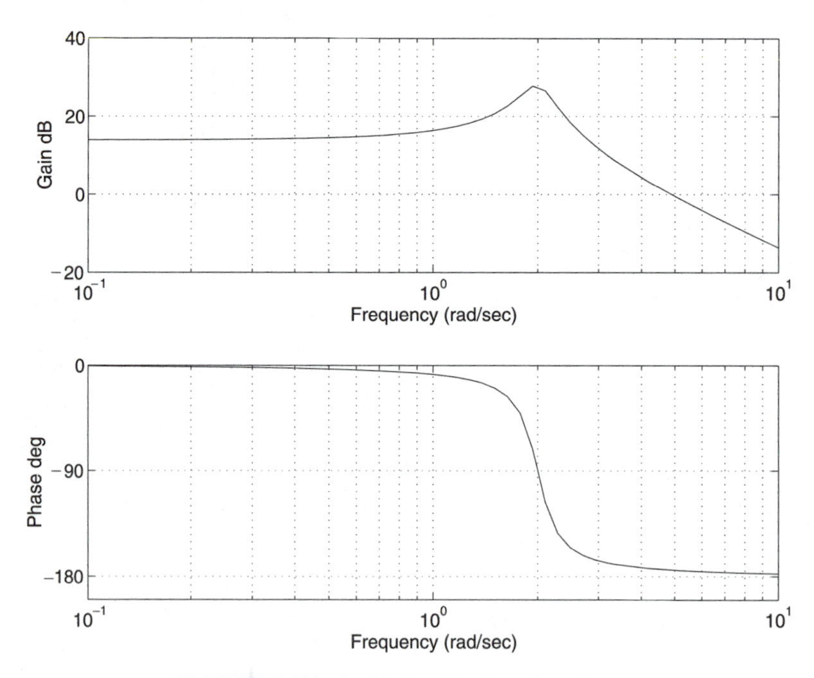

FIGURE 2.27 Bode plots for Exercise 14 (e).

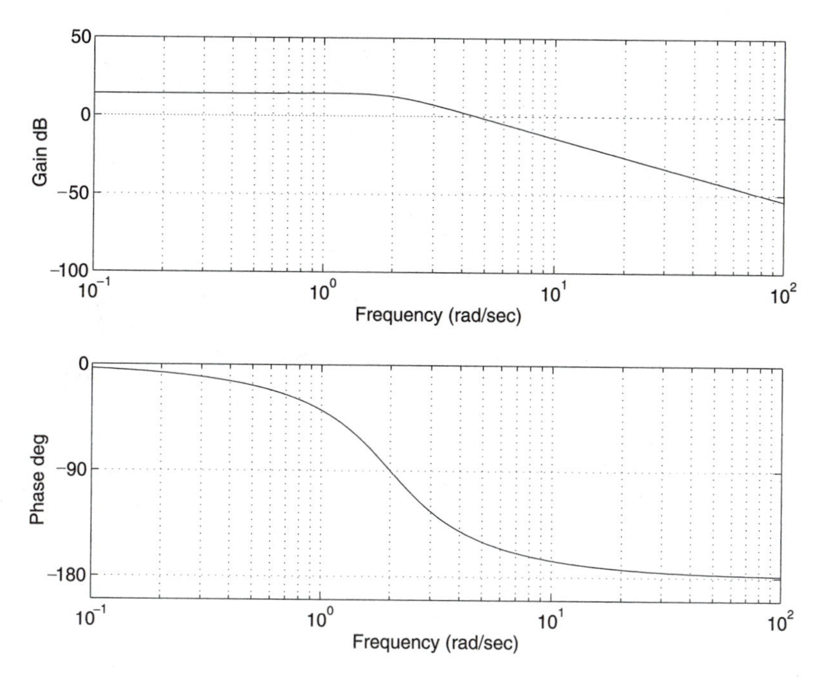

FIGURE 2.28 Bode plots for Exercise 14 (f).

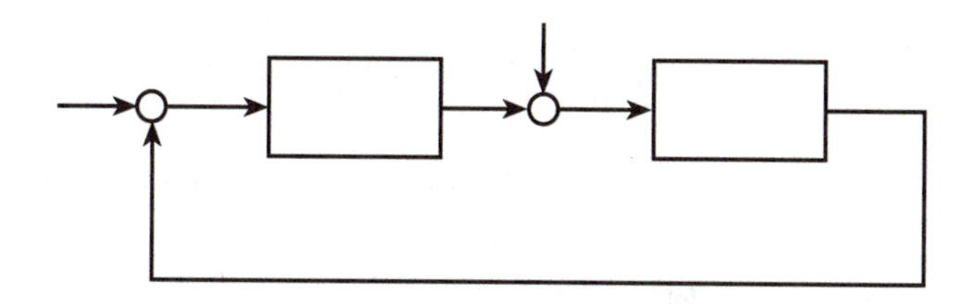

III

STABILITY

One way to quantify the performance of a system is to define the size of signals of interest. A real-valued signal u is, mathematically, a function mapping $(0, \infty)$ to R. The size of a signal is thus the function measured in an appropriate norm. For example, the performance of a tracking system is determined by the size of the error signal. Different norms lead to different performance criterion and may lead to different definitions of stability. We will be concerned with the important L_2 norm defined in Section 3.1.

Recall that for a system with a realization (A, B, C, E) and zero initial condition, the output $y(t)$ is related to the input $u(t)$ by

$$y(t) = \int_0^t C \exp(A(t-\tau))Bu(\tau)d\tau + Eu(t). \tag{3.1}$$

We may write this concisely as $y(t) = (\mathcal{G}u)(t)$ where $\mathcal{G}$ is defined by (3.1).

3.1 L_2 STABILITY

Assume that u is square integrable,

$$\int_0^\infty \|u(t)\|^2 dt < \infty,$$

where $\|z\|^2 = z^*z$ is the usual Euclidean vector norm, and define

$$\|u\|_2 = \left[\int_0^\infty \|u(t)\|^2 dt\right]^{1/2}.$$

The L_2 norm often has a physical interpretation as the energy of the signal. The normed linear space of all R^m-valued square integrable functions with this norm is denoted $L_2(0, \infty; R^m)$. When $m = 1$ we usually abbreviate this to $L_2(0, \infty)$.

Definition 3.1: A system is L_2 *stable* or *externally stable* if, for every $u \in L_2(0, \infty; R^m)$ the output y is in $L_2(0, \infty; R^p)$. In this case $\mathcal{G}$ maps inputs in $L_2(0, \infty; R^m)$ to outputs in $L_2(0, \infty; R^p)$.

We now consider the question of how the norm of the output $\mathcal{G}u$ is related to the norm of the input u.

Definition 3.2: The L_2 *gain* of an externally stable system is

$$\|\mathcal{G}\|_{2,2} = \sup_{\substack{u \in L_2, \\ u \neq \theta}} \frac{\|y\|_2}{\|u\|_2}. \tag{3.2}$$

(Here θ indicates the zero function.) The L_2 gain $\|\mathcal{G}\|_{2,2}$ is the maximum ratio of $\|y\|_2/\|u\|_2$.

We now develop a characterization of the L_2 gain in terms of the system transfer function.

For integrable $u \in L_2(0, \infty, R^m)$ define its Fourier transform

$$\hat{u}(j\omega) = \int_0^\infty u(t)e^{-j\omega t} dt.$$

(This is the Laplace transform on the imaginary axis, or the system frequency response.) Define $\bar{L}_2^m$ to be the space of all C^m-valued functions with finite norm

$$\|\hat{u}\|_2 = \left[\frac{1}{2\pi}\int_{-\infty}^\infty \|\hat{u}(j\omega)\|^2 d\omega\right]^{1/2}.$$

If the dimension of the signal is 1 we often abbreviate this as $\bar{L}_2$. The same norm symbol is used for $\bar{L}_2^m$ and $L_2(0, \infty; R^m)$. Context determines which is intended. In fact, by Parseval's Theorem,

$$\|\hat{u}\|_2 = \|u\|_2$$

and so the L_2 norm of $u(t)$ and the $\bar{L}_2$ norm of its transform are identical. Since $\hat{y}(j\omega) = G(j\omega)\hat{u}(j\omega)$, for a SISO system

$$\|\hat{y}\|_2 \leq \sup_{\omega} |G(j\omega)| \|\hat{u}\|_2$$

and it is possible to show that

$$\sup_{\omega} |G(j\omega)|$$

is the least upper bound.

EXAMPLE 3.3: $G(s) = 1/(s+1)$. Here

$$\begin{aligned}
\sup_{\omega} |G(j\omega)| &= \sup_{\omega} \frac{1}{|j\omega + 1|} \\
&= \sup_{\omega} \frac{1}{\sqrt{\omega^2 + 1}} \\
&= 1.
\end{aligned}$$

This development ignores some complexities. Consider the following slightly different example.

EXAMPLE 3.4: Consider $G(s) = 1/(s-1)$. In this case the frequency response is

$$G(j\omega) = \frac{1}{j\omega - 1}$$

and as before

$$\sup_{\omega} |G(j\omega)| = 1.$$

However, G is the transfer function of the system defined by

$$y(t) = \int_0^t e^{(t-\tau)} u(\tau) d\tau.$$

Any input $u(t) = e^{\alpha t}$ where $\alpha < 0$, $\alpha \neq 1$, is in $L_2(0, \infty)$. However, the output

$$y(t) = \frac{1}{\alpha - 1}(e^{\alpha t} - e^{t}).$$

This function is not in $L_2(0, \infty)$ and so the system is not L_2-stable.

The difference between these two examples is that one system has a transfer function that is analytic in the closed right half-plane $\text{Re}(s) \geq 0$, whereas the other transfer function (for the unstable system) is analytic in the left half-plane $\text{Re}(s) < 0$.

Define the normed linear space $\mathcal{H}_\infty$ consisting of functions G, analytic in the right half-plane $\text{Re}(s) > 0$ and with norm

$$\|G\|_\infty = \sup_{\text{Re}(s)>0} |G(s)| = \sup_\omega |G(j\omega)| < \infty.$$

As mentioned in Section 2.2, the transfer function of any system modeled by a set of linear time-invariant ordinary differential equations is rational with real coefficients. Denote the subset of $\mathcal{H}_\infty$ consisting of rational functions with real coefficients by $R\mathcal{H}_\infty$. Notice that $\mathcal{G}$ is the operator mapping inputs to outputs while multiplication by G is the mapping

$$\hat{y}(s) = G(s)\hat{u}(s)$$

between Laplace transforms of inputs and outputs.

For a multivariable system, $G(j\omega)$ is a matrix. We will define a matrix norm that generalizes magnitude of a complex scalar. Let $\lambda_i(A)$ indicate the ith eigenvalue of a matrix A. The singular values of a matrix A, $\sigma_i(A)$, are defined to be

$$\sigma_i = [\lambda_i(A^*A)]^{1/2},$$

where A^* indicates complex conjugate transpose. We define the norm of a matrix A to be

$$\|A\| = \max_i \sigma_i(A).$$

(See Appendix B for details.) Notice that if A is a scalar, this definition of norm reduces to the magnitude of A. The only change in generalizing the above discussion from SISO to multi-input/multi-output (MIMO) systems is that the magnitude of $G(j\omega)$ is replaced by this matrix norm. Define $M(R\mathcal{H}_\infty)$ to be the space of matrices with entries in $R\mathcal{H}_\infty$ and define

$$\begin{aligned}\|G\|_\infty &= \sup_\omega \|G(j\omega)\| \\ &= \sup_\omega \max_i \sigma_i(G(j\omega)).\end{aligned}$$

Lemma 3.5: Let A be any $n \times n$ matrix and define

$$\alpha = \max_{1\leq i\leq n} \text{Re}(\lambda_i(A)).$$

For any $\epsilon > 0$, there is $M \geq 1$ such that the matrix norm

$$\|\exp(At)\| \leq Me^{(\alpha+\epsilon)t}.$$

□ *Proof:* The proof relies on the Jordan form of A and is omitted here. The proof for the case where A can be diagonalized is straightforward. (In this case the result holds with $\epsilon = 0$.) ■

THEOREM 3.6: A system (2.1) is L_2-stable if and only if $G \in M(R\mathcal{H}_\infty)$. If so, $\|\mathcal{G}\|_{2,2} = \|G\|_\infty$.

□ *Proof:* We will only give the proof for a SISO system. Suppose that $G \in R\mathcal{H}_\infty$ and hence it has no poles in the closed right half-plane. Then, any minimal realization (A, B, C, E) of G has the property that

$$\max_{1 \leq i \leq n} \mathrm{Re}\,(\lambda_i(A)) < 0.$$

Lemma 3.5 implies then that for some $M > 0$ and $p > 0$,

$$Ce^{At}B$$

satisfies

$$|Ce^{At}B| \leq Me^{-pt}.$$

It follows that the system with realization $(A, B, C, 0)$ is L_2-stable. Since the addition of E does not affect stability, it follows that this system is L_2-stable.

Suppose now that $G \notin \mathcal{H}_\infty$. Since G is the transfer function of a system of the form (2.1), it is proper and analytic except at a finite number of poles. Hence, the assumption $G \notin \mathcal{H}_\infty$ implies that G has at least one pole in the closed right half-plane. Choose the pole p with the largest real part. Consider any input $u \in L_2(0, \infty)$ with Laplace transform $\hat{u}$ such that $\hat{u}(p) \neq 0$. The Laplace transform of the corresponding output is

$$G(s)\hat{u}(s) = G_1(s) + \frac{K}{s-p}$$

where $G_1(s)$ is analytic in a neighborhood of p and K is the residue of $G(s)\hat{u}(s)$ at $s = p$. Inverting the Laplace transform, we see that if $\mathrm{Re}(p) > 0$, this output will grow exponentially with time. If p is imaginary, the corresponding output will have a periodic component. In both cases, the output is not in $L_2(0, \infty)$. We conclude that if $G \notin R\mathcal{H}_\infty$, the system is not L_2-stable.

We now consider the gain of an L_2-stable system. The earlier discussion showed that if a system G is L_2-stable then the L_2 gain

$$\|\mathcal{G}\|_{2,2} = \sup_{\omega} |G(j\omega)|. \qquad \blacksquare$$

3.1.1 Computation of the $\mathcal{H}_\infty$ Norm

Exact computation of $\|G\|_\infty$ requires calculation of a supremum of the frequency response, and this is only feasible for very simple systems. An approximation may be obtained by evaluating $|G(j\omega)|$ at a number of points $\{\omega_1, \ldots, \omega_n\}$. This procedure, however, does not yield an error bound on the final calculated norm. The following result leads to an algorithm to calculate $\|G\|_\infty$ with an upper bound on the error.

THEOREM 3.7: Choose A, B, C, E so that

$$G(s) = C(s - A)^{-1}B + E$$

and define, for real $\gamma > 0$, $R = (E^*E - \gamma^2 I)$, $S = (EE^* - \gamma^2 I)$ and

$$Z_\gamma = \begin{bmatrix} A - BR^{-1}E^*C & -BR^{-1}B^* \\ \gamma^2 C^* S^{-1} C & -A^* + C^*ER^{-1}B^* \end{bmatrix}.$$

Assume that A has no imaginary axis eigenvalues, $\gamma > \|E\|$, and $\omega \in R$. Then γ is a singular value of $G(j\omega)$, that is, there is a nonzero v so

$$G(j\omega)^* G(j\omega) v = \gamma^2 v,$$

if and only if $j\omega$ is an eigenvalue of Z_γ.

□ *Proof:* Let γ be a singular value of $G(j\omega)$. This is equivalent to the existence of $u, v \neq 0$ such that

$$\begin{aligned} G(j\omega)u &= v \\ G(j\omega)^* v &= \gamma^2 u. \end{aligned} \tag{3.3}$$

Or, defining

$$\begin{aligned} r &= (j\omega I - A)^{-1} Bu \\ s &= (-j\omega I - A^*)^{-1} C^* v, \end{aligned}$$

we can rewrite (3.3) as

$$\begin{aligned} Cr + Eu &= v \\ B^*s + E^*v &= \gamma^2 u. \end{aligned}$$

We can rewrite the preceding equation so that u and v are given in terms of r, s:

$$\begin{bmatrix} u \\ v \end{bmatrix} = \begin{bmatrix} -E & I \\ \gamma^2 I & -E^* \end{bmatrix}^{-1} \begin{bmatrix} C & 0 \\ 0 & B^* \end{bmatrix} \begin{bmatrix} r \\ s \end{bmatrix}. \tag{3.4}$$

By definition of r and s:

$$(j\omega I - A)r = Bu$$
$$(-j\omega I - A^*)s = C^* v.$$

Using (3.4) to eliminate u and v we obtain

$$j\omega \begin{bmatrix} r \\ s \end{bmatrix} = \cdots$$

$$\left(\begin{bmatrix} A & 0 \\ 0 & -A^* \end{bmatrix} + \begin{bmatrix} B & 0 \\ 0 & -C^* \end{bmatrix} \begin{bmatrix} -E & I \\ \gamma^2 I & -E^* \end{bmatrix}^{-1} \begin{bmatrix} C & 0 \\ 0 & B^* \end{bmatrix} \right) \begin{bmatrix} r \\ s \end{bmatrix}.$$

That is,

$$Z_\gamma \begin{bmatrix} r \\ s \end{bmatrix} = j\omega \begin{bmatrix} r \\ s \end{bmatrix}. \tag{3.5}$$

To summarize, we have shown that (3.3) is equivalent to (3.5). But (3.3) is the statement that γ is a singular value of $G(j\omega)$ and (3.5) states that $Z_\gamma - j\omega I$ is singular. Thus we have proven the lemma. ■

Corollary 3.8: Let (A, B, C, E) be a state-space realization with a transfer function $G \in M(R\mathcal{H}_\infty)$. Choose A so that it has no imaginary eigenvalues. Choose $\gamma \geq \|E\|$. Then $\|G\|_\infty \geq \gamma$ if and only if Z_γ has at least one imaginary eigenvalue.

□ *Proof:* Since $\gamma \geq \lim_{\omega\to\infty} |G(j\omega)|$ and $|G(j\omega)|$ is a continuous function of ω, $\|G\|_\infty \geq \gamma$ if and only if $|G(j\omega)| = \gamma$ for some ω. The result then follows from the preceding lemma. ■

This result implies that the $\mathcal{H}_\infty$ norm of a rational transfer function can be found to arbitrary accuracy via a search: The simplest method is a bisection algorithm.

0. Find $\gamma_{\max}$ so that $Z_{\gamma_{\max}}$ has no imaginary eigenvalues and $\gamma_{\min} \geq |E|$ so that $Z_{\gamma_{\min}}$ has at least one imaginary eigenvalue. Set $\gamma = \frac{1}{2}(\gamma_{\max} + \gamma_{\min})$.
1. Check if Z_γ has any imaginary eigenvalues.
2. Adjust $\gamma_{\min}$, $\gamma_{\max}$ accordingly and define $\gamma = \frac{1}{2}(\gamma_{\max} + \gamma_{\min})$.

3. Return to (1) and repeat until $\gamma_{\max} - \gamma_{\min}$ are within the required tolerance.
4. $|\|G\|_\infty - \gamma| < \frac{1}{2}(\gamma_{\max} - \gamma_{\min})$.

A major advantage of this method over an estimate found by evaluating $G(j\omega)$ at a number of points is that an error bound is obtained. Furthermore, in general, much less computation time is required.

EXAMPLE 3.9: The transfer function

$$G(s) = \frac{1}{s^2 + 0.1s + 25}$$

has a realization

$$A = \begin{bmatrix} 0 & 1 \\ -25 & -0.1 \end{bmatrix}, B = \begin{bmatrix} 0 \\ 1 \end{bmatrix}, C = [1 \quad 0], E = 0.$$

The magnitude plot of G (not in decibels) is shown in Fig. 3.1. We want

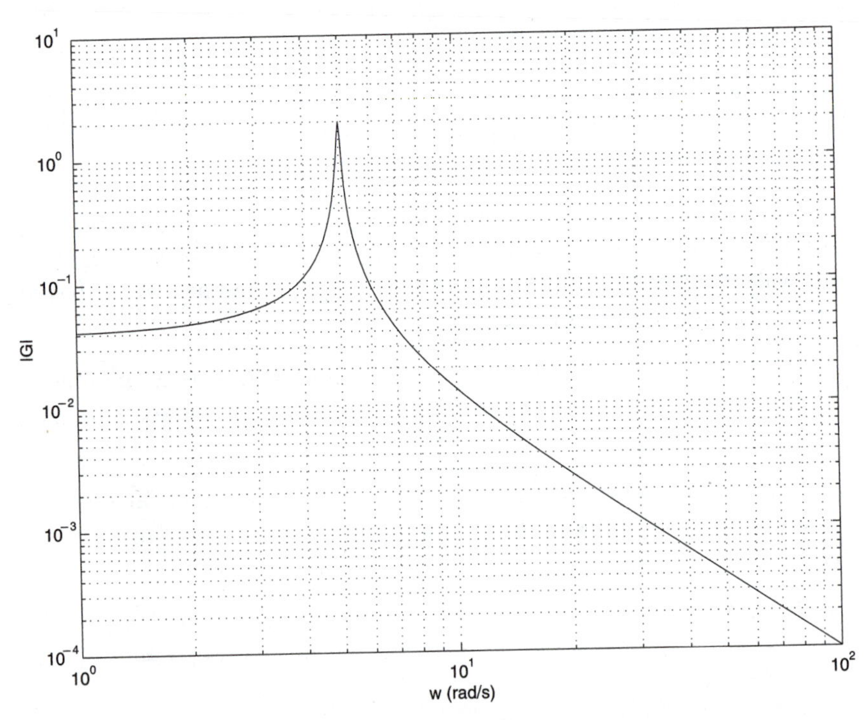

FIGURE 3.1 Magnitude plot of $1/(s^2 + 0.1s + 25)$.

to determine $\|G\|_\infty$ to within an error of 0.01. Use an initial estimate of $\gamma_{\min} = 1$, $\gamma_{\max} = 5$. The following sequence is obtained:

$\gamma_{\max}$	$\gamma_{\min}$	γ	Z_γ has imaginary eigenvalue?
5	1	3	No
3	1	2	Yes
3	2	2.5	No
2.5	2	2.25	No
2.25	2	2.125	No
2.125	2	2.0625	No
2.0625	2	2.0312	No
2.0312	2	2.0156	No

Thus, $\|G\|_\infty \approx 2.0078$ with an error less than 0.0078.

In the previous example, the maximum magnitude is achieved near $\omega = 5$, a natural frequency of G. For many systems, the maximum magnitude is achieved near $\omega = p$ where p is a pole of G, or at $\omega = 0$. Evaluating the transfer function at these frequencies initially can lead to a faster convergence procedure. In the exercises, an algorithm that improves on the bisection procedure is developed.

3.2 *L_∞ STABILITY

This norm is useful when the maximum value of the signal is of concern. We will only consider SISO systems in this section. The extension to MIMO systems is straightforward. A set A has measure zero if $\int_A dx = 0$. If a statement holds for all t in some interval, except on some set of measure zero, we say that the statement holds *almost everywhere* (a.e.). The essential supremum of $|u|$ is

$$\operatorname*{ess\,sup}_{t \geq 0} |u(t)| = \inf\{a | \, |u(t)| \leq a \text{ almost everywhere}\}.$$

That is, $|u(t)| \leq a$ except on a set with measure zero and a is the smallest number with this property.

Define the norm

$$\|u\|_\infty = \text{ess sup}_{0<t<\infty} |u(t)|.$$

$L_\infty(0, \infty)$ is the normed linear space consisting of all functions for which this quantity is finite, with norm $\| \, \|_\infty$.

DEFINITION 3.10: If the operator $\mathcal{G}$ is bounded from $L_\infty(0, \infty)$ to $L_\infty(0, \infty)$ then the system is L_∞-*stable*. The operator norm of $\mathcal{G}$, $\|\mathcal{G}\|_{\infty,\infty}$ is called the L_∞ *gain*.

Assume that $u \in L_\infty(0, \infty)$. Then,

$$|y(t)| \leq \int_0^t |C \exp(A(t-r))B| \, |u(r)| dr + |E| \, \|u\|_\infty$$

$$\leq \left[\int_0^\infty |C \exp(Ar)B| dr + |E| \right] \|u\|_\infty. \tag{3.6}$$

Thus, if $Ce^{At}B \in L_1(0, \infty)$, $y \in L_\infty(0, \infty)$. The following theorem can be proven.

THEOREM 3.11:

$$\|\mathcal{G}\|_{\infty,\infty} = \|Ce^{At}B\|_1 + |E|$$

where $\|g\|_1 = \int_0^\infty |g(r)| dr$.

Computation of the L_∞-gain. The integration required for calculation of the L_1-norm can be approximated by quadrature. Since for $\mathrm{Re}(s) > 0$,

$$\left| \int_0^\infty g_a(t) e^{-st} dt \right| \leq \int_0^\infty |g_a(t)| dt,$$

the $\mathcal{H}_\infty$-norm of the transfer function yields a lower bound.

The preceding statement implies that a system (2.1) is L_2-stable if it is L_∞-stable. In fact, L_2-stability also implies L_∞-stability for this class of systems.

THEOREM 3.12: A system (2.1) is L_2-stable if and only if it is L_∞-stable.

□ *Proof:* Recall that a system is L_2-stable if and only if its transfer function $G \in R\mathcal{H}_\infty$. A transfer function is in $R\mathcal{H}_\infty$ if and only if all its poles p_i have $\mathrm{Re}(p_i) < 0$. Inverting the Laplace transform we see that $|g_a(t)| < Me^{-\alpha t}$, $\gamma < 0$. Alternatively, one could argue that $Ce^{At}B \in L_1(0, \infty)$ if and only if $|Ce^{At}B| < Me^{-\alpha t}$ where $\alpha < 0$. ■

3.3 INTERNAL AND EXTERNAL STABILITY

There is another definition of stability useful for linear systems. This definition does not involve the norm of an operator.

DEFINITION 3.13: If $\max_{1 \leq i \leq n} \mathrm{Re}\,(\lambda_i(A)) < 0$, A is *Hurwitz*.

DEFINITION 3.14: If $\max_{1 \leq i \leq n} \mathrm{Re}\,(\lambda_i(A)) < 0$ (that is, A is Hurwitz) then the system is *internally stable*.

LEMMA 3.15: The matrix A is Hurwitz if and only if $(sI - A)^{-1}$ is a matrix with entries in $R\mathcal{H}_\infty$.

□ *Proof:* By Cramer's Rule

$$(sI - A)^{-1} = \frac{1}{\det(sI - A)} \operatorname{adj}(sI - A)$$

where adj M indicates the adjoint matrix of M. Thus, $(sI - A)^{-1}$ is always a matrix with entries proper and rational in s. A matrix A is Hurwitz if and only if $\det(sI - A)$ has no solutions with $\mathrm{Re}(s) \geq 0$. Thus, $(sI - A)^{-1}$ has entries in $R\mathcal{H}_\infty$ if and only if A is Hurwitz. ■

The following theorem now follows easily.

THEOREM 3.16: A system with state-space realization (A, B, C, E) is externally stable if it is internally stable.

□ *Proof:* The poles of G are a subset of the eigenvalues of A. Thus, if A is Hurwitz then G is analytic in the closed right-half plane. Since all transfer functions of the form $C(sI - A)^{-1}B + E$ are proper, $G \in M(R\mathcal{H}_\infty)$. ■

Thus, every internally stable system is externally stable. Is the converse statement true?

EXAMPLE 3.17:

$$A = \begin{bmatrix} 2 & 0 \\ 0 & -1 \end{bmatrix}, \quad B = \begin{bmatrix} 0 \\ 1 \end{bmatrix}, \quad C = [1 \quad 1], \quad E = 0.$$

This system is not internally stable and an initial condition such as

$$\begin{bmatrix} 1 \\ 0 \end{bmatrix}$$

will grow without bound.

The transfer function is

$$\begin{aligned} G(s) &= \frac{C\operatorname{adj}(sI - A)B}{\det(sI - A)} \\ &= [1 \quad 1] \begin{bmatrix} 0 \\ (s-2) \end{bmatrix} \frac{1}{(s-2)(s+1)} \\ &= \frac{(s-2)}{(s-2)(s+1)} \\ &= \frac{1}{s+1}. \end{aligned}$$

The transfer function is in $R\mathcal{H}_\infty$ (the impulse response is e^{-t}) and so the system is externally stable.

The problem in the preceding example is that the input matrix B is such that the unstable pole at 2 is "canceled" by a zero in the transfer function at 2.

Consider the system

$$\dot{x}(t) = Ax(t) + Bu(t)$$
$$x(0) = x_0.$$

If we can find a control u of the form $u(t) = -Kx(t)$ where K is a matrix of appropriate dimensions so that the trajectory $x(t) \to 0$ for any initial condition $x(0)$, then we say that the system is *stabilizable*. Since the foregoing system with control $u(t) = -Kx(t)$ can be written as

$$\dot{x}(t) = (A - BK)x(t),$$

the following concise definition is usually used.

Definition 3.18: The pair (A, B) is *stabilizable* if there exists K such that $A - BK$ is Hurwitz.

It can easily be shown that the system in the preceding example is not stabilizable.

Let us now consider a similar problem. This one is related to the output operator C.

Example 3.19:

$$A = \begin{bmatrix} 2 & 0 \\ 0 & -1 \end{bmatrix}, \quad B = \begin{bmatrix} 1 \\ 1 \end{bmatrix}, \quad C = [0 \quad 1], \quad E = 0.$$

This system is not internally stable and an initial condition such as

$$\begin{bmatrix} 1 \\ 0 \end{bmatrix}$$

will grow without bound.

The transfer function is

$$\begin{aligned} G(s) &= \frac{C\text{adj}(sI - A)B}{\det(sI - A)} \\ &= \frac{[0 \quad (s-2)]\begin{bmatrix} 1 \\ 1 \end{bmatrix}}{(s-2)(s+1)} \\ &= \frac{(s-2)}{(s-2)(s+1)} \\ &= \frac{1}{s+1}. \end{aligned}$$

The transfer function is in $\mathcal{H}_\infty$ (the impulse response is e^{-t}) and so the system is externally stable.

In the above example, the map from the state to the output if $u(t) = 0$, $t > 0$, is

$$y(t) = Ce^{At}x(0) = [0 \quad e^{-t}]x(0).$$

Because of the nature of C and A here, the eigenfunction (or mode) corresponding to the eigenvalue $\lambda = 2$ is not "detected."

Consider the system

$$\begin{aligned}\dot{x}(t) &= Ax(t) - Fy(t)\\ y(t) &= Cx(t)\\ x(0) &= x_0\end{aligned}$$

where F is a matrix of appropriate dimensions. If a matrix F exists so that the trajectory $x(t) \to 0$ for any initial condition $x(0)$, then we say that the system is *detectable*. This means that any unstable eigenfunction can be "detected" by the output and corrected by the output feedback $Fy(t)$. The foregoing system can be written as

$$\dot{x}(t) = (A - FC)x(t),$$

and so the following definition is usually used.

DEFINITION 3.20: The pair (A, C) is *detectable* if there exists F so that $A - FC$ is Hurwitz.

The definition of detectability is similar to that of stabilizability. Notice also that (A, C) is detectable if and only if (A^*, C^*) is stabilizable.

THEOREM 3.21: Suppose a given system is stabilizable and detectable. It is internally stable if and only if it is externally stable.

□ *Proof:* Every internally stable system is externally stable. It remains only to show that external stability implies internal stability for stabilizable and detectable systems. The feedthrough term E clearly does not affect stability and so we neglect it in order to simplify the proof. Suppose then that

$$G(s) = C(sI - A)^{-1}B \in M(R\mathcal{H}_\infty),$$

that is, $G(s)$ is a matrix with entries in $R\mathcal{H}_\infty$. Choose F and K so that $A_F = A - FC$ and $A_K = A - BK$ are Hurwitz. Since

$$A_K x(t) = Ax(t) - B(Kx(t)),$$

$$e^{A_K t}x_0 = e^{At}x_0 - \int_0^t e^{A(t-\tau)}B(Ke^{A_K\tau}x_0)d\tau. \tag{3.7}$$

Similarly,

$$Ax(t) = A_F x(t) + F(Cx(t))$$

and so

$$e^{At}x_0 = e^{A_F t}x_0 + \int_0^t e^{A_F(t-\tau)}F(Ce^{A\tau})x_0 d\tau. \tag{3.8}$$

Taking Laplace transforms of (3.7) and multiplying by C,

$$C(sI - A_K)^{-1}x_0 = C(sI - A)^{-1}x_0 - G(s)K(sI - A_K)^{-1}x_0.$$

Since A_K is Hurwitz, $C(sI - A_K)^{-1}$ is a vector with entries in $R\mathcal{H}_\infty$. Similarly, so is $K(sI - A_K)^{-1}$. Therefore, $C(sI - A)^{-1}$ is a vector with entries in $R\mathcal{H}_\infty$.

Now, take the Laplace transform of (3.8),

$$(sI - A)^{-1}x_0 = (s - A_F)^{-1}x_0 + (sI - A_F)^{-1}FC(sI - A)^{-1}x_0.$$

The matrix A_F has no eigenvalues in the closed right half-plane and so $(sI - A_F)^{-1}$ is a matrix with entries in $R\mathcal{H}_\infty$. It follows that $(sI - A)^{-1}$ is a matrix with entries in $R\mathcal{H}_\infty$. Therefore, A is Hurwitz. By definition, the system is internally stable. ■

3.3.1 Test for Stabilizability

We have the following simple test for stabilizability (and by the earlier remark on (A^*, C^*), for detectability, as well). Define

$$C_{+e} = \{s|\text{Re}(s) \geq 0\}.$$

Theorem 3.22: Suppose A is an $n \times n$ matrix and B is $n \times m$. The pair (A, B) is stabilizable if and only if

$$\text{rank}[(sI - A) \quad B] = n \tag{3.9}$$

for all $s \in C_{+e}$.

□ *Proof:* We will show that Range $[(sI - A) \quad B] = R^n$ for all $s \in C_{+e}$ if and only if A is stabilizable. This will prove the theorem.

First, suppose that (A, B) is stabilizable. Then there exists K so that for all $s \in C_{+e}$, and all $z \in R^n$,

$$\begin{aligned} z &= (sI - A + BK)(sI - A + BK)^{-1}z \\ &= (sI - A)\tilde{z} + B\tilde{u} \end{aligned}$$

where $\tilde{z} = (sI - A + BK)^{-1}z$ and $\tilde{u} = K(sI - A + BK)^{-1}z$. In other words, there is

$$\begin{bmatrix} \tilde{z} \\ \tilde{u} \end{bmatrix}$$

such that

$$[(sI - A) \quad B]\begin{bmatrix} \tilde{z} \\ \tilde{u} \end{bmatrix} = z.$$

Thus, Range $[(sI - A) \quad B] = R^n$ as required.

Now suppose that (A, B) is not stabilizable. That is, for every K there exists $s \in C_{+e}$, and a nonzero $z \in R^n$ such that $(sI - A + BK)z = 0$. Rewriting, we obtain

$$[(sI - A) \quad B]\begin{bmatrix} z \\ Kz \end{bmatrix} = 0.$$

Thus, $[(sI - A) \quad B]$ has a nontrivial nullspace and so the range of this operator is not all of R^n. This implies that rank $[(sI - A) \quad B] < n$, completing the proof. ■

Corollary 3.23: Suppose A is an $n \times n$ matrix and B is $n \times m$. The pair (A, B) is stabilizable if and only if

$$\text{rank}[(\lambda I - A) \quad B] = n \tag{3.10}$$

for all eigenvalues $\lambda \in C_{+e}$ of A.

□ *Proof:* The matrix $(\lambda I - A)$ has rank n if λ is in the resolvant of A. Thus, it is only necessary to verify (3.9) for eigenvalues $\lambda \in C_{+e}$. ■

Theorem 3.24: Every controllable system is stabilizable.

□ *Proof:* This follows immediately by comparing this test for stabilizability with the PBH test (Thm. 2.7) for controllability. ■

The following is now immediate.

Corollary 3.25: Suppose A is an $n \times n$ matrix and C is $p \times n$.

1. The pair (A, C) is detectable if and only if

$$\text{rank}\begin{bmatrix} (\lambda I - A) \\ C \end{bmatrix} = n \tag{3.11}$$

for all eigenvalues $\lambda \in C_{+e}$ of A.
2. Every observable system is detectable.

3.4 CLOSED-LOOP STABILITY

The standard feedback control system has two components: the object to be controlled, called the plant, and a controller to generate the plant's input. Any sensor and actuator dynamics are included in the plant model. A generic closed-loop system is shown in Fig. 3.2. Note that the plus (+) signs at summing junctions are omitted. Only the minus (−) sign is explicitly indicated. Also, when the context is only the frequency domain, the hats (ˆ) on Laplace transforms are dropped as in Fig. 3.1. We will often say P is stable instead of "$P \in R\mathcal{H}_\infty$" or "the map from e_2 to y_2 is a bounded operator on $L_2(0, \infty)$."

Any closed-loop system must, aside from any other performance objectives, be stable. Stability is so fundamental that it is often not given explicitly as an objective, but it is an important objective even for stable plants. Feedback can be used to stabilize any unstable system. Unfortunately, it can also destabilize a stable system.

Example 3.26: $P(s) = 1/(s+1)$. This is a stable system. Suppose a controller is chosen,

$$H(s) = \frac{-3}{s+2},$$

which is also stable. However, if $d \equiv 0$,

$$\begin{aligned} e_1 &= r - PHe_1 \\ e_1 &= (1+PH)^{-1}r \\ \frac{e_1}{r} &= \frac{(s+1)(s+2)}{s^2+3s-1}. \end{aligned}$$

The zeros of $s^2 + 3s - 1$ are 0.303, −3.30. The preceding transfer function is not in $R\mathcal{H}_\infty$, and the closed loop is unstable.

This brings us to the question of what we mean by stability of the closed loop, since there are a number of maps to consider. We will say that

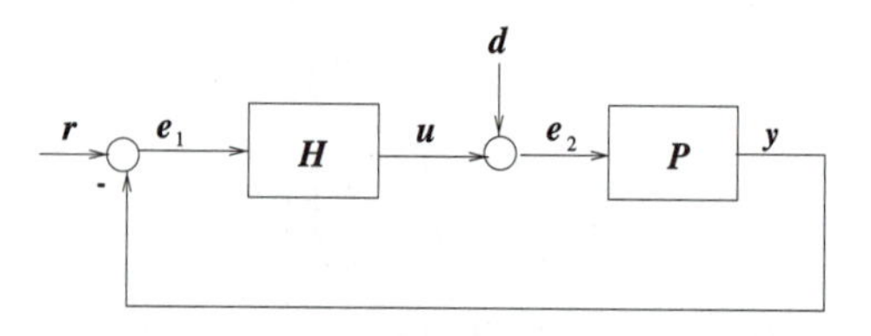

FIGURE 3.2 Standard closed-loop system.

the closed-loop system (Fig. 3.2) is stable if all four maps from (r, d) to (e_1, e_2) are well defined and L_2-stable. That is, these four maps have transfer functions in $M(R\mathcal{H}_\infty)$. We now derive the various transfer functions of a closed loop:

$$e_1 = r - Pe_2$$
$$e_2 = d + He_1$$
$$\begin{bmatrix} I & P \\ -H & I \end{bmatrix} \begin{bmatrix} e_1 \\ e_2 \end{bmatrix} = \begin{bmatrix} r \\ d \end{bmatrix}.$$

The foregoing matrix must be invertible for the various maps to be well defined. That is, we require that

$$\det(I + PH)(s) = \det(I + HP)(s)$$

not be identically 0. We also require that

$$\lim_{|s|\to\infty} \det(I + PH)(s) \neq 0.$$

This second condition is required for the closed loop to be proper. If this second condition is satisfied, then we also have that the determinant is not identically zero.

DEFINITION 3.27: If P and H are such that $\lim_{s\to\infty} \det(I + PH)(s) \neq 0$, then we say the closed loop is *well-posed.*

In general at least one of P or H is strictly proper. In these cases well-posedness is guaranteed.

THEOREM 3.28: Let P have a realization (A_P, B_P, C_P, E_P) and H a realization (A_H, B_H, C_H, E_H). The closed loop is well-posed if -1 is not an eigenvalue of $E_P E_H$.

□ *Proof:* $\lim_{s\to\infty} \det(I + PH)(s) = \det(I + E_P E_H)$. The result follows. ■

If the closed loop is well-posed, then it can be verified that

$$\begin{bmatrix} e_1 \\ e_2 \end{bmatrix} = \Delta(P, H) \begin{bmatrix} r \\ d \end{bmatrix} \tag{3.12}$$

where for SISO systems,

$$\Delta(P, H) = (I + PH)^{-1} \begin{bmatrix} 1 & -P \\ H & 1 \end{bmatrix}. \tag{3.13}$$

For multivariable systems we need to be careful in what order we multiply the transfer matrices P and H. We can show that

$$\Delta(P, H) = \begin{bmatrix} (I + PH)^{-1} & -P(I + HP)^{-1} \\ H(I + PH)^{-1} & (I + HP)^{-1} \end{bmatrix}. \tag{3.14}$$

We can rewrite this so that only $(I + HP)^{-1}$ or $(I + PH)^{-1}$ appear. First, from some elementary matrix manipulations,

$$(I + PH)^{-1} = I - P(I + HP)^{-1}H,$$

with a similar expression for $(I + HP)^{-1}$. This can be used to rewrite (3.14) as

$$\Delta(P, H) = \begin{bmatrix} (I + PH)^{-1} & -(I + PH)^{-1}P \\ H(I + PH)^{-1} & I - H(I + PH)^{-1}P \end{bmatrix} \tag{3.15}$$

$$= \begin{bmatrix} I - P(I + HP)^{-1}H & -P(I + HP)^{-1} \\ (I + HP)^{-1}H & (I + HP)^{-1} \end{bmatrix}. \tag{3.16}$$

We have now a characterization of stability for the closed loop.

Definition 3.29: The closed-loop system is *externally stable* if it is well-posed and $\Delta(P, H) \in M(R\mathcal{H}_\infty)$.

Before proceeding further, we will justify design of closed-loop systems based on the inputs and outputs.

Theorem 3.30: Let (A_P, B_P, C_P, E_P) be a stabilizable/detectable realization of P and let (A_H, B_H, C_H, E_H) be a stabilizable/detectable realization of H. Assume the closed loop is well-posed. Indicate the state of P by x_P and the state of H by x_H. Then the closed loop with state (x_P, x_H) is internally stable if and only if it is externally stable.

□ *Proof:* Since the closed loop is clearly a linear, finite-dimensional time-invariant system, the result will follow from earlier theorems if the closed loop is also stabilizable and detectable.

To simplify the proof, assume $E_H = E_P = 0$. (The general proof is entirely similar.) Let x_P and x_H indicate the state of the plant and controller, respectively. It is routine to verify that the interconnected system is described by

$$\begin{bmatrix} \dot{x}_P \\ \dot{x}_H \end{bmatrix} = A \begin{bmatrix} x_P \\ x_c \end{bmatrix} + B \begin{bmatrix} r \\ d \end{bmatrix}$$

$$\begin{bmatrix} y_P \\ y_H \end{bmatrix} = C \begin{bmatrix} x_P \\ x_H \end{bmatrix}$$

where

$$B = \begin{bmatrix} 0 & B_P \\ B_H & 0 \end{bmatrix}, \qquad C = \begin{bmatrix} C_P & 0 \\ 0 & C_H \end{bmatrix}$$

$$A = \begin{bmatrix} A_P & 0 \\ 0 & A_H \end{bmatrix} + \begin{bmatrix} 0 & B_P \\ B_H & 0 \end{bmatrix} \begin{bmatrix} -C_P & 0 \\ 0 & C_H \end{bmatrix}.$$

Choose some K_P, K_H so that $A_P - B_P K_P$ and $A_H - B_H K_H$ are Hurwitz. Defining

$$K = \begin{bmatrix} 0 & K_H \\ K_P & 0 \end{bmatrix} + \begin{bmatrix} -C_P & 0 \\ 0 & C_H \end{bmatrix},$$

$$A - BK = \begin{bmatrix} A_P - B_P K_P & 0 \\ 0 & A_H - B_H K_H \end{bmatrix}$$

is Hurwitz and the closed loop is stabilizable. Similarly, the closed loop is detectable. Therefore, the closed loop is stabilizable and detectable with all four transfer functions in $M(R\mathcal{H}_\infty)$. The result now follows from the equivalence of internal and external stability for stabilizable/detectable systems (Theorem 3.21). ∎

If the original system is not stabilizable and detectable, misleading results may be obtained when designing a controller.

Example 3.31: The system is

$$\dot{x}(t) = \begin{bmatrix} 0 & 0 \\ 0 & -1 \end{bmatrix} x(t) + \begin{bmatrix} 0 \\ 1 \end{bmatrix} v(t)$$
$$y(t) = [1 \quad 1] x(t).$$

The system has eigenvalues $-1, 0$ and for any proposed feedback

$$K = [k_1 \quad k_2],$$

we have

$$A - BK = \begin{bmatrix} 0 & 0 \\ 0 & -1 \end{bmatrix} - \begin{bmatrix} 0 & 0 \\ k_1 & k_2 \end{bmatrix}$$
$$= \begin{bmatrix} 0 & 0 \\ -k_1 & -k_2 - 1 \end{bmatrix}.$$

Regardless of what values of k_1, k_2 are chosen, $A - BK$ will have a zero eigenvalue. The system is not stabilizable.

However, the transfer function is

$$P(s) = \frac{1}{s+1},$$

which is in $\mathcal{RH}_\infty$.

Consider the controller

$$\begin{aligned}\dot{x}_c(t) &= -2x_c(t) + u_c(t)\\ y_c(t) &= x_c(t).\end{aligned}$$

The controller is internally stable with transfer function

$$H(s) = \frac{1}{s+2} \in \mathcal{RH}_\infty.$$

The transfer function matrix for the closed-loop system is

$$\frac{1}{1+(s+1)(s+2)}\begin{bmatrix}(s+1)(s+2) & -(s+2)\\ (s+1) & (s+1)(s+2)\end{bmatrix}.$$

All four maps are in $\mathcal{RH}_\infty$.

However,

$$\begin{bmatrix}\dot{x}(t)\\ \dot{x}_c(t)\end{bmatrix} = \begin{bmatrix}0 & 0 & 0\\ -1 & -2 & 0\\ 0 & 0 & -3\end{bmatrix}\begin{bmatrix}x(t)\\ x_c(t)\end{bmatrix} + \begin{bmatrix}0 & 0\\ 1 & 0\\ 0 & 1\end{bmatrix}[r \quad d]$$

$$\begin{bmatrix}y(t)\\ y_c(t)\end{bmatrix} = \begin{bmatrix}1 & 1 & 0\\ 0 & 0 & 1\end{bmatrix}\begin{bmatrix}x(t)\\ x_c(t)\end{bmatrix}.$$

The closed loop is not internally stable.

If P and H are both stable then closed loop stability follows if the closed loop is well-posed and $(I + PH)^{-1} \in M(\mathcal{RH}_\infty)$. It is always necessary for stability that $(I + PH)^{-1}$ have no r.h.p. poles. However, if P or H are unstable, this is not sufficient for closed loop stability.

Example 3.32: $P = 1/(s^2 - 1)$, $H = (s-1)/(s+1)$. The transfer function

$$(1 + PH)^{-1} = \frac{(s+1)^2}{(s^2+2s+2)} \in \mathcal{RH}_\infty,$$

but

$$P(1 + PH)^{-1} = \frac{(s+1)}{(s-1)(s^2+2s+2)} \notin \mathcal{RH}_\infty.$$

The problem here is that a r.h.p. pole in P corresponds to a r.h.p. zero in H: They cancel in the first transfer function, but not in the second.

3.5 TESTS FOR CLOSED-LOOP STABILITY OF SISO SYSTEMS

To see how to avoid r.h.p. pole/zero cancellations for SISO systems, write P and H as ratios of coprime polynomials:

$$P = \frac{n_P}{d_P}, \qquad H = \frac{n_H}{d_H}.$$

That is, n_P and d_P have no common zeros, and similarly for n_H, d_H. Define the characteristic polynomial

$$\kappa = n_P n_H + d_P d_H.$$

THEOREM 3.33: Let P and H be SISO systems. The closed loop is stable if and only if it is well-posed and κ has no closed right half-plane zeros.

□ *Proof:* Write $P = n_P/d_P$ where n_P and d_P are coprime polynomials and similarly write $H = n_H/d_H$ where n_H and d_H are coprime. The closed-loop operator can be rewritten as

$$\Delta(P, H) = \frac{1}{n_P n_H + d_P d_H} \begin{bmatrix} d_P d_H & -n_P d_C \\ d_P n_H & d_P d_H \end{bmatrix}.$$

It is evident that if the characteristic polynomial κ has no zeros in the closed right half-plane $\text{Re}(s) \geq 0$ then the closed loop is stable.

Suppose now that κ has a zero in the closed right half-plane, say at $s = p$. If the closed loop is stable, then all four numerators must have a zero at $s = p$. Suppose that $d_P(p) = 0$ so that the (1,1) entry in the transfer matrix is 0. (The argument is identical if $d_H(p) = 0$.) Then since n_P and d_P are coprime, we must have $d_H(p) = 0$. But $\kappa(p) = 0$ so this implies that at least one of n_P or n_H is zero at p. This contradicts the coprimeness assumption. Hence, if the characteristic polynomial has any r.h.p. zeros, the system is unstable. ■

EXAMPLE 3.34: $P = 1/(s^2 + as)$, $H = K$, where $a > 0$ and K is any real number. For what values of K is the closed loop stable?

The coprime polynomials are easily obtained,

$$N_P = 1, D_P = s(s + a), N_H = K, D_H = 1,$$

and the characteristic polynomial is

$$\kappa = K + s(s+a) = s^2 + as + K.$$

The real parts of the roots of this equation are negative if and only if $K > 0$.

Example 3.35: Consider the same plant as in the previous example, but with a more complex controller:

$$H = K_1 + K_2/s + K_3 s/(1+\varepsilon s), \qquad 0 < \varepsilon \ll 1.$$

Now,

$$H = \frac{N_H}{D_H}, \text{ where } \qquad \begin{array}{l} N_H = K_1 s(1+\varepsilon s) + K_2(1+\varepsilon s) + K_3 s^2 \\ D_H = s(1+\varepsilon s). \end{array}$$

It is easy to check that N_H, D_H, are coprime polynomials. The characteristic polynomial is

$$\kappa = \varepsilon s^4 + (1+\varepsilon a)s^3 + (K_1\varepsilon + K_3 + a)s^2 + (K_2\varepsilon + K_1)s + K_2.$$

The roots of this polynomial must be determined numerically.

In order to determine stability we do not need to know the locations of the zeros of the characteristic polynomial; we only need to know whether there are any zeros in the closed right half-plane. The following tests, the Routh–Hurwitz Criterion and the Nyquist Criterion, enable us to determine whether a characteristic polynomial has closed right half-plane zeros, without calculating the zeros of the polynomial.

3.5.1 Routh–Hurwitz Criterion

The Routh–Hurwitz Criterion is an simple algebraic criterion for determining if a polynomial has any zeros in the closed right half-plane. The only assumption is that the coefficients must all be real. We will consider polynomials of the form

$$f(s) = a_n s^n + a_{n-1}s^{n-1} + \cdots + a_0 \tag{3.17}$$

where a_i, $i = 1..n$ are real numbers.

Theorem 3.36: For all zeros of the polynomial f (3.17) to have negative real parts it is necessary that

$$\frac{a_0}{a_n} > 0, \frac{a_1}{a_n} > 0 \ldots \frac{a_{n-1}}{a_n} > 0.$$

□ *Proof:* Let z'_j be the real zeros of f and z''_k the complex roots of f. Then

$$f(s) = a_o \prod_j (s - z'_j) \prod_k (s^2 - 2\mathrm{Re}(z''_k)s + |z''_k|^2).$$

If all the numbers z'_j and $\mathrm{Re}(z''_k)$ are negative we obtain only positive coefficients for the powers of s when we expand the product. ■

The proof of the following criterion is long, and since it is not illuminating it is not given here.

Theorem 3.37 (*Routh–Hurwitz Criterion*): Consider the polynomial f in (3.17) and assume all the coefficients $\{a_i\}$ have the same sign, and are nonzero.

Construct the following table:

$$\begin{array}{cccl} a_n & a_{n-2} & a_{n-4}\cdots & \text{(Any undefined} \\ a_{n-1} & a_{n-3} & \cdots & \text{entries are set to zero.)} \\ b_1 & c_1 & d_1 & \cdots \\ b_2 & c_2 & \vdots & \\ \vdots & \vdots & & \end{array}$$

where

$$b_1 = -\frac{1}{a_{n-1}} \begin{vmatrix} a_n & a_{n-2} \\ a_{n-1} & a_{n-3} \end{vmatrix}, \quad c_1 = -\frac{1}{a_{n-1}} \begin{vmatrix} a_n & a_{n-4} \\ a_{n-1} & a_{n-5} \end{vmatrix} \cdots$$

$$b_2 = -\frac{1}{b_1} \begin{vmatrix} a_{n-1} & a_{n-3} \\ b_1 & c_1 \end{vmatrix}, \quad c_2 = -\frac{1}{b_1} \begin{vmatrix} a_{n-1} & a_{n-5} \\ b_1 & d_1 \end{vmatrix} \cdots$$

All zeros of f will have negative real parts if and only if there are $n+1$ nonzero entries in the first (left-hand) column and all these entries have the same sign.

Example 3.34 (*Cont.*): With $P = 1/s(s+a)$, $H = K$, the characteristic polynomial $\kappa = s^2 + as + K$. The Routh–Hurwitz table is

$$\begin{array}{cc} 1 & K \\ a & 0 \\ K & \end{array}$$

We require $a > 0$, $K > 0$.

EXAMPLE 3.38: Consider

$$\kappa(s) = s^3 + 2s^2 + 5s + 3.$$

The Routh–Hurwitz array is

$$\begin{array}{llll} 1 & 5 & 0 & \dots \\ 2 & 3 & 0 & \dots \\ \frac{7}{2} & 0 & \dots & \\ 3 & 0 & \dots & \end{array}$$

Since all entries in the first column are positive, this polynomial has no zeros in the closed right half-plane. (The zeros can be calculated: -0.737, $-0.631 \pm j1.92$.)

EXAMPLE 3.35 (*Cont.*): Consider the same plant as before, but with the PID controller $H = K_1 + K_2/s + K_3 s/(1 + \varepsilon s)$. Assume $K_1 > 0$, $K_2 > 0$, $K_3 > 0$, and let a_4, a_3, a_2, a_1 be the coefficients of s^4, s^3, s^2, s, respectively. Constructing the Routh–Hurwitz array:

$$\begin{array}{lll} \epsilon & a_2 & K_2 \\ a_3 & a_1 & 0 \\ b_1 & K_2 & 0 \\ b_2 & 0 & \\ K_2 & & \end{array}$$

$$b_1 = -\tfrac{1}{a_3}(a_4 a_1 - a_2 a_3) = a_2 - a_4 \tfrac{a_1}{a_3}$$

$$b_2 = -\tfrac{1}{b_1}(a_3 a_0 - b_1 a_1) = a_1 - a_3 a_0 / b_1.$$

Both b_1 and b_2 become negative as K_2 is increased. The system is stable for certain values of the parameters.

EXAMPLE 3.39: Consider the plant $P(s) = 3/(s^3 + 2s^2 + 5s + 3)$ and controller $H(s) = K/(s^3 + s^2 + 4s + 1)$. For what values of the real number K (if any) is the closed loop stable? The characteristic polynomial is

$$\kappa(s) = s^6 + 3s^5 + 11s^4 + 17s^3 + 25s^2 + 17s + 3K + 3.$$

Finding all the zeros of κ for each value of K is not feasible. To solve this, we will construct the Routh–Hurwitz array, and find for what values of K all entries in the first column are positive:

$$\begin{array}{ccccc}
1 & 11 & 25 & 3K+3 & 0\dots \\
3 & 17 & 17 & 0 & \dots \\
\frac{16}{3} & \frac{58}{3} & 3K+3 & 0 & \dots \\
\frac{49}{8} & \frac{245-27K}{16} & 0 & 0 & \dots \\
\frac{72K}{49}+6 & 3K+3 & 0 & 0 & \dots \\
\frac{-81K^2-196K+2401}{4(12K+49)} & 0 & 0 & 0 & \dots \\
3K+3 & 0 & 0 & 0 & \dots
\end{array}$$

We require that the last three entries in the first column be positive. The simplest condition is $3K + 3 > 0$ or $K > -1$. The next condition is

$$\frac{72K}{49} + 6 > 0$$

or $K > -4.08$. This is implied by $K > -1$. For such K, $12K + 49 > 0$ and so the remaining condition is

$$-81K^2 - 196K + 2401 > 0.$$

Calculating the roots of this quadratic equation, we obtain

$$-6.78 < K < 4.37.$$

Thus, the closed loop is stable for

$$-1 < K < 4.37.$$

3.5.2 Nyquist Criterion

We first review a result from complex analysis, Cauchy's Principle of the Argument. Let C be any simple closed curve.

Definition 3.40: Let C be a closed curve in the complex plane and let s_o be any point not on C. The *winding number* (or *index*) of C with respect to s_o is the number of times that C encircles s_o in the *clockwise* sense. Formally,

$$\text{wind}(C, s_o) = \frac{1}{2\pi j} \int_C \frac{ds}{s - s_o},$$

where the integral is evaluated around C clockwise.

Notice that the usual definition in mathematics textbooks is that counterclockwise encirclements are positive. Here the convention that clockwise encirclements are positive is used in order to be consistent with common usage in engineering.

Now let F be a transfer function. It is analytic except at a (finite) number of poles. Assume that F has no poles or zeros on a simple closed curve $\mathcal{C}$. As s is varied along the contour $\mathcal{C}$ will obtain a set of values of $F(s)$. Denote the curve formed by the set of values of $F(s)$ as s travels $\mathcal{C}$ by $\mathcal{F} = F \circ \mathcal{C}$. Since $\mathcal{C}$ is closed, the curve $\mathcal{F}$ forms a closed curve in the complex plane. We can define the winding number of this curve.

Definition 3.41: Let F be function analytic in a region containing a simple closed curve $\mathcal{C}$. Assume also that F has no zeros on $\mathcal{C}$. The *winding number of* $F(s)$ *on* $\mathcal{C}$ is

$$\text{wind}_{\mathcal{C}} F = \text{wind}(F \circ \mathcal{C}, 0).$$

Essentially, the winding number of $\mathcal{F}$ is the number of times that the origin is encircled clockwise. If the origin is encircled twice counterclockwise, the winding number is -2.

We need the following famous result from complex analysis.

Theorem 3.42 (*Cauchy's Principle of the Argument*): Let F be a function analytic on and within a closed contour $\mathcal{C}$ except at a finite number of poles inside $\mathcal{C}$ and assume that F has no poles or zeros on $\mathcal{C}$. Then

$$\text{wind}_{\mathcal{C}} F = Z_F - P_F$$

where

$$\left.\begin{aligned} Z_F &= \#\,\text{zeros of } F \text{ inside contour} \\ P_F &= \#\,\text{poles of } F \text{ inside contour} \end{aligned}\right\}\text{counted according to their multiplicity.}$$

We now return to our problem of a test that determines whether $1 + P(s)H(s)$ where P and H are scalar functions has any zeros in the closed right half-plane. For our problem, $F(s) = 1 + P(s)H(s)$. The closed loop can only be stable if F has no closed right half-plane (r.h.p.) zeros. Assume for now that P and H have no imaginary axis poles. Consider the clockwise contour $\mathcal{C}$ shown in Fig. 3.3 where the radius α is chosen large enough so that (1) the contour encloses all the r.h.p. poles of F and (2) $F(j\alpha) \approx F(-j\alpha)$. Notice that since the contour is traversed clockwise, the values of the function in question are calculated as ω increases from $-\infty$ to ∞.

Using Cauchy's Principle of the Argument, for a stable system, the number of *counterclockwise* encirclements of the origin must equal the

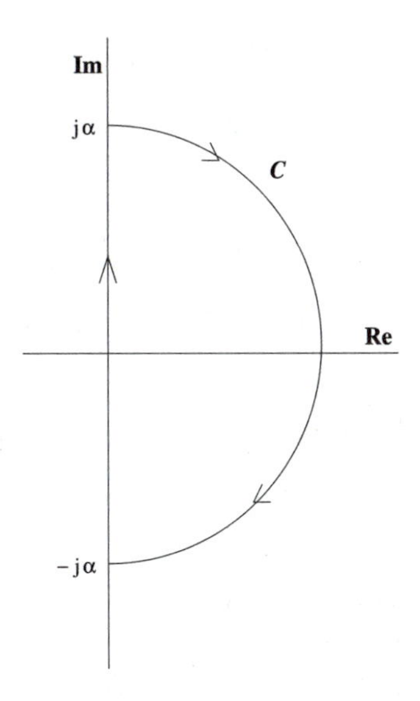

FIGURE 3.3 Contour for Nyquist plots.

number of unstable poles of $P(s)H(s)$. This is now stated and proven formally.

THEOREM 3.43: Let P and H have β_P and β_H poles, respectively, in the right half-plane and assume that neither P or H have poles on the imaginary axis. Considering the clockwise contour $\mathcal{C}$, we obtain a closed curve $\mathcal{K}$ from the set of values of $1 + PH$ as s is varied along this contour. The closed loop is stable if and only if the curve $\mathcal{K}$ (1) does not pass through the origin and (2) encircles the origin $\beta_P + \beta_H$ times counterclockwise.

□ *Proof:* Write $P = n_P/d_P$ where n_P and d_P are coprime polynomials and similarly write $H = n_H/d_H$. The characteristic polynomial of the closed loop is

$$\kappa = n_P n_H + d_P d_H.$$

The closed loop will be stable if and only if κ has no (closed) right half-plane zeros.

First assume that the closed loop is stable, that is, κ has no closed right half-plane zeros. Then $1 + PH$ has no imaginary-axis zeros and the curve $\mathcal{K}$ does not pass through the origin. Also,

$$\kappa = d_P(1 + PH)d_H.$$

For any two complex-valued functions G and H

$$\arg(GH) = \arg(G) + \arg(H),$$

and so

$$\text{wind}_{\mathcal{C}}\,\kappa = \text{wind}_{\mathcal{C}}\,d_P + \text{wind}_{\mathcal{C}}(1 + PH) + \text{wind}_{\mathcal{C}}\,d_H. \tag{3.18}$$

Since κ is a polynomial, with no closed right half-plane zeros $\text{wind}_{\mathcal{C}}\,\kappa = 0$ and it follows that

$$\begin{aligned}\text{wind}_{\mathcal{C}}(1 + PH) &= -\text{wind}_{\mathcal{C}}\,d_P - \text{wind}_{\mathcal{C}}\,d_H\\ &= -\beta_P - \beta_H.\end{aligned}$$

In other words, $\mathcal{K}$ encircles the origin $\beta_P + \beta_H$ times counterclockwise.

Now prove the converse. Assume that the curve $\mathcal{K}$ does not pass through the origin and that $\mathcal{K}$ encircles the origin $\beta_P + \beta_H$ times counterclockwise. The first part of this statement is that $1 + PH$ has no imaginary axis zeros, and since d_P and d_H have no imaginary axis zeros, κ has no imaginary axis zeros. The second part of the statement is that the winding number of $1 + PH$ with respect to $\mathcal{C}$ is $-\beta_P - \beta_H$. Using (3.18), $\text{wind}_{\mathcal{C}}\,\kappa = 0$, and so κ has no closed right half-plane zeros. We conclude that the closed loop is stable. ∎

In short, for a stable system the number of *counterclockwise* encirclements of the origin must equal the number of unstable poles of $P(s)H(s)$. It is easier to plot the loop transfer function $L(s) = P(s)H(s)$ instead of $1 + L(s)$. We plot $L(s)$ as s ranges over $\mathcal{C}$, clockwise, and count encirclements of the point $(-1, 0)$ instead of $(0, 0)$. The curve $\mathcal{L}$ as $L(s)$ traverses the clockwise contour $\mathcal{C}$ is known as the *Nyquist plot* of the function $L(s)$. What if P or H has imaginary axis poles? In this case, we alter the contour $\mathcal{C}$ slightly so that the the imaginary axis poles are not contained in the contour, but not so as to remove other poles and zeros. In this case, the imaginary axis poles are not included when counting unstable poles.

Definition 3.44: Let L be a transfer function. If L has no imaginary axis poles, consider the *clockwise* contour $\mathcal{C}$, in Fig. 3.3. If L has any imaginary axis poles, we alter the contour indenting to the right around imaginary axis poles. We obtain a closed curve $\mathcal{L}$ from the set of values of L as s is varied along the contour. This closed curve is known as the *Nyquist plot* of L.

We now restate the previous theorem using these changes.

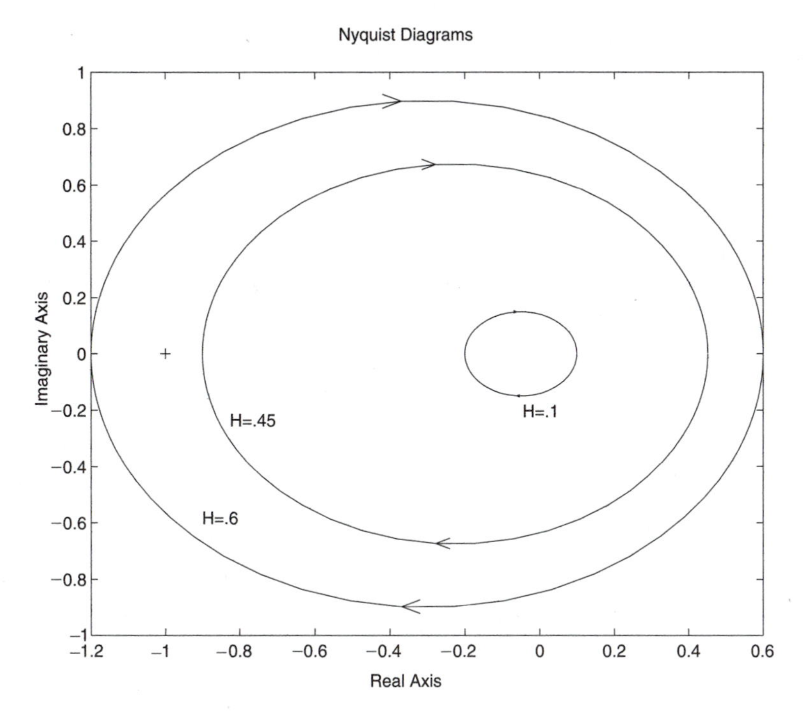

FIGURE 3.4 Nyquist plots for Example 3.46.

Corollary 3.45 (*Nyquist Criterion*): Construct the Nyquist plot of PH. Let β_P and β_H be the number of poles of P and H, respectively, with real part greater than zero. The closed loop system is stable if and only if the Nyquist plot does not pass through -1 and encircles this point $\beta_P + \beta_H$ times counterclockwise.

Example 3.46: The Nyquist diagrams for $P = (s-2)/(s+1)$ and several different gains for a proportional controller $H = K$ are shown in Fig. 3.4. Since both plant and controller are stable, there should be no encirclements of the point $(-1, 0)$. Thus, the closed loop is stable for $H = 0.1, 0.45$ and unstable for $H = 0.6$. Since for $H = 0.6$, the Nyquist plot encircles -1 once, clockwise, $1 + PH$ has one r.h.p. zero and hence the closed loop has one r.h.p. pole.

Note however that when $H = 0.45$ the Nyquist curve passes quite close to the point $(-1, 0)$. This indicates that small changes in system parameters could lead to a Nyquist curve that encircles the point $(-1, 0)$ and thus an unstable closed loop.

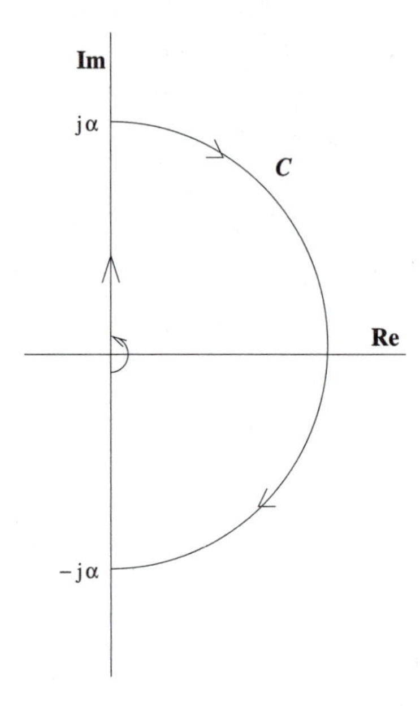

FIGURE 3.5 Nyquist contour for Example 3.47.

Example 3.47: Consider now the plant $P = (s-2)/s(s+1)$ with the constant gain controller $H = 0.5$. The Nyquist contour is shown in Fig. 3.5 and the corresponding Nyquist plot in Fig. 3.6. We see that there is one clockwise encirclement, indicating that the closed loop is unstable.

Notice the importance of the $(-1, 0)$ point. The closed loop will be stable if $L(s)$ is never -1. The closeness of $L(s)$ to -1 is a measure of how "close" the closed system is to being unstable. One way to quantify the "distance" of a system from instability is through *gain* and *phase margins.*

Definition 3.48: The *gain margin* $k_{\max}$ is the reciprocal of the gain at the frequency where the phase is 180°.

The gain margin is a measure of the amount by which the system gain would have to increase for the closed loop to become unstable. It can be estimated from the Nyquist plot of L: $-1/k_{\max}$ is the point where the Nyquist plot has phase angle 180° and is closest to the point $(-1, 0)$. The system gain could be increased by $k_{\max}$ before the loop $L(s)$ would intersect the point $s = -1$. A gain margin greater than 1 tends to indicate

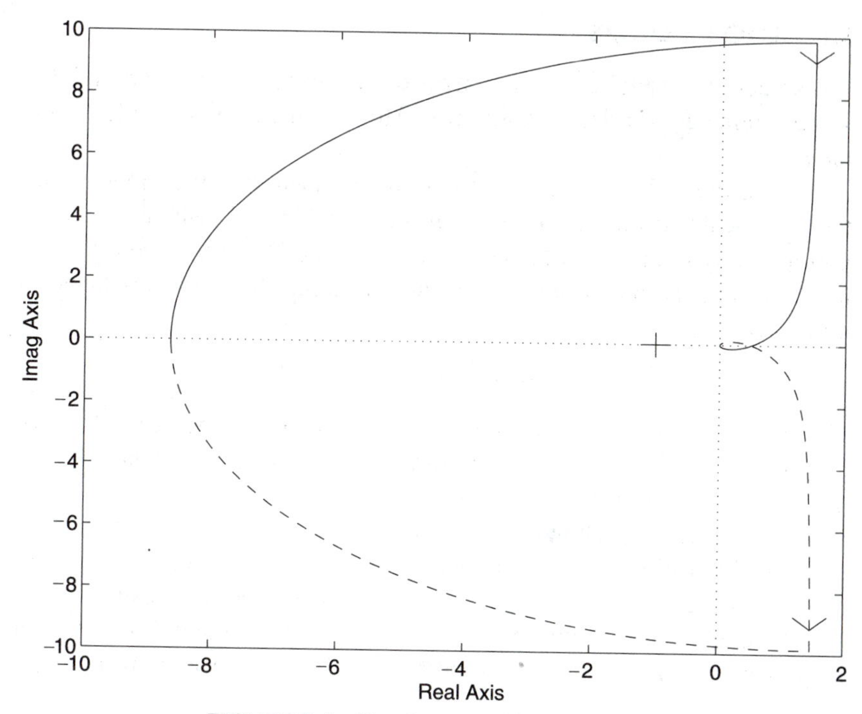

FIGURE 3.6 Nyquist plots for Example 3.47.

that the closed loop is stable, and a gain margin less than 1 that the closed loop is unstable.

DEFINITION 3.49: The *phase margin* $\phi_{\max}$ is the angle that $L(j\omega)$ makes with respect to the negative real axis.

A phase margin is a measure of how much additional delay (phase lag) could be introduced before the system became unstable. An additional phase lag of $\phi_{\max}$ is possible before the loop $L(s)$ would intersect the point $s = -1$. A positive phase margin tends to indicate stability; a negative phase margin tends to indicate instability.

The phase and gain margins can usually be measured more easily using a Bode plot of $L(j\omega)$ than with a Nyquist plot.

Since no model used in controller design is perfectly accurate a controller should stabilize the system with some margin. This will ensure that the closed loop will remain stable in the presence of some modeling errors.

3.6 ROOT LOCUS

It is sometimes useful to plot the zeros of the characteristic polynomial as some parameter is varied (usually from 0 to ∞). Such a plot is called a *root locus*.

For instance, consider the common situation where the parameter of interest is a scale on the control. Write the plant $P = n_P/d_P$ where n_P, d_P are coprime polynomials and the controller as $H = Kn_H/d_H$ where n_H, d_H are coprime polynomials and K is a variable gain. The characteristic polynomial is

$$d_H d_P + K n_H d_P.$$

It is clear that the roots will move from the open loop poles to the open loop zeros as K increases from 0 to infinity. The situation is identical if the parameter K scales the plant.

A root locus plot shows clearly all the possible closed loop pole locations as K is varied. Since all models of physical systems have errors, it is desired that K be chosen so that not only is the closed loop stable, but so that there is a margin between the imaginary axis and the poles. This will help to ensure that errors in the model will not lead to system instability.

EXAMPLE 3.50: Consider the plant $P = (s - 2)/(s + 1)$ with a constant gain controller $H = K$. The characteristic polynomial is $s + 1 + K(s - 2)$. Fig. 3.7 shows how the closed-loop pole moves from the open-loop pole to the open-loop zero as K increases.

EXAMPLE 3.51: Consider the plant $P = 1/(s + 2)$ and the controller $H = K + \frac{4}{s}$. The characteristic polynomial is

$$\kappa(s) = s^2 + 2s + 4 + Ks.$$

A root locus diagram is shown in Fig. 3.8.

3.7 ROBUST STABILITY

In most cases, the plant model used in controller design is not perfect. We say that the system is uncertain, since we do not know the exact model. This uncertainty may arise because of modeling errors, or because the exact model is too complicated to be used in controller design. We usually know that the exact plant belongs to some set of plants $\mathcal{P}$. System uncertainty may be handled by modeling the plant as belonging to this set. The nominal plant P used in controller design may or may not be the actual plant. It is desirable that errors in the nominal plant model P will not lead to an

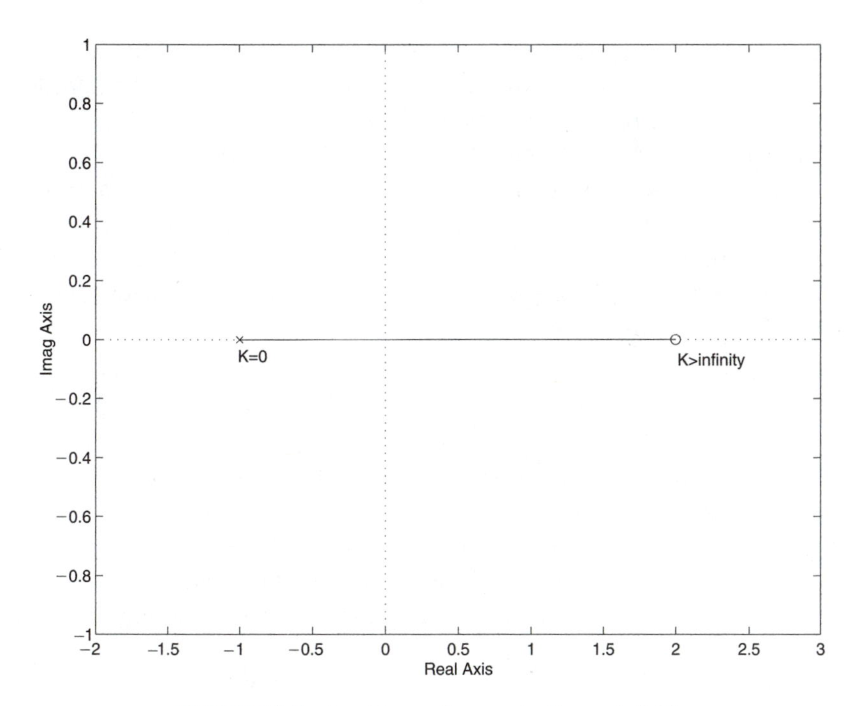

FIGURE 3.7 Root locus diagram for Example 3.50.

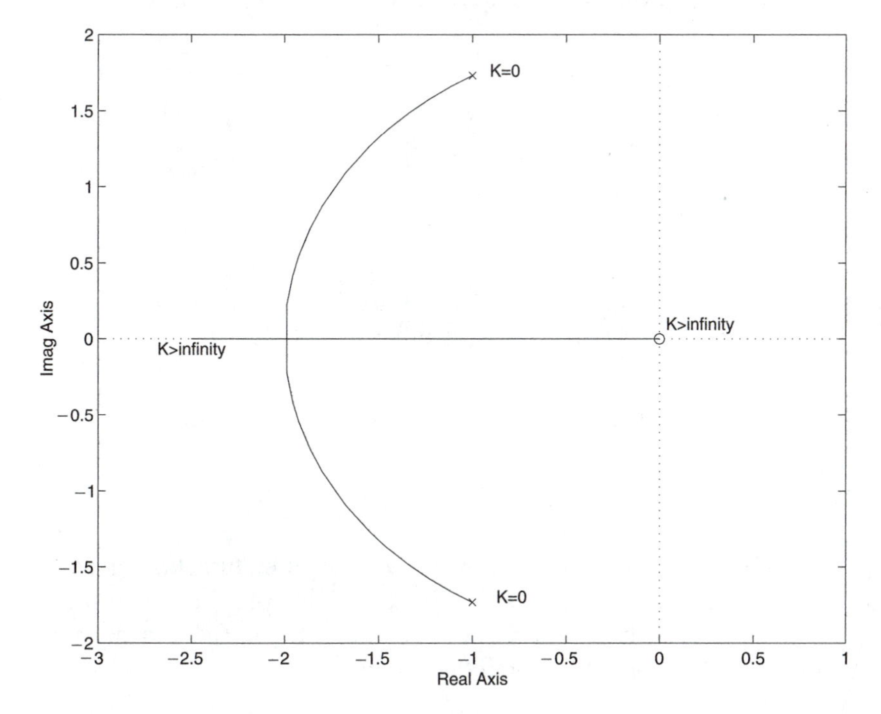

FIGURE 3.8 Root locus diagram for Example 3.51.

unstable closed loop when the controller is implemented. Thus, we look for a controller H that stabilizes every plant in the set $\mathcal{P}$. Such a controller is said to provide *robust stability*. Thus, we would like to have a test for robust stability of a controller with respect to a set of plants $\mathcal{P}$.

Recall that the Nyquist plot gives information about stability margins. Ensuring reasonable gain and phase margins is one way of allowing for plant uncertainty. For instance, if a controller stabilizes a plant P with stability margin $k_{\max}$ then we can say that the controller provides robust stability for all plants in the set

$$\mathcal{P} = \{\tilde{P} \mid \tilde{P} = kP, 1 \leq k < k_{\max}\}.$$

However, both gain and phase margins can be large and yet the Nyquist plot can pass close to the point $(-1, 0)$. Thus, simultaneous small changes in gain and phase can cause the number of encirclements to change, and thus the perturbed system will be unstable. For situations where both the gain and phase are uncertain approaches different from ensuring gain and phase margins may be more appropriate.

For a plant P and controller H write $L = PH$. We have

$$\begin{aligned}
\text{distance from } -1 \text{ to Nyquist plot} &= \inf_{\omega} |-1 - L(j\omega)| \\
&= \inf_{\omega} |1 + L(j\omega)| \\
&= \left[\sup_{\omega} \frac{1}{1 + L(j\omega)}\right]^{-1} \\
&= [\|S\|_{\infty}]^{-1}
\end{aligned}$$

where $S(s) = (1 + L(s))^{-1}$, the system sensitivity. Suppose that P is perturbed to $\tilde{P}$, having the same number of unstable poles. If

$$|\tilde{P}(j\omega)H(j\omega) - P(j\omega)H(j\omega)| < \|S\|_{\infty}^{-1} \quad \forall \omega.$$

then the number of encirclements of -1 does not change and stability is preserved.

This bound is usually very conservative: Larger perturbations could be allowed at frequencies where $P(j\omega)H(j\omega)$ is far from -1. Better robust stability criteria can be obtained by considering particular uncertainty models that include information about the uncertainty.

Example 3.52: $P(s) = 1/(s^2 + as + 1)$. This could represent the transfer function of a mass–spring–dashpot system, an RLC circuit, etc. Suppose we are sure of the general form of the model (a second-order system) but that the constant a is known only to lie in some interval $[a_{\min}, a_{\max}]$. The plant belongs to the set

$$\left\{ \frac{1}{s^2 + as + 1} \; ; \; a_{\min} \leq a \leq a_{\max} \right\}.$$

Example 3.53: Assume that the plant is stable and its transfer function is obtained by identifying the frequency response. By repeating the experiment a number of times we arrive at a nominal transfer function ("best fit"), P_0, and an upper bound on the error at each frequency, $W_2(j\omega)$. The plant belongs to the set

$$\mathcal{P} = \{\tilde{P} \in \mathcal{H}_\infty ; |\tilde{P}(j\omega) - P(j\omega)| \leq |W_2(j\omega)|\}.$$

The first example involves a *structured uncertainty*: It is parametrized by a finite number of scalar parameters. The second example contains *unstructured uncertainty*: The family of possible plants contains a ball in $\mathcal{H}_\infty$. Unstructured uncertainty is characterized by a nominal plant P together with a weighting function W_2. Trial and error using Bode plots is usually required to select an appropriate weight W_2. Typically, $|W_2(j\omega)|$ is an increasing but proper function of ω since uncertainty increases with frequency.

We will concentrate on unstructured uncertainty for two reasons:

1. Most models neglect, or are otherwise incorrect in their treatment of, high-frequency dynamics, and there are usually other unmodeled "unstructured" dynamics.
2. There are general analysis and synthesis methods to handle unstructured uncertainty.

The price we pay for the simple description is conservativeness: The set $\mathcal{P}$ usually contains many plants that never occur.

3.7.1 Multiplicative Uncertainty

Define for some function W_2, with no imaginary axis poles,

$$\mathcal{M}(P, W_2) = \{\tilde{P} \text{ has the same number of unstable poles as } P \text{ and } \tilde{P} = (I + \Delta)P, \; \|\Delta(j\omega)\| \leq |W_2(j\omega)|\}.$$

Note that perturbed plants are of the form

$$\tilde{P} = (1 + \Delta)P.$$

Let H be some controller that stabilizes P. What conditions must Δ satisfy so that H stabilizes $\tilde{P}$ for all allowable perturbations Δ?

THEOREM 3.54: Assume H stabilizes P. The controller H provides robust stability for $\mathcal{M}(P, W_2)$ if and only if

$$\sup_{\omega} \|[W_2 PH(I + PH)^{-1}](j\omega)\| < 1.$$

□ *Proof:* We only give the proof for SISO plants. Construct the Nyquist plot of

$$\tilde{P}H = (1 + \Delta)PH = (1 + \Delta)L$$

where $L = PH$. Assume that $\tilde{P}$ has the same $j\omega$-axis poles as P and the same number of unstable poles as P. Then, H will stabilize $\tilde{P}$ if and only if the Nyquist plot of $\tilde{P}H$ does not pass through -1 and it encircles -1 exactly as many times as does the Nyquist plot of L. The number of encirclements will not change for all possible perturbations Δ, $|\Delta(j\omega)| \leq |W_2(j\omega)|$ if and only if for every ω the point -1 lies outside the disk with center $L(j\omega)$ and radius $|W_2(j\omega)L(j\omega)|$. (See Fig. 3.9.) In other words, the perturbed system is stable if and only if

$$|W_2(j\omega)L(j\omega)| < |1 + L(j\omega)|$$

or,

$$\left| W_2(j\omega)\frac{L(j\omega)}{1 + L(j\omega)} \right| < 1$$

for all ω. The result follows. ■

EXAMPLE 3.55: Many systems have a time delay. For example, a time delay of 0.1 seconds is present in the idle speed control example (Example 2.21). Since most controller design methods apply to rational transfer functions, it is usual to approximate a time delay by a rational function. Any proper rational function $A(s)$ has a limit as $s \to \infty$ in any direction while $e^{-j\omega T}$ oscillates as $\omega \to \infty$. Thus, any rational approximation to an exponential is increasingly inaccurate at higher frequencies and has an error when measured in the $\mathcal{H}_\infty$ norm.

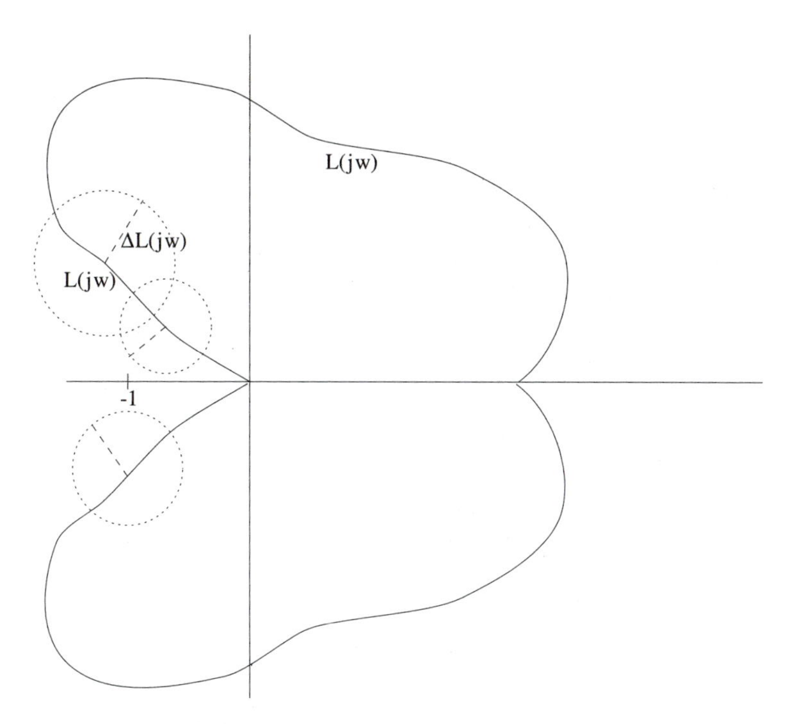

FIGURE 3.9 Nyquist diagram with loop perturbations.

In this example we will approximate e^{-Ts} by $1 - Ts$. Let G indicate the purely rational part of a given transfer function P and assume that G is strictly proper. Then $P = e^{-Ts}G$ and the nominal plant used in design is $P_o = G(1 - Ts)$. We want to find a weighting function W_2 so that P is included in the multiplicative uncertainty set $\mathcal{M}(P_o, W_2)$. We need a function W_2 so that

$$|W_2(j\omega)| > \left| \frac{P(j\omega)}{P_o(j\omega)} - 1 \right|$$

for all ω. In this case,

$$\frac{P(j\omega)}{P_o(j\omega)} - 1 = \frac{\exp(-0.1j\omega)}{1 - 0.1j\omega} - 1.$$

A plot of this function for $T = 0.1$ yields a magnitude plot that increases from 0 at $\omega = 0$ to 1 as $\omega \to \infty$. (See Fig. 3.10.) A "corner" in the overall increase occurs around $\omega_0 = 40$. The graph suggests that the weighting

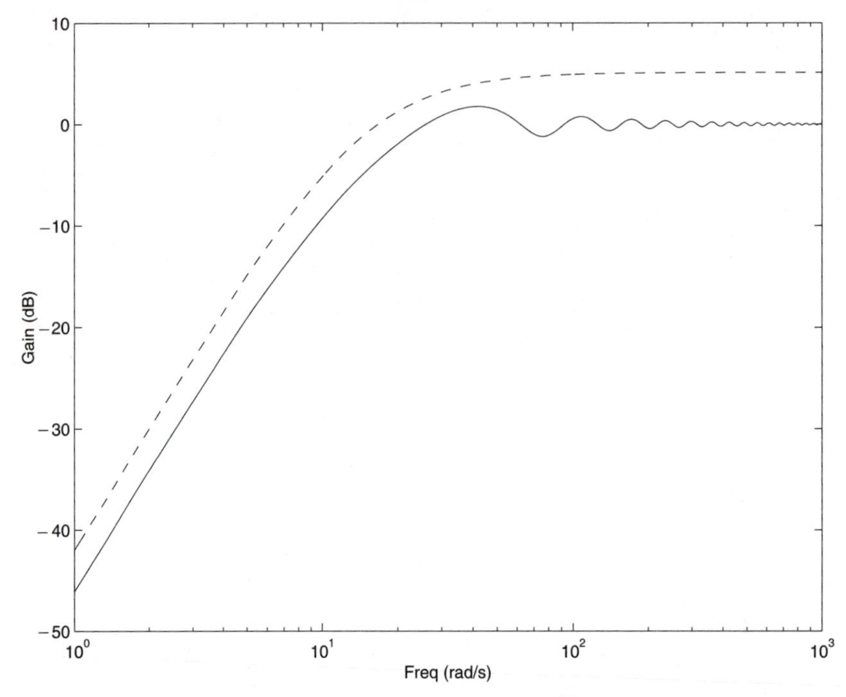

FIGURE 3.10 Magnitudes of frequency response of $e^{-0.1s}/(1-0.1s)-1$ (solid line) and $1.8s^2/(s+15)^2$ (dashed line).

function should have a pole at about −40, a zero at 0, and a limiting value above 1 at infinity. The sharp increase for low frequencies suggests a double zero at 0. Since we want the graph to be flat at high frequencies, the numerator and denominator must have the same degree. An initial guess is

$$W_2(s) = \frac{s^2}{(s+40)^2}.$$

Some trial and error yields

$$W_2(s) = \frac{1.8s^2}{(s+15)^2},$$

shown in Fig. 3.10. Since $|W_2(s)| > 1$ at infinity, some uncertainty due to other modeling errors is included in the uncertainty model.

3.7.2 Additive Uncertainty

Now consider additive uncertainty. Define for a function W_2 with no poles on the imaginary axis

$$\mathcal{A}(P, W_2) = \{\tilde{P} \text{ has the same number of unstable poles as } P \\ \text{and } \|\tilde{P}(j\omega) - P(j\omega)\| \leq |W_2(j\omega)|\}.$$

Perturbed plants are of the form

$$\tilde{P} = P + \Delta$$

where $\|\Delta(j\omega)| \leq |W_2(j\omega)\|$. Example 3.53 involves additive uncertainty.

Theorem 3.56: Assume H stabilizes P. The controller H robustly stabilizes $\mathcal{A}(P, W_2)$ if and only if

$$\sup_{\omega} \|[W_2 H S](j\omega)\| < 1.$$

□ *Proof:* As in Theorem 3.54, we only prove the SISO case. In the case of additive uncertainty, the perturbed loop is

$$\tilde{L} = \tilde{P}H = PH + \Delta H.$$

In order that H stabilizes $\tilde{P}$ as well as P, the number of encirclements of -1 of the two Nyquist plots must be the same. This means that for each frequency ω, the point -1 must lie outside the disk with center $L(j\omega)$ and radius $|W_2(j\omega)H(j\omega)|$. In other words, we require

$$|W_2(j\omega)H(j\omega)| < |1 + L(j\omega)|.$$

Or,

$$|W_2(j\omega)H(j\omega)S(j\omega)| < 1.$$

Thus, if

$$\|W_2 H S\|_\infty < 1,$$

then H stabilizes $\tilde{P}$ as well as P. ■

Example 3.57: Consider the nominal model

$$P = \frac{1.43}{(1 + 0.1s)(s + 1.3)(s + 1.1)}$$

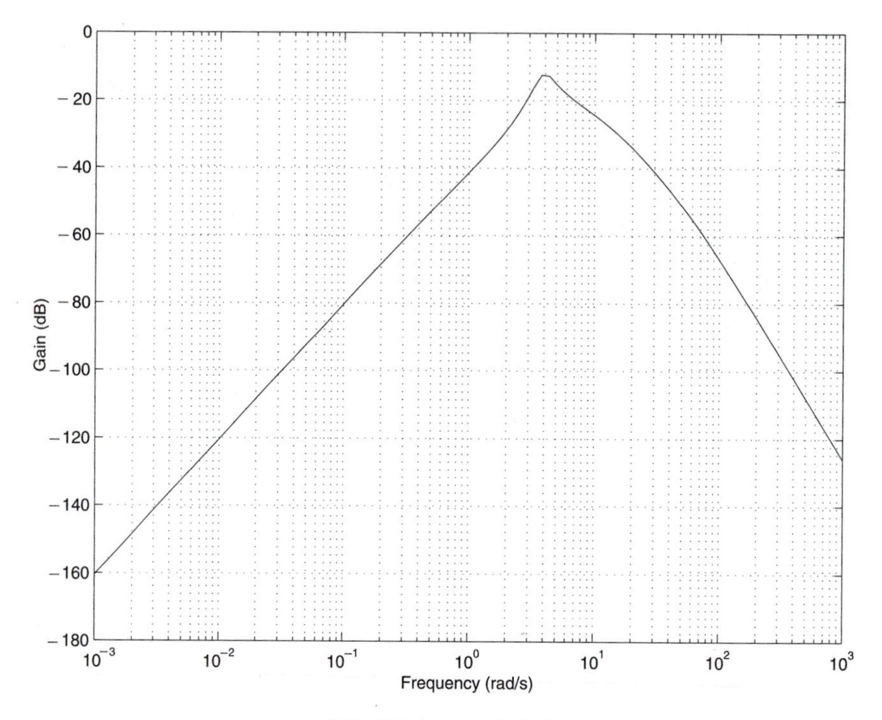

FIGURE 3.11 $|W_2HS|$.

with an additive uncertainty model with

$$W_2 = 5\frac{s^2}{(s+1)^2(s+10)(s+50)}.$$

The gain of W_2HS for

$$H = \frac{20}{1+s/5}$$

is in Fig. 3.11. It is well below 1 at all frequencies. The Nyquist plot is in Fig. 3.12. Since both P and H are stable, the encirclement of -1 shows that this controller is not stabilizing. The robust stability theorem given earlier states that the controller stabilizes all plants in the set if $\|W_2HS\|_\infty < 1$ and the controller stabilizes the nominal plant. The inequality $\|W_2HS\|_\infty < 1$ implies that the number of encirclements of -1 in the Nyquist plot is the same for all plants in the uncertainty set. So, in this example, the controller not only does not stabilize the nominal plant, it does not stabilize any plant in the uncertainty set.

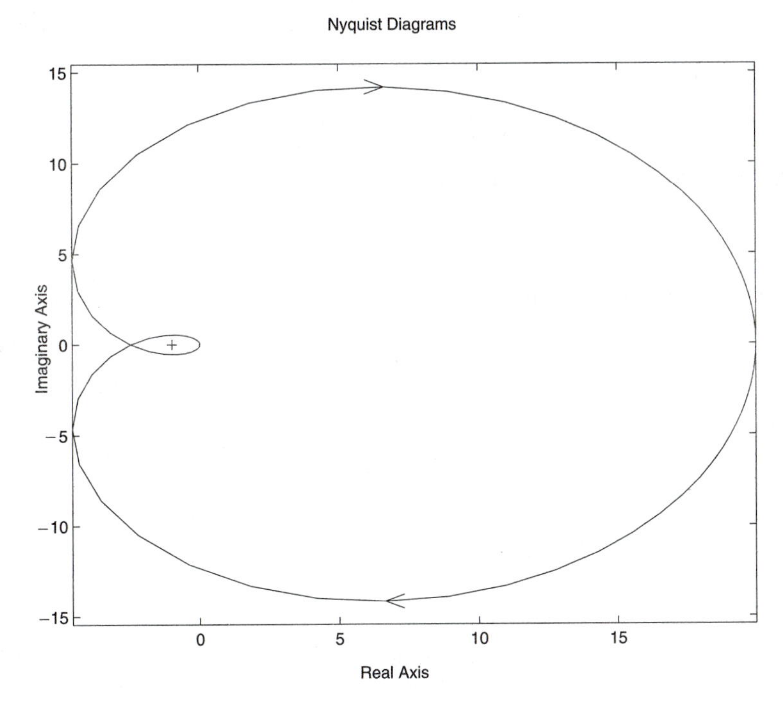

FIGURE 3.12 Nyquist plot of PH.

An identical observation can be made about multiplicative uncertainty: The controller must stabilize the nominal plant and satisfy the inequality $\|W_2T\|_\infty < 1$.

It is desirable, whenever possible, to use an uncertainty model that reflects the actual plant uncertainty.

For instance, suppose only the gain is uncertain:

$$\tilde{P} = kP, \; k > 1.$$

This is a *very* structured uncertainty. For what range of k is there a single stabilizing controller? Assume that P is unstable. (If P is stable, the zero controller will stabilize P and the answer is $k_{\max} = \infty$.)

Suppose we imbed this uncertainty into a multiplicative uncertainty model

$$\begin{aligned} W_2(s) &= \frac{\tilde{P}}{P} - 1 \\ &= k - 1. \end{aligned}$$

Let $\mathcal{S}(P)$ indicate the set of all controllers that stabilize P. Consider the problem

$$\inf_{H \in \mathcal{S}(P)} \|(k-1)T\|_\infty$$

where the complementary sensitivity $T = PH/(1+PH)$. Define

$$\gamma = \inf_{H \in \mathcal{S}(P)} \|T\|_\infty.$$

Then, for stability we must have

$$k - 1 < \tfrac{1}{\gamma}$$

$$k < 1 + \tfrac{1}{\gamma}.$$

Write $P = n_P/d_P$ where n_P and d_P are coprime polynomials, and similarly $H = n_H/d_H$. Then T can be written

$$T = \frac{n_P n_H}{n_P n_H + d_P d_H}.$$

Since d_P has r.h.p. zeros

$$\inf \|T\|_\infty \geq 1,$$

and so

$$\gamma > 1.$$

Thus, the allowable range of k is

$$k_{\max} < 2.$$

Using an additive model of uncertainty yields a similar result.

However, it can be shown that

$$k_{\max} = \left(\frac{\gamma+1}{\gamma-1}\right)^2, \tag{3.19}$$

which is a much better result.

NOTES AND REFERENCES

Theorem 3.6 generalizes to systems with nonrational transfer functions. Under extremely general conditions, a linear time-invariant system is L_2-stable if and only if its transfer function is in $\mathcal{H}_\infty$ [38]. Theorem 3.8 was first given in [3] and a more efficient refinement was published in [2]. A general proof of Theorem 3.30 is in [44]. Proofs of the Routh–Hurwitz Criterion can be found in, for example, [8, Sect. 8.5], [22, Sect. II.6]. Generalizations of the Nyquist criterion (Corollary 3.45) and of the robust stability theorems in

Section 3.7 to multivariable systems can also be found in [44]. They depend on the stable coprime factorization theory explained later in this book. Some other unstructured uncertainty models and robust stability results may be found in [13]. The robust stability result (3.19) is due to [26].

EXERCISES

1. *Show that $\|\mathcal{G}\|_{2,2} < \infty$ if for all $u \in L_2(0, \infty)$ the output y is in $L_2(0, \infty)$. (Hint: Prove first that $\mathcal{G}$ is a closed operator. Then use the Closed Graph Theorem and the fact that L_2 is complete.)
2. *Prove Theorem 3.11 when $E = 0$. That is, show that "$\leq$" in the inequality (3.6) cannot be replaced by "$<$."
3. Prove Theorem 3.6 for a system with m inputs and p outputs.
4. Let G be the transfer matrix of an externally stable system with m inputs and p outputs y_g and let H be the transfer matrix of another externally stable system with p inputs u_h and q outputs. Suppose the two systems are connected in series so that $u_h = y_g$. Show that the series connection has transfer matrix HG and that the L_2 gain of the series connection is less than $\|G\|_\infty \|H\|_\infty$.
5. A consequence of Theorem 3.7 is that if Z_a has an imaginary eigenvalue $j\omega_o$, $|G(j\omega_o)| = a$. This may be used in a method to calculate the $\mathcal{H}_\infty$ norm of a transfer function that is more efficient than the simple bisection algorithm given in the notes. (See Fig. 3.13.)

   (a) Write an algorithm for strictly proper transfer functions in $R\mathcal{H}_\infty$ that uses the above fact.

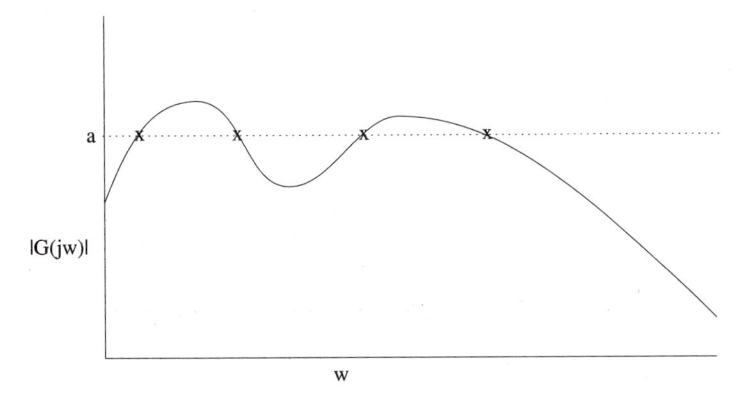

FIGURE 3.13 Figure for exercise on calculation of $\mathcal{H}_\infty$ norm.

(b) Use your algorithm to calculate the transfer function in Example 3.9 with an error less than 0.01.

6. *Generalize Theorem 3.12 to multivariable systems. (Hint: What matrix norm should replace absolute value in the integral?)

7. Determine whether the closed loop is externally stable for the following transfer functions.

(a) $P(s) = \dfrac{1}{s+4}, \; H(s) = \dfrac{10}{s+7}$

(b) $P(s) = \dfrac{1}{s^2-4}, \; H(s) = \dfrac{s-2}{s+2}$

(c) $P(s) = \dfrac{1-s}{s}, \; H(s) = 1$

(d) $P(s) = \dfrac{s-1}{s^2+s-2}, \; H(s) = \dfrac{4}{s-1}$

(e) $P(s) = \dfrac{2}{s^3+3s^2+16s+3}, \; H(s) = \dfrac{s+1}{s}$

8. Which systems in Chapter 2, Exercise 3, are externally stable and which are internally stable? Determine the $\mathcal{H}_\infty$ norm of the externally stable systems.

9. Verify expressions (3.14)–(3.16).

10. A more general feedback system than that used in this chapter (Fig. 3.2) is shown in Fig. 3.14.
Let

$$P = \tfrac{n_P}{d_P}$$
$$H = \tfrac{n_H}{d_H}$$
$$F = \tfrac{n_F}{d_F}.$$

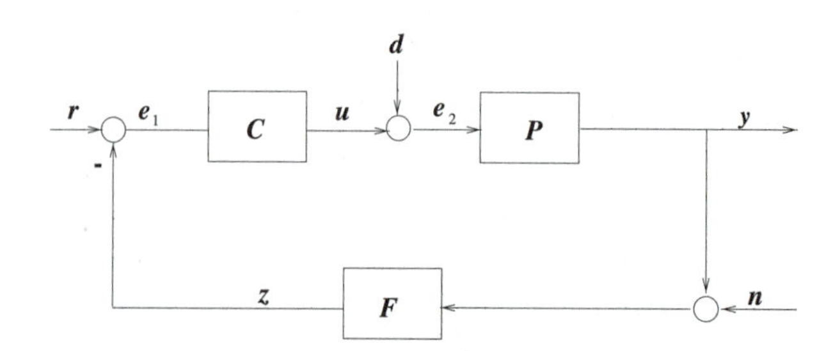

FIGURE 3.14 General feedback system.

(a) Calculate the nine maps from (r, d, n) to (u, y, z).

(b) Write a definition of external stability that is analogous to Definition 3.1.

(c) The characteristic equation is

$$\kappa = n_P n_H n_F + d_P d_H d_F.$$

Prove the system is externally stable if and only if κ has no r.h.p. zeros.

11. Check whether the following pairs are stabilizable:

(a)

$$A = \begin{bmatrix} 2.2 & -.8 \\ -3.2 & -.2 \end{bmatrix}, B = \begin{bmatrix} 2 \\ 8 \end{bmatrix}$$

(b)

$$A = \begin{bmatrix} 1 & 0 \\ 0 & -1. \times 10^{-9} \end{bmatrix}, B = \begin{bmatrix} 1. \times 10^{-9} \\ 0 \end{bmatrix}$$

(c)

$$A = \begin{bmatrix} 1 & 0 \\ 0 & -10 \end{bmatrix}, B = \begin{bmatrix} 1. \times 10^{-9} \\ 0 \end{bmatrix}$$

12. Fill in the details of the proof of Thereom 3.30 by showing that the closed-loop system is detectable.

13. Which of the following functions are transfer functions of an externally stable system? Find a stabilizable/detectable realization of each transfer function. (The easiest way to do this is to construct a minimal realization.)

(a) $$\frac{1}{2s^2 + 5s + 5}$$

(b) $$\frac{8s}{s - 11}$$

(c) $$\frac{1}{s+3} + \frac{2}{s+10}$$

14. **(a)** For the usual feedback configuration (Fig. 3.2) derive the transfer function matrix Δ_y describing the maps from (r, d) to (u, y).

(b) Define $e_1 = r - y,\ e_2 = d + u$. Show that $\Delta_y \in M(R\mathcal{H}_\infty)$ if and only if $\Delta \in M(R\mathcal{H}_\infty)$. Hence feedback stability may be defined using Δ or Δ_y.

15. Consider a SISO plant $P = n_P/d_P$ where n_P, d_P are coprime polynomials, and a controller $H = n_H/d_H$ where n_H, d_H are coprime polynomials. Show that if the closed loop (P, H) is stable then no r.h.p. pole of P can be a r.h.p. zero of H and vice versa. Hence, in a stable feedback system there can be no unstable pole–zero cancellation between P and H.

16. Use the quadratic formula to show that both zeros of $s^2 + as + K$ are in the open l.h.p. if and only if $a > 0$ and $K > 0$.

17. Let $P(s) = 1/[s(1 + 2s)^2]$, $H(s) = K$. For what range K (if any) is the closed loop stable?

18. Consider the plant

$$P(s) = \frac{1}{(s + 1)(s + 2)}$$

and the proportional-integral controller

$$H(s) = K_1 + \frac{K_2}{s}.$$

(a) Set $K_2 = 0$. For what range of K_1 is the closed loop stable?
(b) Determine the range of K_2, in terms of K_1, for which the closed-loop system is stable.
(c) Graph the Nyquist plots of PH for $K_1 = 5$ and several different values of K_2.

19. Prove or disprove the following statement: If a given controller stabilizes two plants, then they have the same number of poles in the closed right half-plane.

20. The gain $k > 0$ of the plant

$$P = \frac{k}{s(s + 1)}$$

is uncertain. Consider the controller

$$H = 1 + 1/s + s/(1 + 0.01s).$$

(a) For what range of k is the system stable? Use a Routh–Hurwitz table. Also provide a root locus plot.
(b) Estimate the gain and phase margins when $k = 1$.

21. Plot the root locus as the gain K is varied for each of the following plants with a proportional controller $H = K$.

(a) $P(s) = \frac{1}{(s+1)(s+10)}$

(b) $P(s) = \frac{(s+2)}{(s+1)(s+10)}$

(c) $P(s) = \frac{(s-2)}{(s+1)(s+10)}$

22. Graph the Nyquist plots for the plants in the previous question for $K = 4$. Is the closed loop stable? If so, determine the gain and phase margins using Bode plots.

23. For a strictly proper plant, the root locus will have asymptotes tending to ∞. (This can be intepreted as the root locus moving from poles to zeros "at infinity.")

(a) Prove that the angles of the asymptote(s) are

$$\pm\frac{(2m+1)\pi}{p-z}$$

where p is the number of poles, z is the number of (finite) zeros, and $0 \leq m < p - z$.

(b) Prove that if a transfer function G has relative degree 1, and all its poles and zeros are in the open left half-plane, then the closed-loop system is stable for all sufficiently large constant feedback $K > 0$.

24. Consider the set of plants

$$\mathcal{P} = \{P \in \mathcal{H}_\infty \ ; \ \mathrm{Re}P(j\omega) \geq 0\}.$$

Find a sufficient condition for a stable controller to robustly stabilize all plants in P. Is it necessary in some sense?

25. Suppose the controller for a plant P is known only to be an additive uncertainty set $\mathcal{A}(H_o, W_2)$ around a nominal controller H_o. The nominal controller stabilizes P.

(a) Give an example of a situation of this type.

(b) Prove that the closed loop is stable for all possible controllers if and only if

$$\|W_2(I + PH)^{-1}P\|_\infty < 1.$$

26. Let the loop gain $L(s)$ be

$$L(s) = \frac{K}{s(1 + s/2)^2}$$

for some real K.

(a) Give a root locus plot.
(b) Determine the range of K for stability.
(c) Give a Nyquist plot for a value of K where the closed loop is stable, and a Nyquist plot for a value where the closed loop is unstable.

27. Let

$$P(s) = \frac{k}{s}$$

where the gain k is known to lie in the interval [1,10].

(a) Model the plant uncertainty using multiplicative perturbations.
(b) Prove that a controller with constant gain provides robust stability for this set.

28. Consider the multiplicative uncertainty set $\mathcal{M}(P_0, W_3)$ where W_3, P_0 have no imaginary axis poles. Find a W_2 so that $\mathcal{M}(P_0, W_3) \subset \mathcal{A}(P_0, W_2)$.

Useful MATLAB routines: "nyquist," "rlocus," "rlocfind," "bode," "eig."

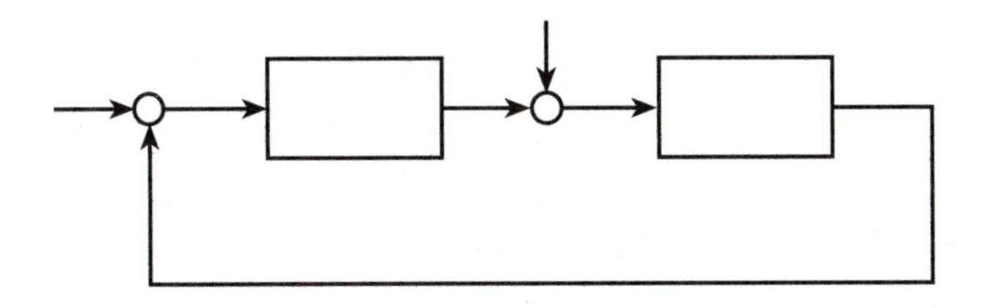

IV
BASIC LOOPSHAPING

Clearly any closed-loop system must be stable. However, since most systems are already stable, the purpose of controller design is generally to improve performance (without destroying stability). In this chapter we discuss some of the more common design criteria in the context of SISO systems.

Choice of performance specifications is a very important step in controller design. Once the performance criteria are specified precisely, that is, what we mean by "good behavior," the problem of controller synthesis is often routine. However, the choice of performance is not a trivial one. Desired performance often conflicts with stability requirements. Furthermore, many performance criteria conflict with other criteria and a trade-off must be made. Finally, a design engineer might choose a simple controller with acceptable performance over one that has better performance, but is considerably more difficult to implement.

An important function is the transfer function from the input r to the error e:

$$S = \frac{1}{1 + PH},$$

where we assume here that H stabilizes P. This transfer function is called the *sensitivity function*. The reason for the name sensitivity function is as follows: Let T indicate the transfer function from an input r to the output y as in Fig. 3.2:

$$T = \frac{PH}{1 + PH}.$$

One way to quantify how sensitive T is to changes in P is to take the limiting ratio of a relative perturbation in T to one in P:

$$\lim_{\Delta P \to 0} \frac{\Delta T/T}{\Delta P/P} = \frac{dT}{dP} \cdot \frac{P}{T} = \frac{1}{(1 + PH)} = S.$$

Thus, S is the sensitivity of the closed-loop transfer function to an infinitesimal perturbation in P. If $|S(s)| \leq 1$, then errors in the model will have less effect on the closed loop than they will on the open loop. The sensitivity of an open loop is 1. Thus, to the extent that feedback reduces sensitivity, it reduces the need for plant identification.

Example 4.1 (*Noninverting Operational Amplifier*): The operational amplifier (op amp) is a crucial component in electronic devices. Essentially, it multiplies an input voltage $(V_i - V_a)$ by some factor A. Since the multiplication factor A is difficult to manufacture and maintain at a precise level, other configurations are used to improve accuracy. (The desire to improve amplifier performance lay behind Bode's fundamental work on feedback control.)

The simple noninverting configuration is shown in Fig. 4.1. The input impedance is very high, and so we can neglect current flowing through the op amp. A current balance at node a yields

$$\frac{V_a}{R_1} = \frac{V_o - V_a}{R_2}.$$

Rearranging and using $V_o = A(V_i - V_a)$ to eliminate V_a, we obtain

$$\frac{V_o}{V_i} = \frac{k}{1 + k/A}$$

where $k = 1 + R_2/R_1$. The op amp gain A is generally very large and so the compensated op amp gain is close to k.

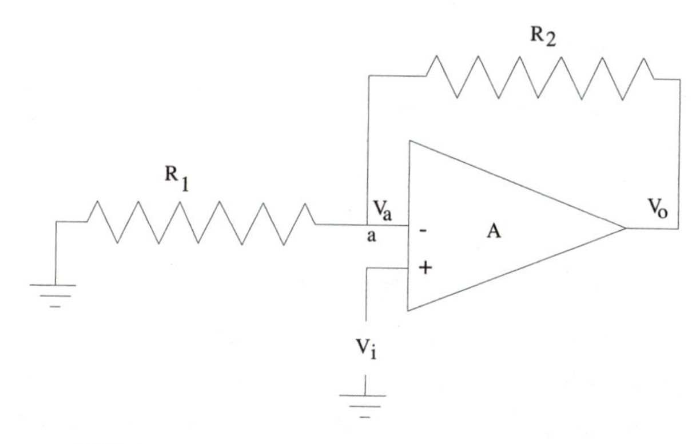

FIGURE 4.1 Noninverting operational amplifier.

The sensitivity of the uncompensated op amp to changes in the parameter A is 1. The sensitivity of the compensated system transfer function

$$G = \frac{V_o}{V_i}$$

to changes in A is

$$\begin{aligned} S &= \frac{\partial G}{\partial A}\frac{A}{G} \\ &= \frac{k}{1+\frac{k}{A}}\frac{1}{A} \\ &= \frac{k}{A+k}. \end{aligned}$$

The sensitivity of the compensated op amp is always less than 1. Since A is very large, the sensitivity is very small.

This discussion assumed that both the original and the perturbed closed loops are stable. The controller needs to be chosen so that it stabilizes all possible plants. The less information that is available about a plant, the less possible it is in general to select a controller to reduce sensitivity. For example, increasing the loop gain $L = PH$ will tend to decrease the sensitivity. However, unless the designer is careful about the phase, the plot of $L(j\omega)$ may pass through the -1 point and the closed loop will be unstable.

4.1 COMMON CONTROLLERS

Before describing some common performance criteria, we describe several classes of controllers that are used to stabilize systems and achieve performance criteria.

4.1.1 PID Controllers

We now discuss a class of controllers commonly used in control, PID controllers.

Proportional

$$H = K.$$

This is the simplest type of controller and only the gain of the system is affected. Typically, increasing K will improve the closed-loop performance. However, increasing K generally decreases stability margins. A more complicated controller is often required.

PI

It is shown later that the inclusion of an integral term (K_I/s) in the controller will reduce the steady-state error in response to a step to zero, provided that the controller is stabilizing. This removes the need for a large proportional gain to reduce the error. The pole at $s = 0$ also helps attenuation at other low frequencies. However, transient response is generally slower than with purely proportional feedback.

PD

The inclusion of a "derivative" term $(K_d s/(1 + \epsilon s))$ where $\epsilon > 0$ is small often smooths the system response by penalizing sharp changes in the response. For a second-order system, the damping is increased. A PD controller tends to lead to a faster system response. Too large a value of K_d, however, leads to a closed loop with a large gain at high frequencies. This is undesirable since it causes amplification of any undesirable high-frequency "noise" in the input signal.

PID

Commonly, a PID controller is used:

$$H = K_p + K_I/s + K_d s/(1 + \epsilon s).$$

Such a controller can incorporate the beneficial features of proportional, integral, and derivative control. PID controllers are very common in applications, and can be bought from a number of manufacturers. The parameters K_p, K_I, and K_d are generally "tuned" until satisfactory closed-loop performance is obtained.

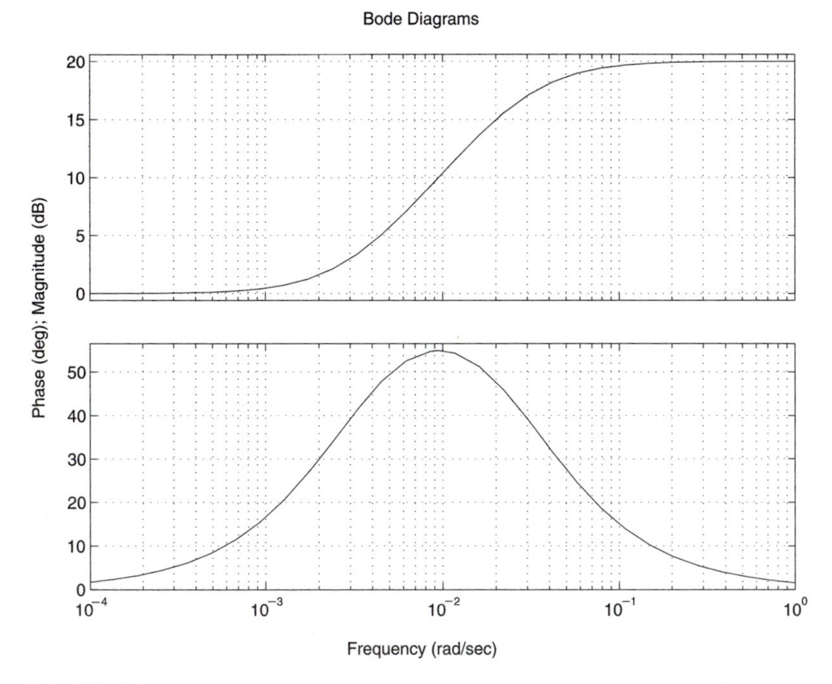

FIGURE 4.2 Bode plot of lead compensator $(1 + \alpha Ts)/(1 + Ts)$, $T = 33$, $\alpha = 10$.

4.1.2 Phase Lead/Lag Compensation

Lead and lag compensators are of the form

$$K\frac{1 + s/z}{1 + s/p}$$

where $z < p$ for a lead compensator and $z > p$ for a lag compensator. (See Fig. 4.2 and 4.3.)

Note that although we are discussing lead/lag compensation independently of PID controllers, a lead compensator with $z \ll p$ is essentially a differentiator, whereas a lag compensator with $z \gg p$ is an integrator. PID controllers tend to be used to attain transient performance goals; lag/lead controllers are used primarily to shape the loop PH in order to improve stability margins and other frequency domain performance goals.

Phase Lead

$$H(s) = \frac{1 + \alpha Ts}{1 + Ts}, \qquad \alpha > 1.$$

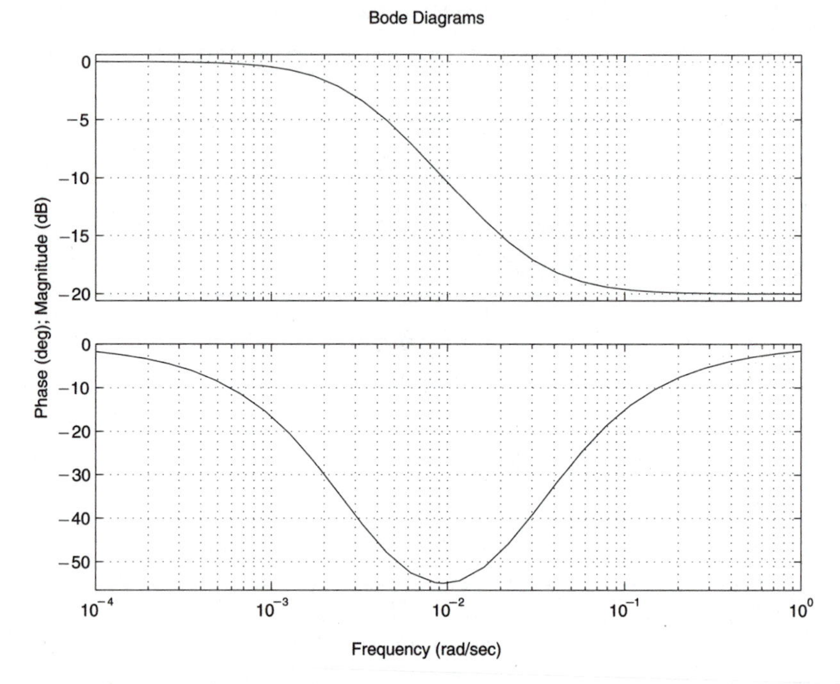

FIGURE 4.3 Bode plot of lag compensator $(1 + Ts)/(1 + \alpha Ts)$, $T = 33$, $\alpha = 10$.

As its name suggests, this controller introduces additional phase in the system. The maximum phase lead occurs at

$$\omega_{\max} = \frac{1}{T\sqrt{\alpha}}$$

and its value is

$$\phi_{\max} = \sin^{-1}\left(\frac{\alpha - 1}{\alpha + 1}\right).$$

The gain at the frequency $\omega_{\max}$ is

$$10 \log_{10} \alpha \, dB.$$

Increasing the phase lead at an appropriate point improves the system phase margin. A phase lead controller also increases the dominant pole of the system. (This is known as increasing bandwidth, since the frequency response is large until the dominant pole is reached.) This leads to a faster system response. Unfortunately, it also means that the system is also affected by higher frequency noise.

Phase Lag

$$H(s) = \frac{1 + Ts}{1 + \alpha Ts}, \qquad \alpha > 1.$$

This controller introduces additional lag in the phase. However, phase lag is generally a destabilizing influence. The attenuation in this type of controller increases the gain margin. The formulas for the maximum phase lag and its location are the same as for the phase lead compensator except that $\phi_{\max}$ is negative:

$$\omega_{\max} = \frac{1}{T\sqrt{\alpha}}$$

$$\phi_{\max} = -\sin^{-1}\left(\frac{\alpha - 1}{\alpha + 1}\right).$$

Design is generally so that the phase angle peaks at a frequency below where the stability margin is determined.

Lag compensators are most often used to reduce the response to high-frequency output disturbance such as sensor noise. A disadvantage, as with *PI* control, is that since the bandwidth is reduced, the closed-loop response may be sluggish.

Phase lag and lead compensators are often used in series with a constant gain controller. As for *PID* controllers, a phase lead and a phase lag controller are sometimes used in series in order to combine the benefits of each.

In the following sections some performance criteria are described.

4.2 TRACKING

In this section both time and frequency domain functions are used. The frequency domain functions (i.e., Laplace transforms) are indicated explicitly in this section by $\hat{y}(s)$. Suppose we wish our closed loop to track a given signal asymptotically. Assume that the closed loop is stable. The key tool to determine whether a closed loop system tracks a signal is the Final Value Theorem.

Theorem 4.2 (*Final Value Theorem*): If $\hat{y}(s)$ is the Laplace transform of an integrable function such that $\lim_{t\to\infty} y(t)$ exists, then

$$\lim_{t\to\infty} y(t) = \lim_{s\to 0} s\hat{y}(s)$$

where s on the right-hand side is real.

□ *Proof:* Write $\lim_{t\to\infty} y(t) = \alpha$. For real s, $s > 0$,

$$\int_0^\infty e^{-st} dt = \frac{1}{s}$$

and so for $s > 0$,

$$\begin{aligned} |s\hat{y}(s) - \alpha| &= \left| s \int_0^\infty e^{-st}(y(t) - \alpha) dt \right| \\ &\leq s \int_0^\infty e^{-st} |y(t) - \alpha| dt. \end{aligned}$$

Therefore, for any $T > 0$

$$\begin{aligned} |s\hat{y}(s) - \alpha| &\leq s \int_0^T |y(t) - \alpha| dt + \int_T^\infty e^{-st} |y(t) - \alpha| s dt \\ &\leq s \int_0^T |y(t) - \alpha| dt + \sup_{t>T} |y(t) - \alpha| s \int_T^\infty e^{-st} dt \\ &\leq s \int_0^T |y(t) - \alpha| dt + \sup_{t>T} |y(t) - \alpha|. \end{aligned}$$

For any $\epsilon > 0$, choose T so that the second term on the right-hand side is less than $\epsilon/2$. Then choose R small enough that the first term is less than $\epsilon/2$ for $s < R$. Thus, for any $\epsilon > 0$ we can choose R so that for $s < R$

$$|s\hat{y}(s) - \alpha| < \epsilon.$$

It follows by definition of the limit that

$$\lim_{s\to 0} s\hat{y}(s) = \lim_{t\to\infty} y(t). \qquad \blacksquare$$

Let r be the reference input or ideal response and y the actual response. Referring to Fig. 3.2, the signal e is the *tracking error*. Defining $\hat{S} = (1 + \hat{P}\hat{H})^{-1}$,

$$\hat{e} = \hat{S}\hat{r}. \tag{4.1}$$

Consider first a step input:

$$r(t) = \begin{cases} r_0 & t \geq 0 \\ 0 & t < 0. \end{cases}$$

An application of this type of problem is a temperature-control thermostat in a room: When the setting is changed, the room temperature should eventually change to the new setting. The Laplace transform of a unit step is $\hat{r}(s) = \frac{1}{s}$. Therefore, using (4.1), the steady-state response to a unit step is

$$\lim_{t\to\infty} e_{\text{step}}(t) = \lim_{s\to 0} s\hat{S}(s)\frac{1}{s} = \lim_{s\to 0} \hat{S}(s).$$

The steady-state error has magnitude $|\hat{S}(0)|$. The closed loop tracks a step if and only if $\hat{S}$ has at least one zero at 0.

Another common reference input is a ramp,

$$r(t) = \begin{cases} r_0 t & t \geq 0 \\ 0 & t < 0, \end{cases}$$

where r_0 is some constant. This situation arises in, for example, a radar dish designed to track a satellite with constant orbital velocity. The Laplace transform of a unit ramp is $1/s^2$. Using the Final Value Theorem again, the closed loop tracks a ramp if and only if $\hat{S}$ has at least two zeros at 0.

Corollary 4.3 (*Corollary to Final Value Theorem*): A stable closed loop tracks a step if $\hat{S}$ has at least one zero at 0. It tracks a ramp if $\hat{S}$ has at least two zeros at 0.

Notice that in both cases an open-loop pole at $s = 0$ cancels a pole at $s = 0$ in the reference input $\hat{r}$, resulting in no r.h.p. poles for $\hat{e}$. More generally, write

$$P = \frac{n_P}{d_P}, \quad H = \frac{n_H}{d_H}$$

where n_P, etc., are coprime polynomials. Since

$$\hat{S} = \frac{d_P d_H}{n_P n_H + d_P d_H},$$

the loop transfer function $\hat{P}\hat{H}$ must have poles at the r.h.p. poles of the reference input $\hat{r}$ in order that the error $\hat{e}$ have no r.h.p. poles. This is known as the *Internal Model Principle*.

Example 4.4: $\hat{P}(s) = 1/s(s+a)$, $\hat{H} = K$, $\quad a > 0$.

$$\hat{S}(s) = \frac{1}{1+\hat{P}K}(s) = \frac{(s)(s+a)}{(s)(s+a)+K}$$

We have

$$\lim_{s\to 0} s\hat{S}(s) \cdot \frac{1}{s} = 0, \quad \lim_{s\to 0} s\hat{S}(s)\frac{1}{s^2} \neq 0.$$

The closed loop will track a step but not a ramp.

EXAMPLE 4.5: Consider the same plant as in Example 4.4, but with a controller

$$\hat{H}(s) = K_1 + K_2/s + K_3 s/(1 + \varepsilon s), \qquad 0 < \varepsilon \ll 1.$$

We have shown in the previous chapter (Example 3.35) that the closed loop is stable for some values of K_1, K_2, K_3. In this case,

$$\hat{S}(s) = \frac{s^2(s + a)(s + \varepsilon)}{\kappa(s)},$$

where κ is the closed-loop characteristic polynomial. This function has a double zero at 0, so the closed loop (if stable) will asymptotically track a step and a ramp.

EXAMPLE 4.6: A model for heating of water in a tank was derived in Example 2.1. The transfer function is

$$P(s) = \frac{K}{s + \tau}.$$

where $K = 1/C_P M = 0.256$, and $\tau = (C_P m + k)/C_P M$. The temperature of the tank is in degrees Celsius above ambient. For this problem the parameters are such that $K = 1/C_P M = 0.256$, and $\tau = 0.0013$. This means that the 98% settling time is 50 minutes. The objective is to design a controller so that the closed loop has a 98% settling time of 15 minutes and the steady-state response to a step input has zero error. This will lead to a tank where the water temperature will converge to a desired temperature in 15 minutes. The uncontrolled response is shown in Fig. 4.4.

Since the plant does not have a pole at 0, in order that the closed loop track a step, the controller must have a pole at 0. Try first the simplest such controller,

$$H(s) = \frac{K_I}{s},$$

where the constant K_I needs to be determined. The resulting closed loop will have characteristic polynomial

$$\kappa(s) = s^2 + \tau s + K K_I$$

and the closed-loop system poles will be

$$-\frac{\tau}{2} \pm \frac{1}{2}\sqrt{\tau^2 - 4KK_I}.$$

The closed loop will be stable, and track a step, for all values of $K_I > 0$. (See Fig. 4.5.) In order to improve the settling time, the real part of the

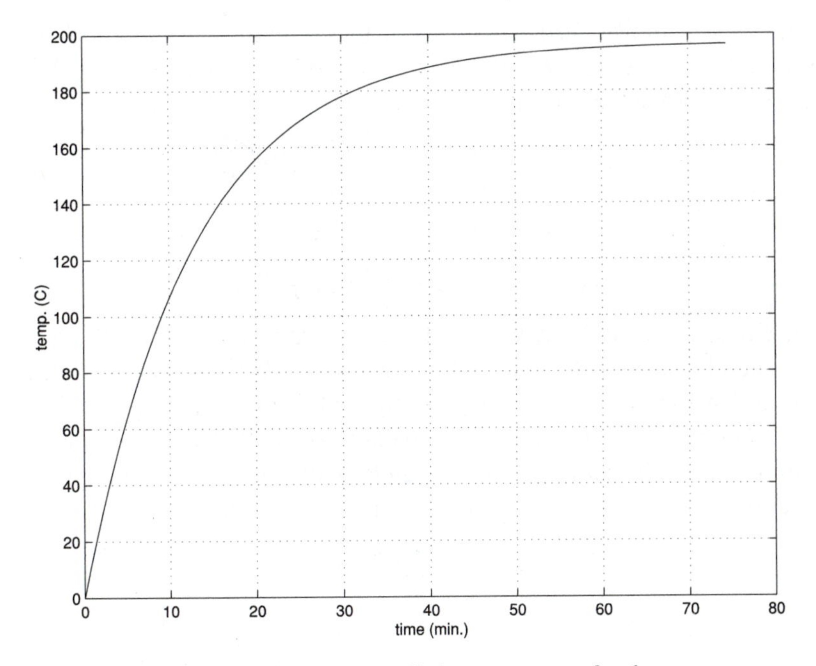

FIGURE 4.4 Uncontrolled step response of tank.

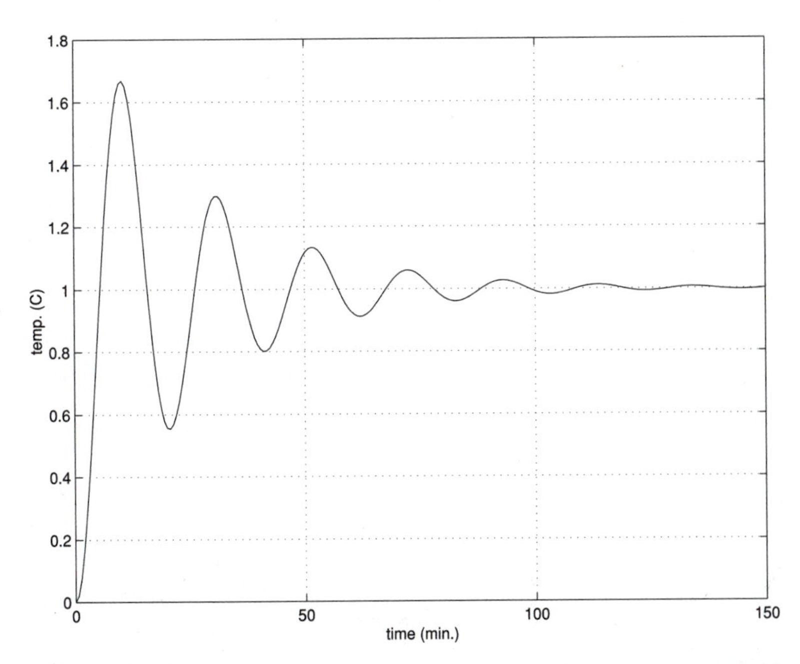

FIGURE 4.5 Closed-loop step response with integral controller ($K_I = 0.0001$).

poles must be moved further to the left in the complex plane. It is impossible to do this by adjusting K_I. We now try a more complicated controller in order to improve the settling time. We try

$$H(s) = \frac{K_I}{s} + K_P = \frac{(K_p s + K_I)}{s}$$

where K_P is also a constant. The characteristic polynomial is now

$$\kappa(s) = s^2 + (\tau + K_P K)s + K K_I.$$

The closed loop will be stable for all positive values of K_P and K_I. Notice that the closed loop is a second-order system. We will choose K_P and K_I so that the closed loop has the required settling time and is slightly underdamped so that errors in the parameters will affect overshoot, not settling time. The settling time of a underdamped second-order system is determined by the coefficient of s. We need to choose K_P so that

$$\exp - \frac{(\tau + K_P K)}{2} T \leq .02$$

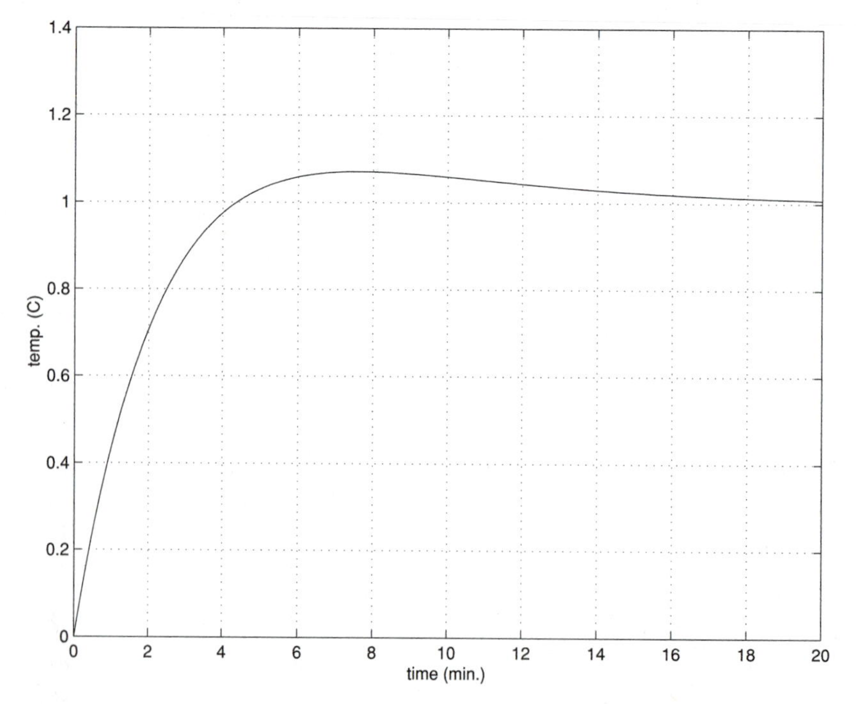

FIGURE 4.6 Closed-loop step response with PI controller ($K_P = 0.0350$, $K_I = 0.0001$).

where $T = 15 \times 60$. Solving, we obtain $K_P \geq 0.03$. Set $K_P = 0.0350$. We will now choose K_I so that the poles are complex, but with a small imaginary part. In other words, the system has damping slightly under critical damping. For critical damping, the discriminant in the quadratic formula is zero:

$$(\tau + K_P K)^2 - 4K_I K = 0.$$

Solving for K_I,we obtain $K_I = 1.023 \times 10^{-4}$. We set $K_I = 1.02 \times 10^{-4}$ to obtain a slightly underdamped system. (See Fig. 4.6.)

4.3 SENSITIVITY REDUCTION

Sensitivity reduction is an important performance goal in many controller design problems. Reducing the sensitivity at 0 reduces the tracking error in response to a step.

Another situation where sensitivity reduction is desirable is in the rejection of an output disturbance, such as sensor noise. The measured output due to the output disturbance r is (see Fig. 3.2)

$$\frac{1}{1 + \hat{P}\hat{H}}\hat{r}.$$

Reducing the sensitivity is equivalent to improving output disturbance rejection. In the idle speed example (Example 2.21), the output disturbances are load changes. We do not want the engine speed to be sensitive to these disturbances. Hence, the sensitivity S should be small at frequencies where output disturbances are present.

For frequencies where $|L| = |PH| \gg 1$,

$$\begin{aligned} |(1 + PH)^{-1}| &= \frac{1}{|1 + PH|} \\ &< \frac{1}{|PH| - 1} \\ &\approx \frac{1}{|PH|} = \frac{1}{|L|}. \end{aligned}$$

Thus, in order to achieve small sensitivity in a frequency range, one needs to design a controller that leads to large loop gain in this range. Loopshaping by varying the controller can be advantageous for a variety of reasons. Two of the most important are these:

1. A reasonable controller may be much simpler than the loop L. It often makes sense to begin controller design with $H =$ constant so that $L =$ constant $\cdot P$, and then add dynamic elements to get the right loop shape.
2. The general form of the controller may be predetermined. For instance, PID, phase-lag, and phase-lead controllers are all very popular in applications.

With this approach the controller design procedure is to choose the parameters in the particular type of controller so that the loopshape satisfies given inequalities. The closed loop is then checked for stability and satisfaction of the performance and robust stability criteria.

EXAMPLE 4.7 (*Control of Loudspeaker Response*): Loudspeakers are often used as inputs to acoustic control systems. In this context, it is useful if the loudspeaker behaves as a constant-volume velocity source. A constant-volume velocity source provides a fixed velocity that is independent of the frequency of the driving voltage. That is,

$$\frac{\dot{x}}{V_a} \quad \text{is constant,}$$

where x is the loudspeaker diaphragm position, and V_a is the applied voltage. Actual loudspeaker response is highly frequency dependent. We will design a controller so that the compensated loudspeaker response will be close to constant over the frequency range of interest.

The dynamics of a loudspeaker can be modeled by the system of ordinary differential equations

$$\begin{aligned} m\ddot{x}(t) + c\dot{x}(t) + kx(t) &= \psi i(t) \\ L\dot{i}(t) + R_d i(t) + \psi \dot{x}(t) &= V_a(t) \end{aligned}$$

where i is the driving current. Physical interpretation of the various parameters along with typical values are in Table 4.1.

Choosing the state $z = (x, \dot{x}, i)$ we obtain the state-space equations

$$\dot{z}(t) = \begin{bmatrix} 0 & 1 & 0 \\ -\frac{k}{m} & -\frac{c}{m} & \frac{\psi}{m} \\ 0 & -\frac{\psi}{L} & -\frac{R_d}{L} \end{bmatrix} z(t) + \begin{bmatrix} 0 \\ 0 \\ \frac{1}{L} \end{bmatrix} u(t)$$

where $u(t) = V_a(t)$ if the speaker is uncontrolled. Since we are trying to control velocity, consider the output

$$y(t) = [0 \quad 1 \quad 0] z(t).$$

TABLE 4.1 Table of loudspeaker parameters

Parameter	Description	Value
m	Mass	0.12 kg
c	Damping	3.1 Ns/m
k	Mechanical stiffness	2085 N/m
ψ	Force constant	6 N/A
L	Coil inductance	0.19×10^{-3} H
R_d	Coil resistance	4.9 Ω

Let $P_v(s)$ indicate the transfer function of the uncontrolled loudspeaker with this output. The open loop response of the system is in Fig. 4.7. The loudspeaker is clearly not a pure velocity source. We would like the error

$$e(t) = \frac{1}{K} V_a(t) - y(t)$$

to be small for some constant K. That is, with $r(t) = \frac{1}{K} V_a(t)$, we want small sensitivity.

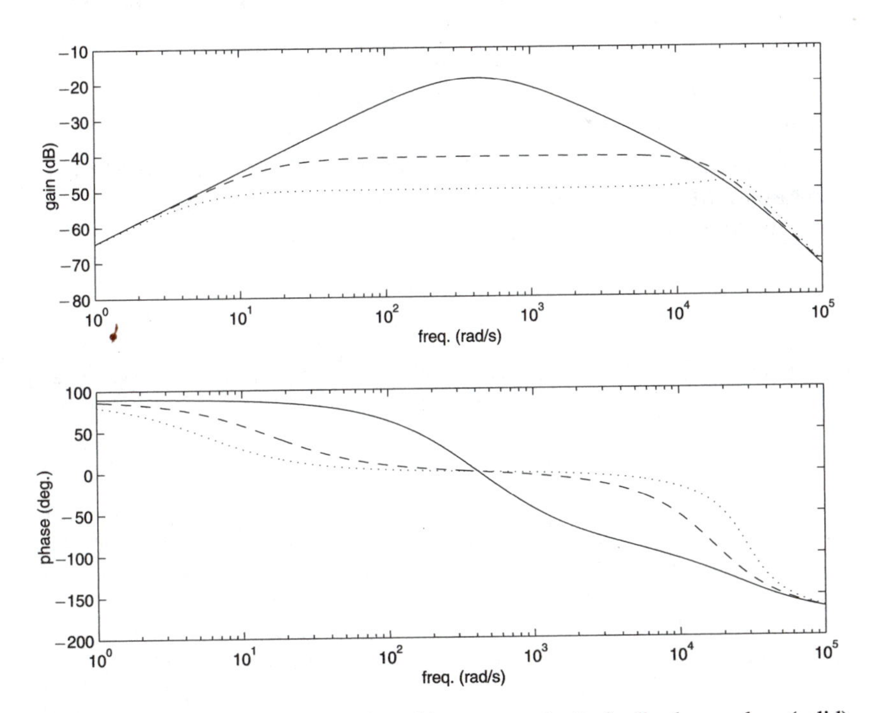

FIGURE 4.7 Bode plot of loudspeaker with constant velocity feedback: open loop (solid), $K = 100$ (dashed), $K = 300$ (dotted).

If, using the standard feedback configuration, we choose $H = K$, $r(t) = \frac{1}{K} V_a(t)$, we obtain that the voltage to the speaker is

$$u(t) = V_a(t) - K y(t).$$

This yields a closed-loop transfer function

$$\frac{y}{V_a} = \frac{\psi s}{(ML)s^3 + (cL + mR_d)s^2 + kL + cR + \psi^2 + \psi K)s + kR}.$$

The frequency response for varying values of K is shown in Fig. 4.7. For $K = 300$ the magnitude response is flat, with almost no gain or phase shift in the range of 10–10^4 rad/s. The transfer function is close to a constant on this frequency range and the loudspeaker will respond as though it were a velocity source.

Example 4.8: Consider the simple plant

$$P(s) = \frac{1}{1 + s/a}$$

where $a > 0$. The proportional controller $H = K$ leads to a sensitivity

$$S(s) = \frac{s + a}{s + a(1 + K)}.$$

A sample frequency response is shown in Fig. 4.8a.

We have $S(0) = 1/(K + 1)$, but because of the zero at $-a$ due to the plant pole, $|S(j\omega)|$ starts to rapidly increase to 1 after $\omega = a$.

Consider now the family of lead compensators,

$$H(s) = K \frac{1 + s/z}{1 + s/(\alpha z)}$$

where $\alpha > 1$. In order to remove the effect of the plant pole, we set $z = a$. Then

$$S(s) = \frac{s + \alpha a}{s + \alpha a(K + 1)}.$$

We still have $S(0) = 1/(K + 1)$ and $\lim_{s \to} S(s) = 1$. However, the bandwidth over which $|S(j\omega)|$ is small increases with increasing α. See Fig. 4.8b.

Example 4.9: Recall that the transfer function of the idle speed control system is

$$P = \frac{k_m k_E}{(1 + Ts)(s + a_m)(s + a_E)} \qquad \begin{array}{l} a_m = k_m = 1.1 \\ a_E = k_E = 1.3, T = 0.1. \end{array}$$

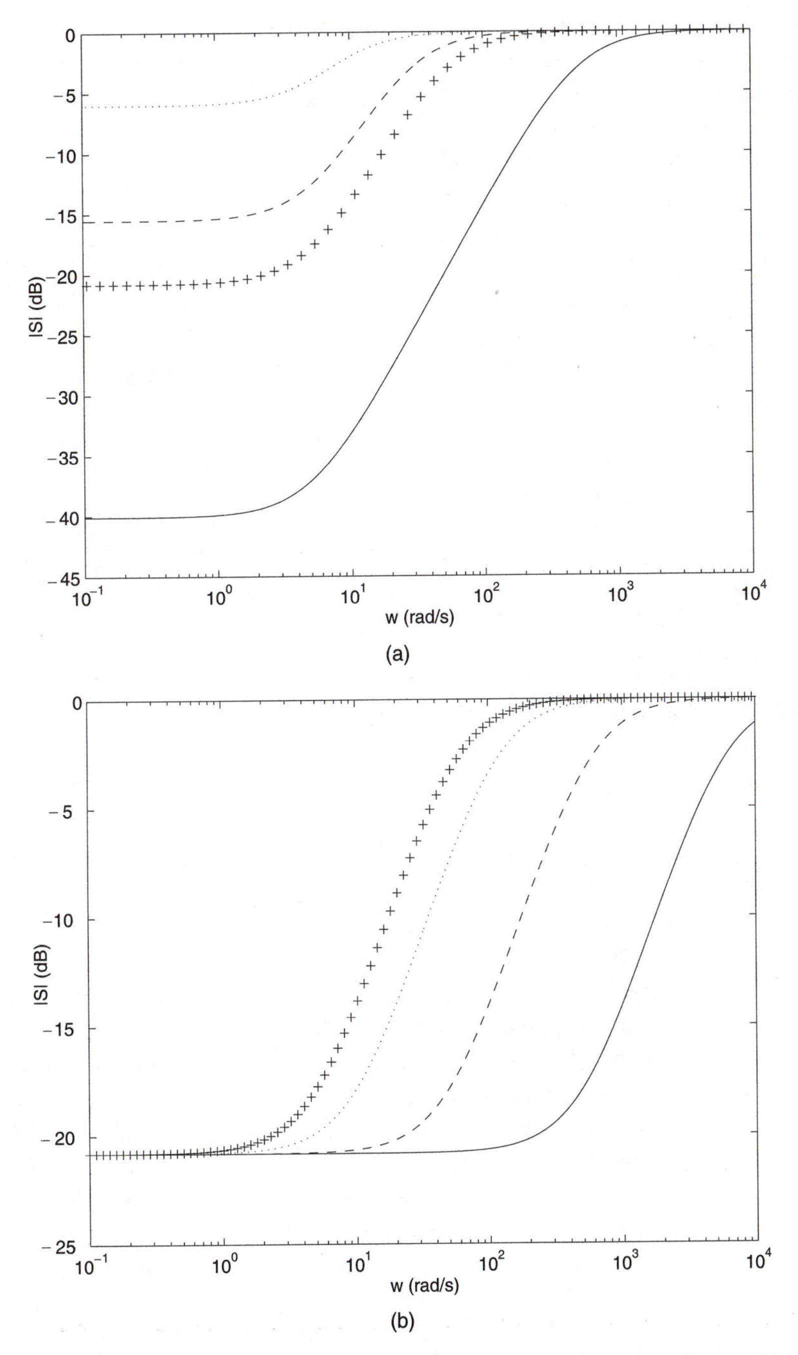

FIGURE 4.8 Sensitivity reduction for $P = \frac{1}{1+s/5}$. (a) H = 1, 5, 10, 100. (b) $H = 10\,(+)$ and lead controller with $K = 10$, $z = a = 5$, and $\alpha = 2, 10, 100$.

Suppose that we want good tracking of a step input, with reasonable gain and phase margins. We also want the system to settle to steady-state within 3 s. (Otherwise the response of the engine could be slow, with persistent oscillations.)

First try a simple proportional controller $H = K$. The characteristic polynomial is

$$\kappa = s^3 + 12.4s^2 + 25.43s + 14.3(1+K),$$

with closed-loop sensitivity

$$S = \frac{1}{1 + P \cdot K} = \frac{s^3 + 12.4s^2 + 25.43s + 14.3}{\kappa(s)}.$$

Routh–Hurwitz: $a_3 = 1$, $a_2 = 12.4$, $a_1 = 25.43$, $a_0 = 14.3(1+K)$,

$$\begin{matrix} a_3 & a_1 \\ a_2 & a_0 \\ b_1 & 0 \\ a_0 & 0 \end{matrix}; \qquad b_1 = -\frac{1}{a_2}\begin{vmatrix} a_3 & a_1 \\ a_2 & a_0 \end{vmatrix} = a_1 - \frac{a_3 a_0}{a_2} = 24. - 1.2K.$$

The closed-loop system is stable for $K \leq 20$. The steady-state error in response to a step is

$$S(0) = \frac{1}{1+K}.$$

Thus, to minimize the steady-state error, we should choose K as large as possible.

However, as K is increased, the gain and phase margins decrease. (See Fig. 4.9.) Also, the settling time increases (Fig. 4.10).

For good tracking of a step input, say, better than 5%, we require that $H(0) \geq 20$. For $K = 20$ the phase margin is only 1.3° and the gain margin is also very small. (See Fig. 4.11 or Fig. 4.12.) This is unacceptable. A more complicated controller than a constant-gain controller is needed.

We now consider a phase lead controller to increase the phase lead so that the closed loop is stable with acceptable gain and phase margins. A phase lead controller is chosen over phase lag since the settling time of the closed loop is more likely to be acceptably fast.

With $H = 20$, the phase margin is 1.3. Add a phase lead controller

$$\frac{1 + \alpha T s}{1 + T s}$$

in series with $H = 20$.

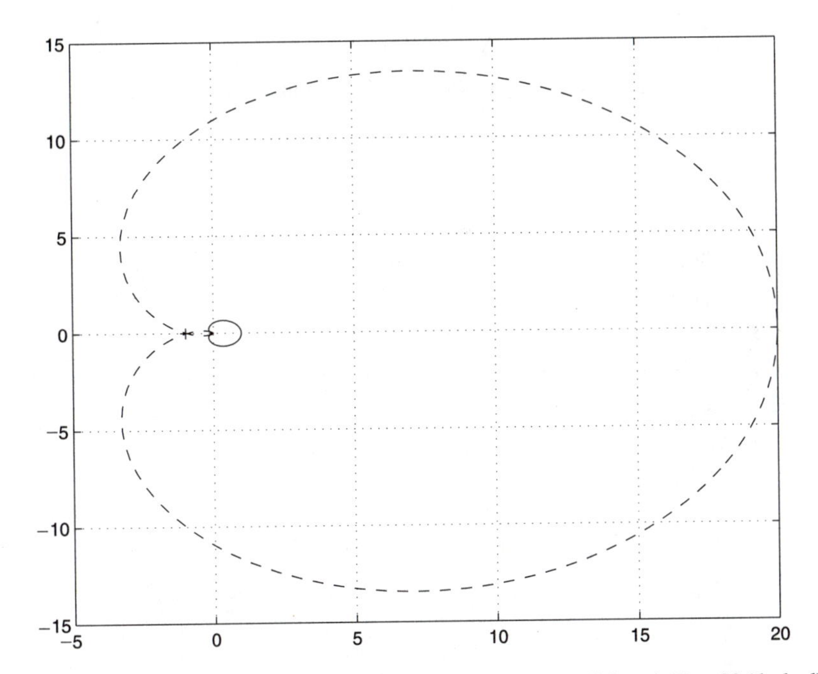

FIGURE 4.9 Nyquist plot for idle speed example, $H = 1$ (solid) and $H = 20$ (dashed).

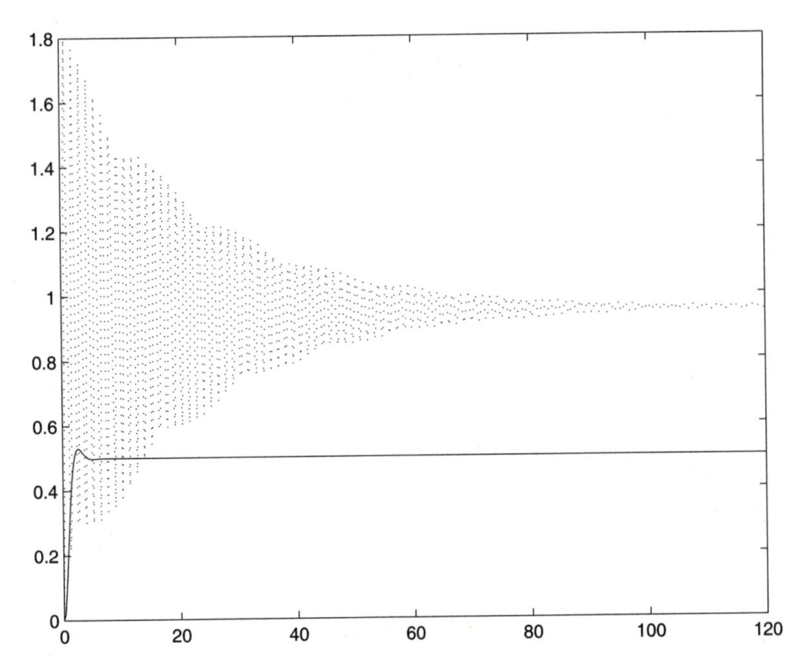

FIGURE 4.10 Closed-loop step response, $H = 1$ (solid) and $H = 20$ (dashed).

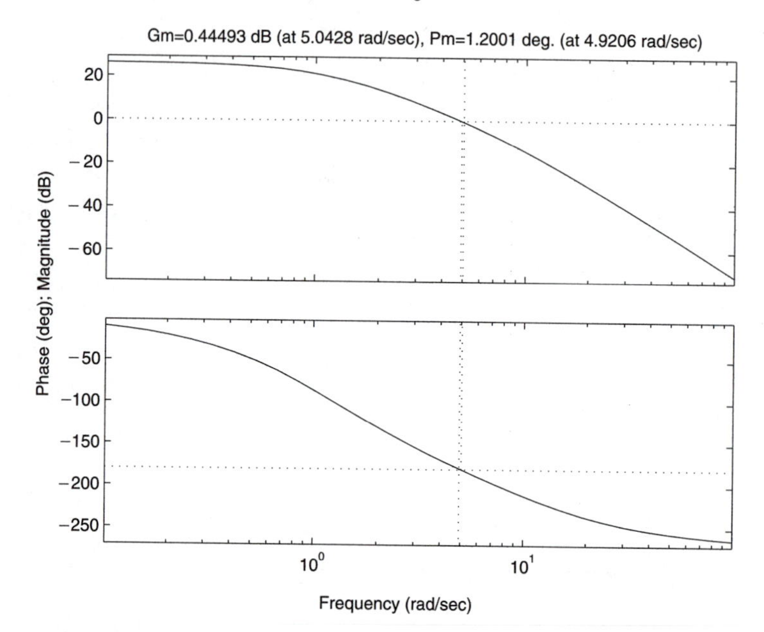

FIGURE 4.11 Bode plot of $20P$ showing phase and gain margins.

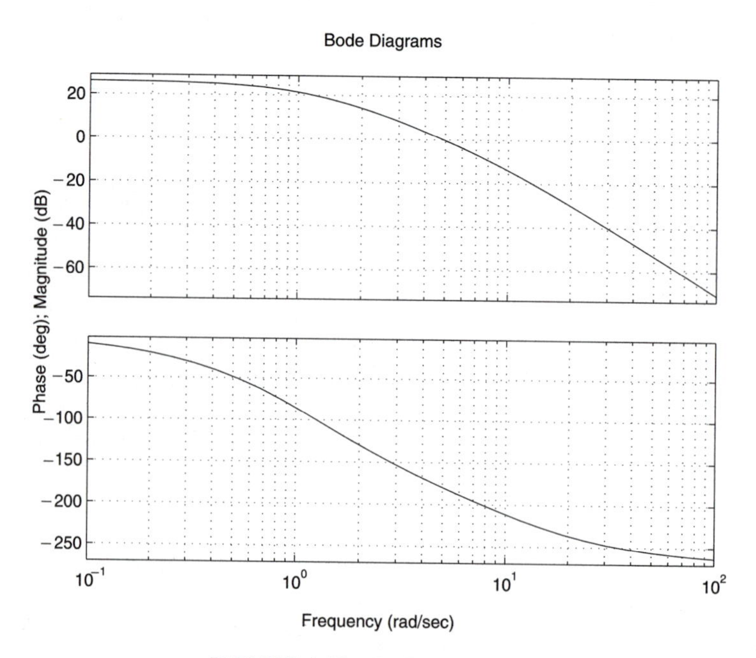

FIGURE 4.12 Bode plot of $20P$.

Design the phase lead controller to add $\phi_m = 60°$ phase lead. Using the formula

$$\sin \phi_m = \frac{\alpha - 1}{\alpha + 1}$$

we obtain that $\alpha = 13.9$.

Recall that at the frequency of maximum phase lead,

$$\omega_m = \frac{1}{T\sqrt{\alpha}},$$

the gain is $10 \log_{10} \alpha = 11.43$ dB. Let us find the frequency where $20P$ has gain -11.43 dB or absolute value 0.2682 and choose T so that ω_m is at this frequency. Then maximum phase lead will occur at approximately the point where the gain is 0 dB in the compensated system. For this system, $|20P(j\omega)| = 0.2682$ at about 8.75 rad/s. The frequency $\omega_m = 8.75$ and so $T = 0.03$. The controller is

$$H(s) = 20\frac{(1 + \alpha T s)}{(1 + T s)}$$

where $\alpha = 13.9$, $T = 0.03$. The gain and phase margins of this system are acceptable (Fig. 4.13) and the settling time is about 1.5 s (Fig. 4.14).

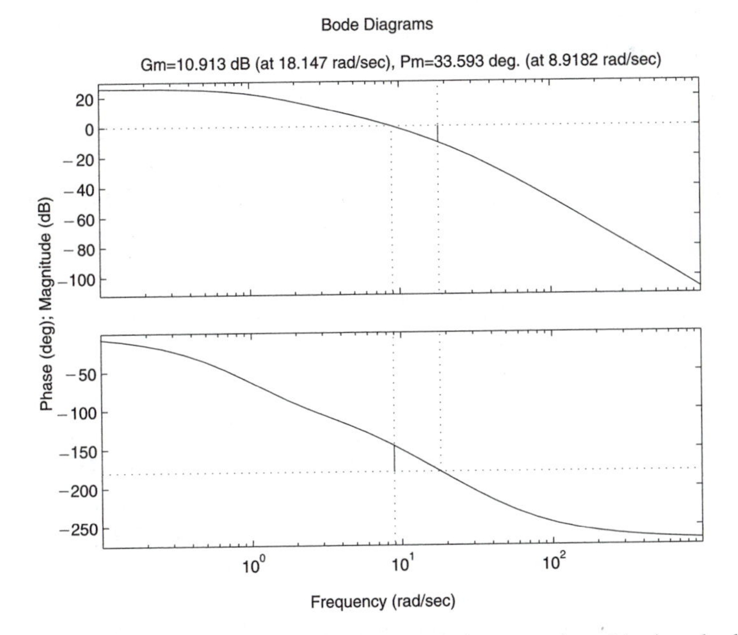

FIGURE 4.13 Bode plot of PH showing gain and phase margins with phase lead controller $H = 20(1 + \alpha T s)/(1 + T s)$, $\alpha = 13.9$, $T = 0.03$.

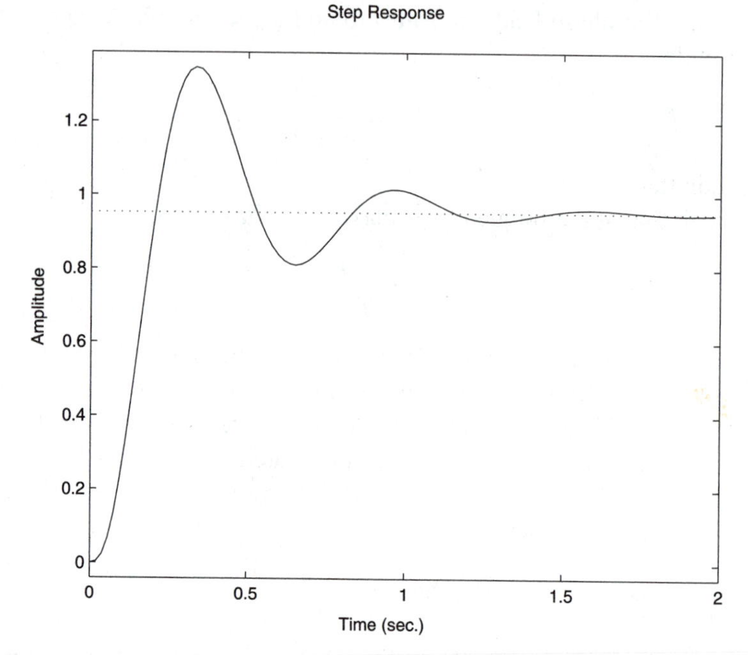

FIGURE 4.14 Closed-loop step response with phase lead controller $H = 20(1 + \alpha T s)/(1 + T s)$, $\alpha = 13.9$, $T = 0.03$.

4.4 $\mathcal{H}_\infty$ PERFORMANCE BOUNDS

Let us now consider more precisely the criteria of regulation and disturbance rejection introduced earlier.

Suppose we want the energy of the error e to be small. We know that

$$\|e\|_2 = \|Sr\|_2 \leq \|S\|_\infty \|r\|_2.$$

Our performance specification is then, for a given tolerance $\varepsilon > 0$, find a stabilizing controller H, so

$$\|S\|_\infty < \varepsilon.$$

Alternatively, we might try to find a stabilizing controller H for which $\|S\|_\infty$ is a minimum. This would be an optimal control problem.

However, for most plants, the feedthrough term E in (2.1), (2.2) is zero and so

$$\lim_{|s| \to \infty} P(s) = 0.$$

Thus, regardless of the choice of H,

$$\lim_{s\to\infty} S(s) = 1$$

and $\|S\|_\infty \geq 1$.

Furthermore, most reference inputs (desirable inputs) or "disturbances" (undesirable inputs) have a certain energy spectrum. Let H^2 indicate the set of Laplace transforms of functions in $L_2(0, \infty)$. Define

$$\Omega = \{r \in H^2 \mid r = W_1 r_0, \|r_0\|_2 \leq 1\},$$

where W_1 is some function. The energy spectrum of r is weighted by W_1,

$$\frac{1}{2\pi}\int_{-\infty}^{\infty} |r(j\omega)/W_1(j\omega)|^2 d\omega \leq 1,$$

for all r in the above set. For $r \in \Omega$,

$$\sup_{r\in\Omega} \|e\|_2 = \sup_{\|r_0\|_2\leq 1} \|SW_1 r_0\|_2 = \|W_1 S\|_\infty.$$

If the inputs have a certain energy spectrum, a realistic performance objective is to find a stabilizing H so that, for tolerance $\varepsilon > 0$,

$$\|W_1 S\|_\infty < \varepsilon. \tag{4.2}$$

The inequality $\|W_1 S\|_\infty < \varepsilon$ implies that $\|e\|_2 < \varepsilon$ for all signals in the set Ω.

Another situation where a performance objective of the form (4.2) occurs is where the sensitivity is to be reduced more at certain frequencies than at other frequencies. If we let a function W_1 describe the relative weighting of different frequencies, we have the performance objective

$$\sup_{\omega} |W_1(j\omega)S(j\omega)| < \varepsilon,$$

where H stabilizes P. This is exactly inequality (4.2).

A further case where a design objective of the form (4.2) is appropriate is when it is known by previous experience that the sensitivity function should lie below a given curve:

$$|S(j\omega)| < |f(j\omega)|.$$

Setting $W_1 = 1/f$, we obtain (4.2).

A desired shape for the closed-loop sensitivity lay behind the design examples in the previous sections. The difference here is that an upper bound on the magnitude of the desired sensitivity function is now given. Thus, this $\mathcal{H}_\infty$-performance criterion in terms of weighted sensitivity is a more precise statement of the criteria discussed earlier. In all cases, we wish the sensitivity to be small in a certain frequency range in order to provide (1) good tracking and/or (2) rejection of output disturbance. There are two major advantages to this reformulation of the design problem. One is that this approach can be generalized to multivariable systems by extending the $\mathcal{H}_\infty$ norm to the $\mathcal{H}_\infty$ norm for matrices. The second advantage is that the design problem is now described quite specifically and mathematically. This enabled rigorous design procedures to be developed.

In the following example we use an *ad hoc* method to obtain a controller that satisfies performance and robust stability constraints. Formal design methods are developed in later chapters.

Example 4.10 (*Idle-Speed Control Example 2.21 Continued*): Recall that the simple model developed for an idle speed system had transfer function

$$P = \frac{k_m k_E}{(1+Ts)(s+a_m)(s+a_E)} \qquad \begin{array}{l} a_m = k_m = 1.1 \\ a_E = k_E = 1.3, T = 0.1. \end{array}$$

Here the input is a valve setting and the output is engine speed. The components of the engine model are shown in Fig. 2.4. A controller for a fuel-injected engine should adjust a valve setting so that a constant engine speed is maintained, despite changes in load, such as turning on the air conditioning. In Example 4.9 we designed a phase lead controller for this system. We now design a controller using formal $\mathcal{H}_\infty$ criteria.

We first define the set for which our controller must provide robust stability. The term $1/(1+Ts)$ is an approximation to the delay term $\exp(-Ts)$, and there are other modeling errors in this simple model. Thus, the final controller should be robust with respect to these errors. For an additive model, we require a transfer function W_2 so that

$$\begin{aligned} |W_2(j\omega)| &\geq |P(j\omega) - \tilde{P}(j\omega)| \\ &= \left| \left(\frac{1}{1+T(j\omega)} - \exp(-Tj\omega) \right) \frac{k_m k_e}{(s+a_m)(s+a_e)} \right|. \end{aligned}$$

By trial and error we obtain

$$W_2(s) = \frac{5s^2}{(1+s)^2(1+s/10)(1+s/50)}.$$

The magnitude curves are shown in Fig. 4.15.

Now we define the performance criteria. We want good tracking of a step input (better than 5% error). The settling time should be less than 3 s.

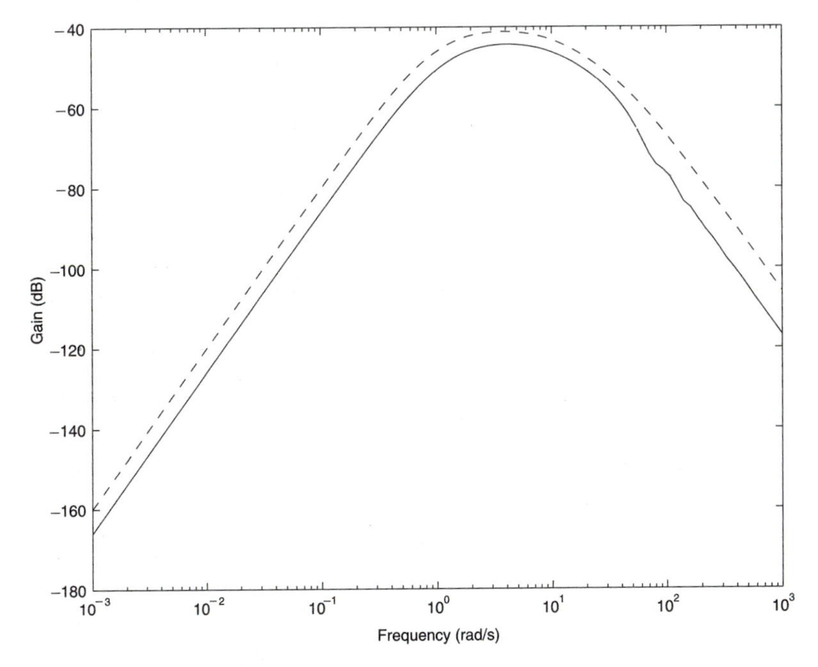

FIGURE 4.15 $|P - \tilde{P}|$ (solid) and $|W_2|$ (dashed).

The closed loop should also be insensitive to load disturbances d. If we use the L_2-norm to measure the relevant signals, this criterion is as follows: Choose H so that $\|SE\|_\infty$ is small. Recall that for good tracking of a step we required that $S(0)$ be small. Since $E(j\omega)$ decreases with ω, this objective is not in contradiction to the criterion that $|S(0)|$ be small. A good choice of the weight W_1 for this problem would be $W_1 \approx E$, but with $W_1(0)$ large. We want a weight W_1 on the sensitivity that has $|W_1(0)| > 20$ in order to satisfy the steady-state tracking error criteria. Also, the disturbance enters through the transfer function (see Fig. 2.4)

$$\frac{k_e}{s + a_e} = \frac{1.3}{s + 1.3},$$

and so it makes sense to design for small sensitivity in the frequency range (0, 1.3). We could choose

$$W_1 = \frac{20 \times 1.3}{s + 1.3}.$$

However, this function drops off quite slowly and is of significant size for

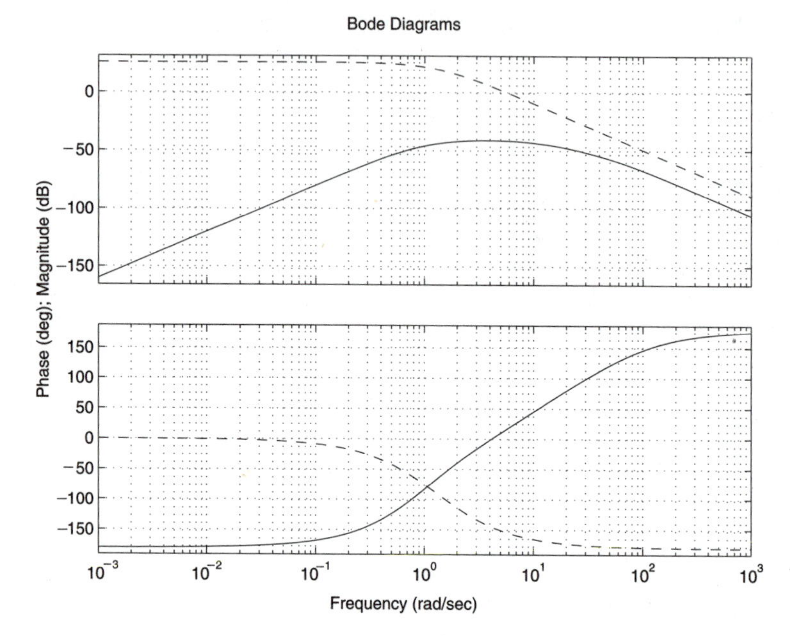

FIGURE 4.16 $|W_2|$ (solid) and $|W_1|$ (dashed).

a large frequency interval above 1.3 rad/s. In this example we use

$$W_1 = \frac{20 \times 1.3^2}{(s + 1.3)^2}.$$

This function has the required gain at 0, but drops off quite sharply after 1.3 (see Fig. 4.16).

We would like to obtain a controller that satisfies

$$\|W_1 S\|_\infty < 1, \qquad \|W_2 H S\|_\infty < 1,$$

and that also leads to a suitably fast settling time.

An initial trial controller is $H = 20$. This controller leads to a stable closed loop and also leads to a small enough tracking error. The relevant gains, $|W_1 S|$ and $|W_2 H S|$, are shown in Fig. 4.17. We need to flatten both functions around $\omega = 5$.

We try putting a zero in the controller here:

$$H_1 = 20\frac{1 + s/5}{1 + s/1000}.$$

The pole at 1000 is to ensure a proper transfer function. The Nyquist plot of the loop PH (Fig. 4.18) shows that H is stabilizing. (Notice that it is

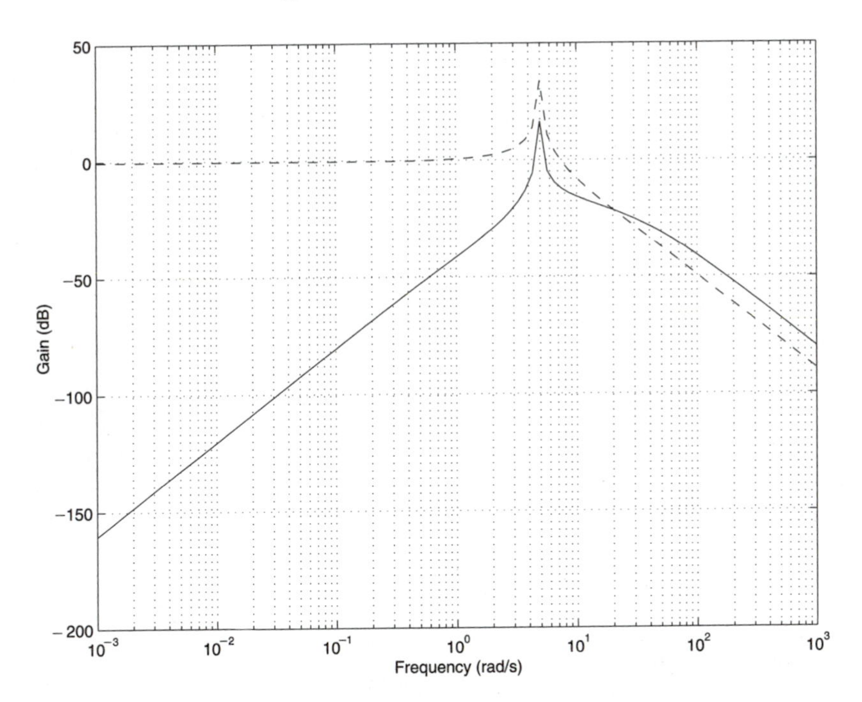

FIGURE 4.17 $H = 20$, $|W_2HS|$ (solid) and W_1S (dashed).

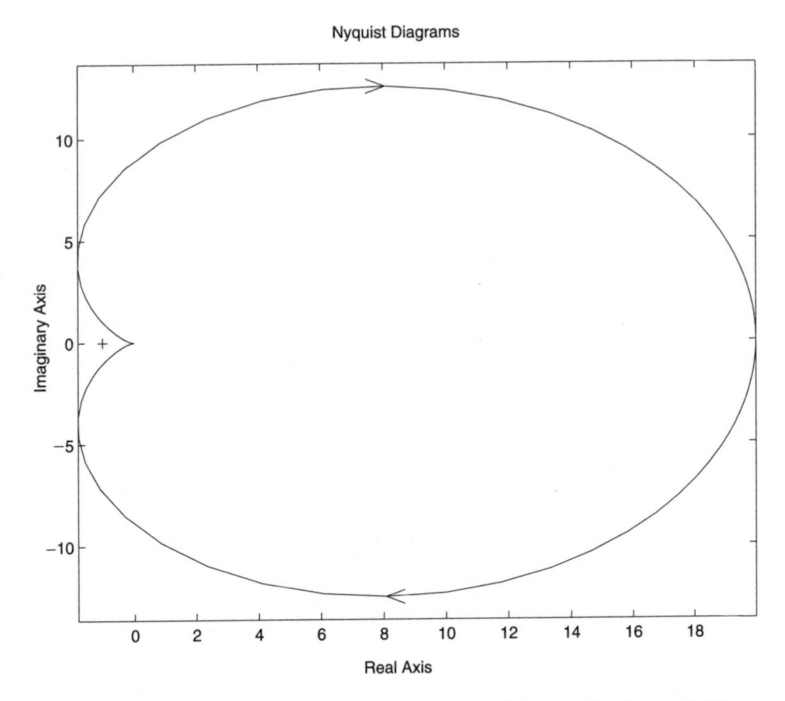

FIGURE 4.18 Nyquist plot of PH with $H = 20(1 + s/5)/(1 + s/1000)$.

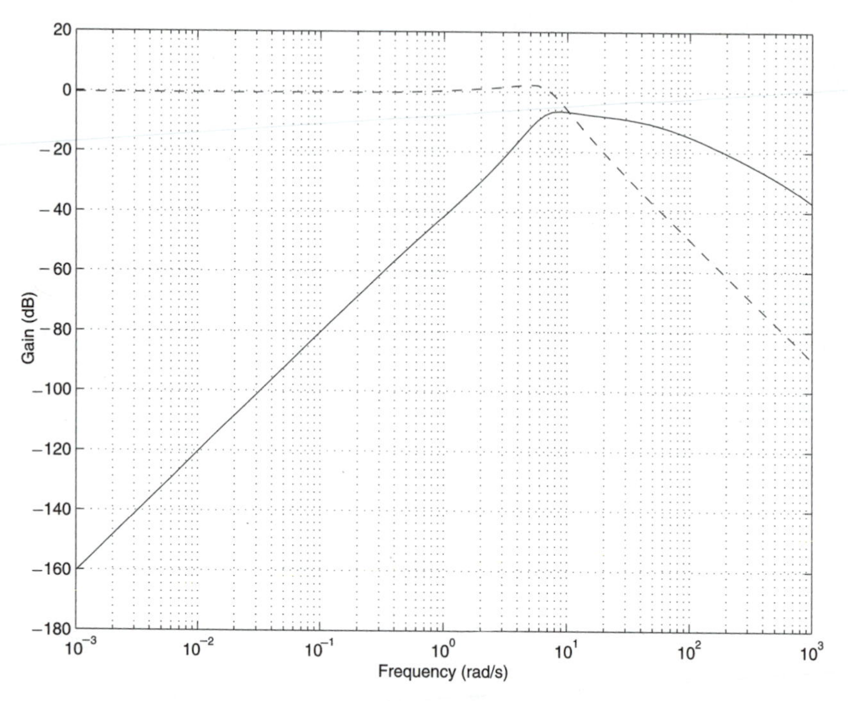

FIGURE 4.19 $H = 20(1 + s/5)/(1 + s/1000)$, $|W_2HS|$ (solid) and W_1S (dashed).

essentially a PD controller.) The relevant transfer functions are shown in Fig. 4.19. The nominal performance is fine. The robust stability function W_2HS is satisfactory.

The step response is fine (Fig. 4.20). The gain and phase margins are in Fig. 4.21. Note that both are quite large.

Another possible controller is the previously designed phase lead controller:

$$H = 20\frac{1 + s/2.4}{1 + s/33}$$

($T = 0.03$, $\alpha = 13.9$). The relevant transfer functions are in Fig. 4.22. Again, the performance measure is fine, but W_2HS is a little large at one point. We try to improve the controller by moving the zero:

$$H = 20\frac{1 + s/10}{1 + s/33}.$$

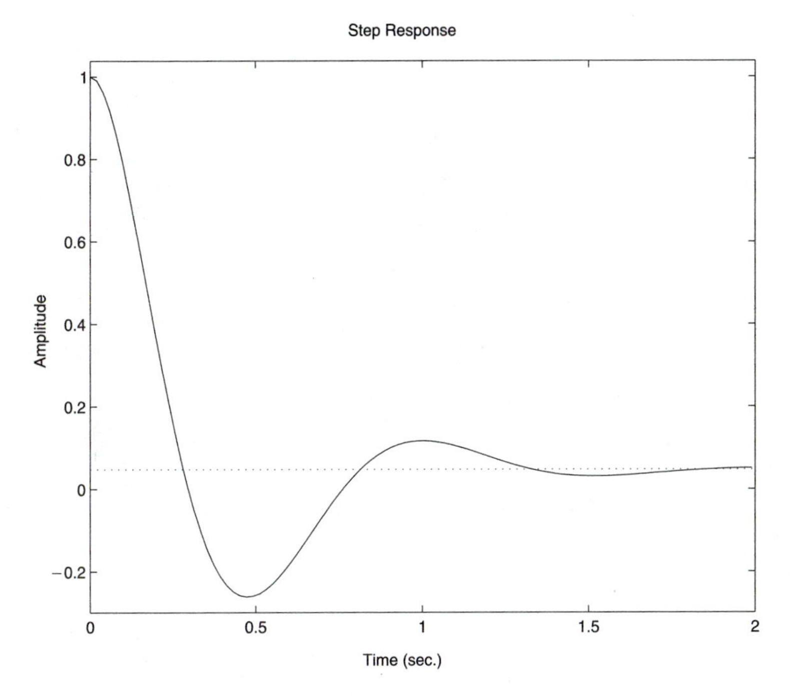

FIGURE 4.20 Step response $H = 20(1 + s/5)/(1 + s/1000)$.

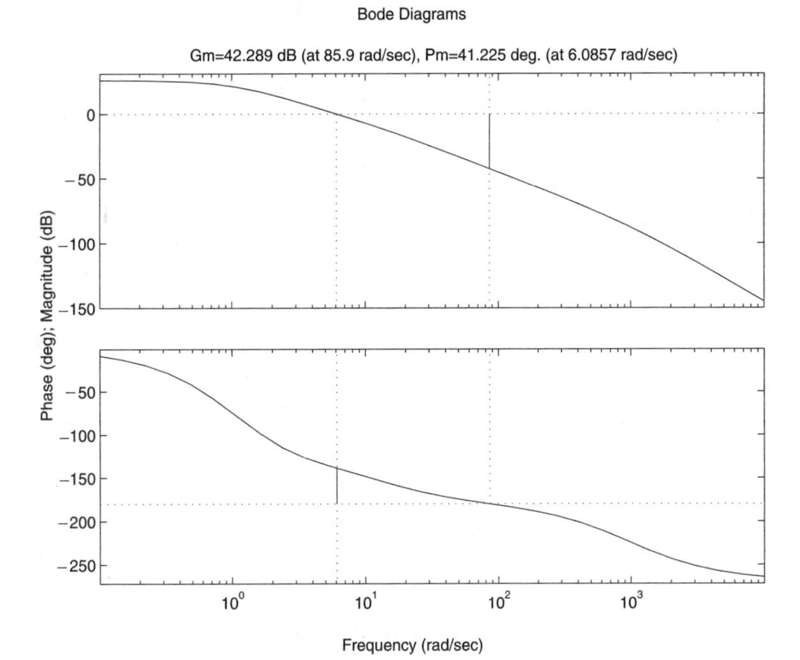

FIGURE 4.21 Gain and phase margins, $H = 20(1 + s/5)/(1 + s/1000)$.

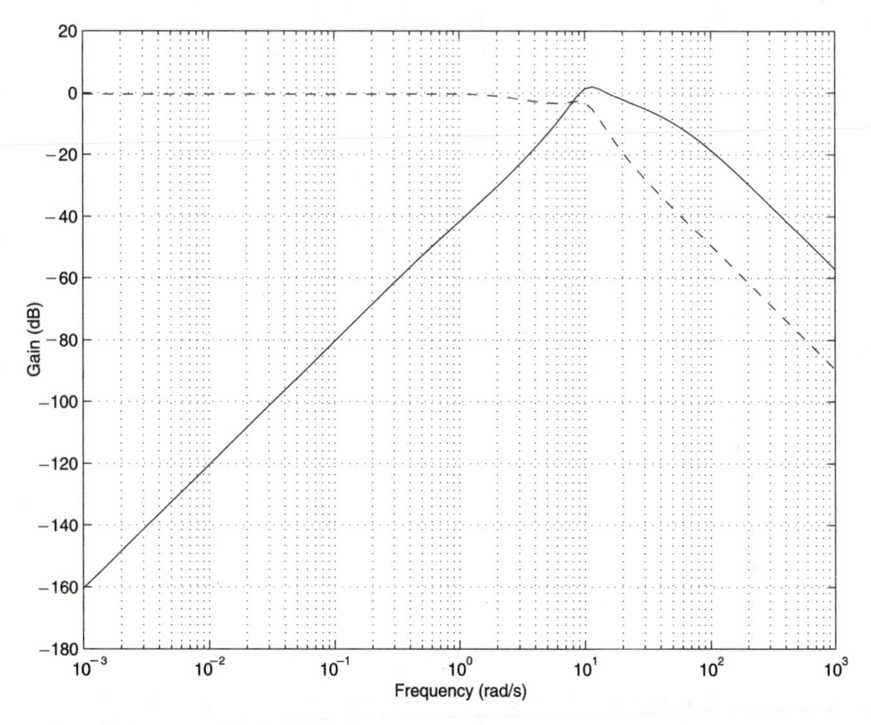

FIGURE 4.22 Phase lead controller, $|W_2 H S|$ (solid) and $W_1 S$ (dashed).

This controller is stabilizing (Fig. 4.23). The relevant transfer functions are in Fig. 4.24. The robust stability criterion is now satisfied, although at some cost to the weighted performance. Unfortunately, the step response is now unacceptable (Fig. 4.25). We move the zero down:

$$H = 20\frac{1 + s/5}{1 + s/33}.$$

This controller is stabilizing (Fig. 4.26). The relevant weighted gains are in Fig. 4.27 and the step response is in Fig. 4.28. The weighted gains look similar to the previous controller. However, the step response is now satisfactory.

For comparison, the magnitude of the unweighted sensitivity is shown for this controller, the first "PD" controller, and the phase lead controller in Fig. 4.29. They are very similar. For most problems there are a number of equally satisfactory (or unsatisfactory) controllers.

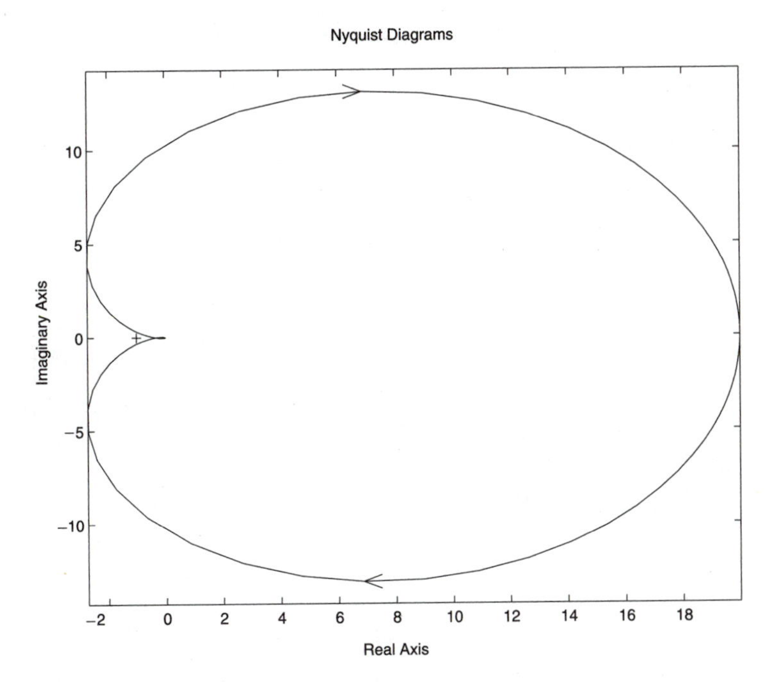

FIGURE 4.23 Nyquist plot of PH with $H = 20(1 + s/5)/(1 + s/1000)$.

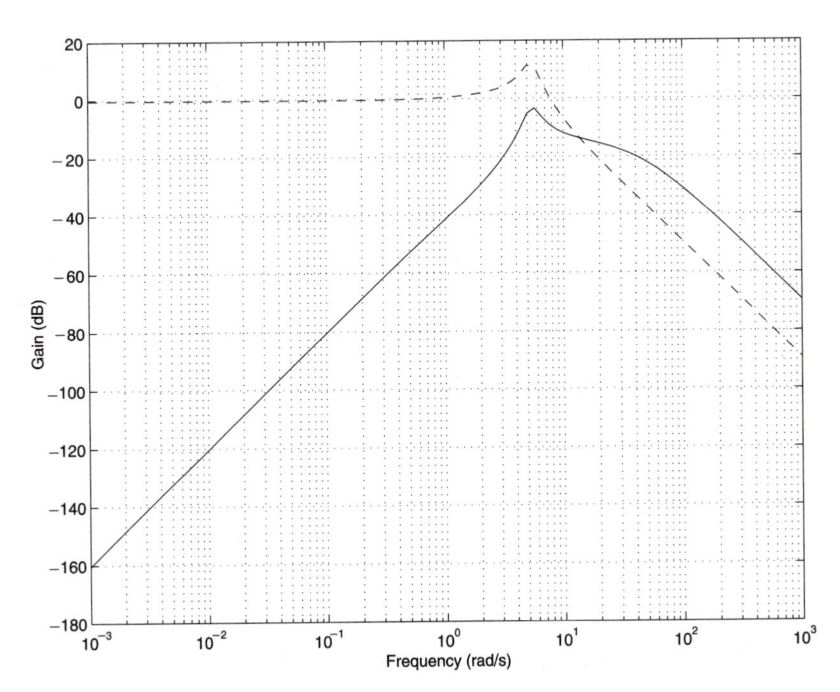

FIGURE 4.24 $H = 20(1 + s/10)/(1 + s/33)$, $|W_2 H S|$ (solid) and $W_1 S$ (dashed).

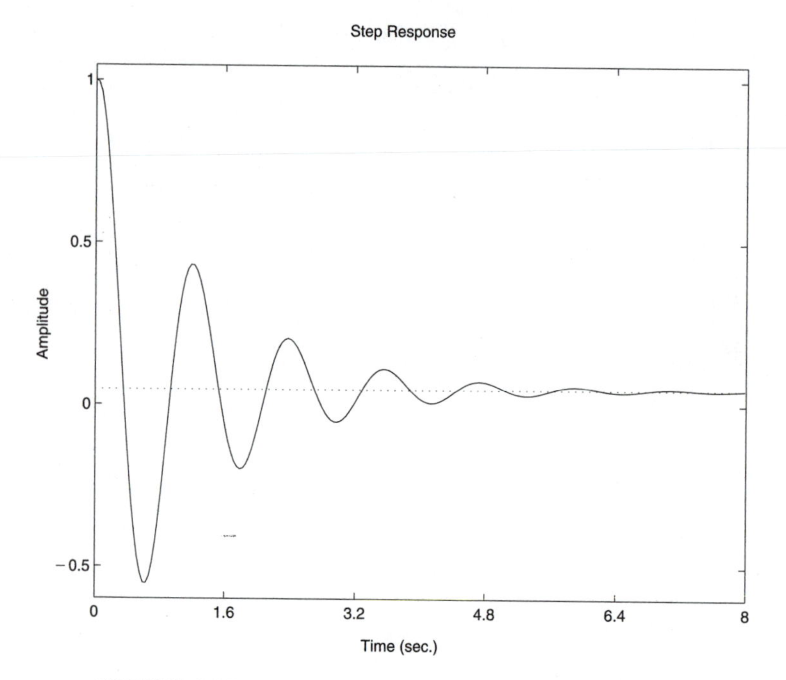

FIGURE 4.25 Step response $H = 20(1 + s/10)/(1 + s/33)$.

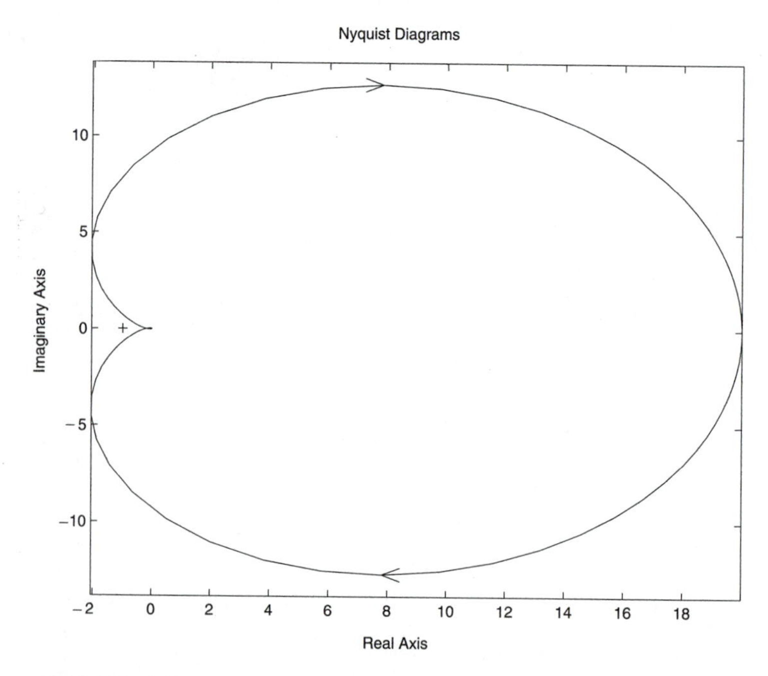

FIGURE 4.26 Nyquist plot of PH with $H = 20(1 + s/5)/(1 + s/33)$.

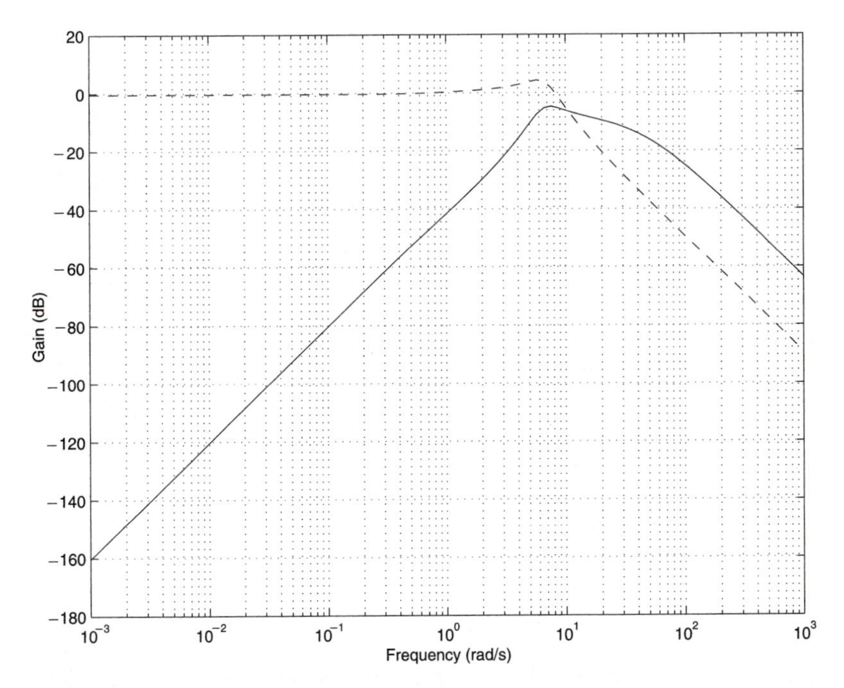

FIGURE 4.27 $H = 20(1 + s/5)/(1 + s/33)$, $|W_2HS|$ (solid) and W_1S (dashed).

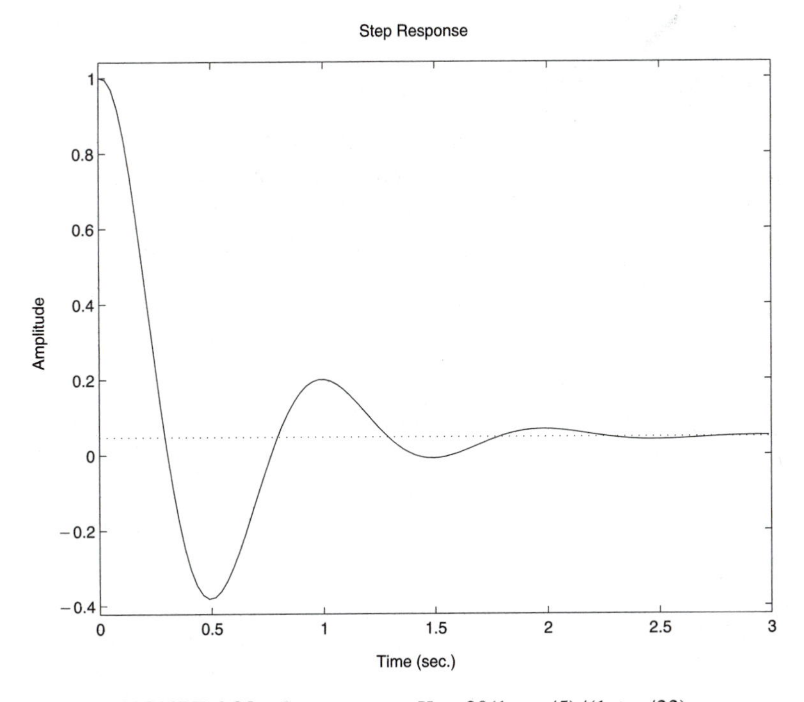

FIGURE 4.28 Step response $H = 20(1 + s/5)/(1 + s/33)$.

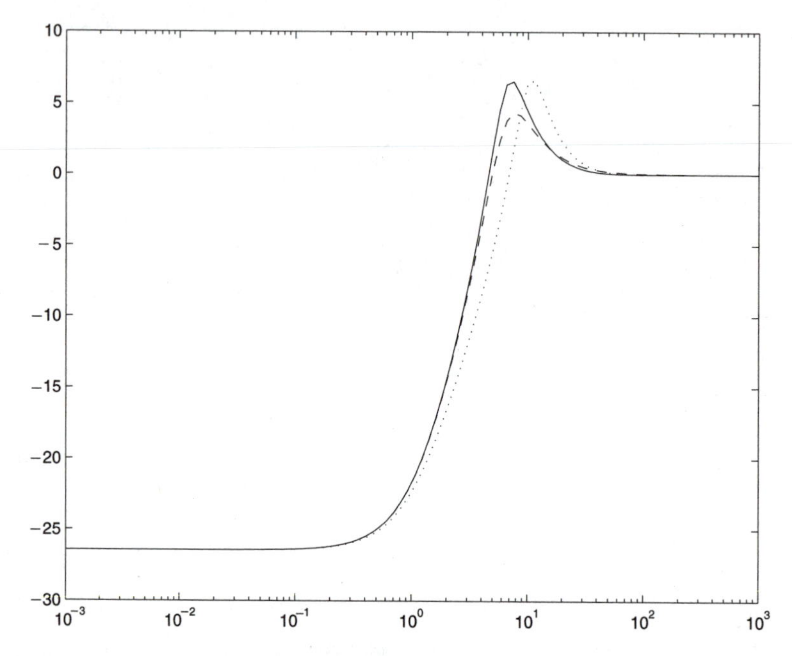

FIGURE 4.29 Sensitivity with last controller $H = 20(1 + s/5)/(1 + s/33)$ (solid), phase lead controller $H = 20(1 + s/2.4)/(1 + s/33)$(dots), and first controller $H = 20(1 + s/5)/(1 + s/1000)$ (dashes).

4.5 ROBUST PERFORMANCE

Suppose the plant belongs to some set $\mathcal{P}$. A controller H provides *robust performance* if it stabilizes every plant in $\mathcal{P}$, and furthermore, performance, of a specified type, holds for all plants in $\mathcal{P}$.

Assume that we are considering multiplicative uncertainty: $\mathcal{P} = \mathcal{M}(P, W_2)$ for some function W_2.

Clearly, we require at least that

$$\|W_2T\|_\infty < 1, \qquad \|W_1S\|_\infty < 1, \tag{4.3}$$

so that robust stability is obtained and satisfactory performance is obtained with the nominal plant P. If a given controller satisfies these two conditions, we say that it provides *nominal performance*.

If we want robust performance, that is, the performance objective

$$\|W_1\tilde{S}\|_\infty < 1$$

is satisfied for every $\tilde{P} \in \mathcal{M}(P, W_2)$, a stronger condition than (4.3) is required.

For example, suppose the performance condition is

$$\|W_1\tilde{S}\|_\infty < 1,$$

where $\tilde{S} = (1 + \tilde{P}H)^{-1}$ is the sensitivity. Consider multiplicative uncertainty for some nominal plant P and weight W_2. A controller H stabilizes every $\tilde{P} \in \mathcal{M}(P, W_2)$ if and only if

$$\|W_2T\|_\infty < 1,$$

where

$$T = PH(1 + PH)^{-1}.$$

Suppose H satisfies this condition. For any $\tilde{P} \in \mathcal{M}(P, W_2)$

$$\tilde{P} = (1 + \Delta)P$$

for some Δ with $|\Delta(j\omega)| < |W_2(j\omega)|$. The performance is

$$\left\|\frac{W_1}{1 + (1 + \Delta)L}\right\|_\infty = \left\|\frac{W_1S}{1 + \Delta T}\right\|_\infty = \left\|\frac{W_1}{(1 + L) + \Delta L}\right\|_\infty.$$

The robust performance condition is

$$\|W_2T\|_\infty < 1, \qquad \left\|\frac{W_1S}{1 + \Delta T}\right\|_\infty < 1, \tag{4.4}$$

for all allowable perturbations Δ, $|\Delta(j\omega)| < |W_2(j\omega)|$.

Theorem 4.11: The controller H stabilizes all plants in the family $\mathcal{M}(P, W_2)$ with performance

$$\|W_1\tilde{S}\|_\infty < 1$$

if and only if

$$\||W_1S| + |W_2T|\|_\infty < 1 \tag{4.5}$$

where S and T are the nominal sensitivity and complementary sensitivity, respectively.

□ *Proof:* From the preceding discussion it is clear that we must have (4.4) for all Δ with $|\Delta(j\omega)| < |W_2(j\omega)|$. Thus, we have robust performance if and only if this condition is satisfied with the "worst" choice of Δ:

$$\|W_2T\|_\infty < 1, \qquad \left\|\frac{W_1S}{1 - |W_2T|}\right\|_\infty < 1. \tag{4.6}$$

Inequality (4.6) is identical to

$$\frac{|W_1 S(j\omega)|}{1 - |W_2 T(j\omega)|} < 1 \quad \forall \omega.$$

Rearranging

$$|W_2 T|(j\omega) + |W_1 S|(j\omega) < 1 \quad \forall \omega$$

or

$$\| |W_1 S| + |W_2 T| \|_\infty < 1$$

as required. ■

The preceding theorem is concerned with reducing sensitivity in the presence of multiplicative uncertainty. Results similar to this theorem exist for other performance criteria and plant uncertainty models.

The robust performance problem is to design a proper controller H so that (1) the nominal plant P is stabilized and (2)

$$\| |W_1 S| + |W_2 T| \|_\infty < 1.$$

The robust performance problem is not always solvable. Some possible problems are the following:

- $\| W_1 S \|_\infty < 1$ is not achievable for the plant
- $\| W_3 T \|_\infty < 1$ is not achievable for the plant.

Also, since

$$S(j\omega) + T(j\omega) = 1,$$

the functions T and S cannot be made simultaneously less than 1. The two conditions in (4.3) are opposing conditions. There is a trade-off in design between robust stability and system performance. Typically, as mentioned before, we try to design S to be small at low frequencies and T small at high frequencies. However, any uncertainty in the model at low frequencies restricts the attainable performance at low frequencies.

4.5.1 Loopshaping

We now look briefly at approximately solving the robust performance problem via loopshaping. Recall that for frequencies where $|L| = |PH| \gg 1$,

$$|(1 + PH)^{-1}| \approx \frac{1}{|L|}.$$

(For convenience the dependence on ω is not explicitly indicated.)

The complementary sensitivity $T = PH(1+PH)^{-1}$ appears in the criterion for robust stability to multiplicative plant uncertainty. For frequencies where $|L| \ll 1$,

$$\begin{aligned} |T| &= |L(1+L)^{-1}| \\ &= \left|\frac{1}{1/L - 1}\right| \\ &\approx |L|. \end{aligned}$$

Notice that the open-loop requirement for good performance ($|S|$ small) is in conflict with the open-loop robust stability requirement ($|T|$ small). This again demonstrates the trade-off between performance and robust stability.

Since the loop gain L is typically large at low frequencies and small at high frequencies, the design criterion $\| |W_1 S| + |W_2 T| \|_\infty < 1$ is approximately

$$\begin{aligned} |T||W_2| + \left|\frac{1}{L}\right| |W_1| &< 1 \qquad \text{for small } \omega \\ |S||W_1| + |L||W_2| &< 1 \qquad \text{for large } \omega. \end{aligned}$$

If $|L| \gg 1$, then $|T| \approx 1$. Similarly, if $|L| \ll 1$, $|S| \approx 1$. Therefore, the foregoing inequalities can be simplified further to

$$\begin{aligned} \frac{1}{|L|}|W_1| + |W_2| &< 1 \qquad \text{for small } \omega \\ |W_1| + |L||W_2| &< 1 \qquad \text{for large } \omega. \end{aligned}$$

Now,

$$\min(|W_1|, |W_2|) < 1$$

at each frequency ω. Typically, W_1 is large for low frequencies, small at high frequencies. The weight on the complementary sensitivity typically has the opposite behavior. The robust performance criterion is then, usually, approximately,

$$\frac{1}{|L|}|W_1| < 1 \qquad \text{for low frequencies}$$
$$(|PH| \gg 1, |W_2| \ll 1)$$

and

$$|L||W_2| < 1 \quad \text{for high frequencies} \quad (|PH| \ll 1, |W_1| \ll 1).$$

This suggests choosing the controller so that

$$\begin{aligned} |L| &> |W_1| && \text{at low frequencies} \\ |L| &< \frac{1}{|W_2|} && \text{at high frequencies.} \end{aligned} \tag{4.7}$$

The first inequality is more precise version of the earlier objective to increase L for good sensitivity reduction. The second inequality is concerned with robust stability and is similar to the objective of reducing L at frequencies where phase shift is significant, thus avoiding the $(-1, 0)$ point. The frequency that marks the transition from "low" to "high" frequency is the point where the loop gain crosses the 0-dB line.

4.6 DESIGN CONSTRAINTS

It is useful to establish limitations on achievable performance. We assume throughout this section that a given closed loop is stable. We will be concerned with the important transfer functions

$$S = (1 + PH)^{-1}$$

$$T = PH(1 + PH)^{-1}.$$

Recall that T arises in the robust stability criterion for multiplicative uncertainty and S arises in a performance criterion.

4.6.1 Restriction on Weights

Note that $S + T = 1$, so that

$$|S(j\omega)| + |T(j\omega)| \geq 1.$$

1. Recall that the condition for weighted robust performance is

$$\||W_1 S| + |W_2 T|\|_\infty < 1.$$

Theorem 4.12: A necessary condition for the robust performance problem to be solvable is that, for all ω,

$$\min\{|W_1(j\omega)|, |W_2(j\omega)|\} < 1.$$

□ *Proof:* First assume that $|W_1(j\omega)| \leq |W_2(j\omega)|$. We have

$$\begin{aligned}|W_1(j\omega)| &= |W_1(j\omega)[S(j\omega)+T(j\omega)]| \\ &\leq |W_1(j\omega)S(j\omega)| + |W_2(j\omega)T(j\omega)| \\ &\leq \||W_1S| + |W_2T|\|_\infty.\end{aligned}$$

Therefore, if robust performance is achieved ($\||W_1S| + |W_2T|\|_\infty < 1$) then $\min\{|W_1|, |W_2|\} < 1$.

The argument is identical if $|W_2(j\omega)| < |W_1(j\omega)|$. ■

2. If P has any r.h.p. zeros $\{z_i\}$ with $\text{Re}(z_i) \geq 0$, then for any choice of controller

$$(W_1S)(z_i) = W_1(z_i).$$

$W_1(z_i)$ must be less than 1. The presence of r.h.p. zeros reduces the achievable sensitivity reduction. Note that

$$T(z_i) = 0$$

so a large amount of uncertainty is allowed around z_i.

3. Suppose that P is unstable, with some r.h.p. poles $\{p_i\}$. Writing $P = n/d$, $H = n_H/d_H$,

$$S(p_i) = \frac{d(p_i)d_H(p_i)}{n_H(p_i)n(p_i) + d_H(p_i)d(p_i)} = 0$$

and

$$T(p_i) = 1.$$

Thus we require $W_2(p_i) < 1$. The presence of r.h.p. poles restricts the allowable uncertainty, but makes sensitivity reduction easier.

In summary, at every frequency either $|W_1|$ or $|W_2|$ must be less than 1. At r.h.p. zeros $\{z_i\}$ and r.h.p. poles $\{p_i\}$,

$$|W_1(z_i)| < 1, \qquad |W_2(p_i)| < 1.$$

The effect of r.h.p. zeros is to restrict achievable sensitivity reduction, while r.h.p. poles restrict allowable uncertainty.

4.6.2 Inner and Outer Factors

We know that at any r.h.p. zero, z_i, of the plant $S(z_i) = 1$. To proceed further we need some lemmas, and inner/outer factorizations.

A function $g_i \in R\mathcal{H}_\infty$ is *inner* if

$$|g_i(j\omega)| = 1 \quad \forall\omega.$$

A function $g_o \in R\mathcal{H}_\infty$ is *outer* if g_o has no zeros in Re(s) > 0. Engineers sometimes refer to inner functions as *all-pass functions*. Another name for outer functions is *minimum phase functions*.

THEOREM 4.13: Every function g in $R\mathcal{H}_\infty$ can be factored as a product of inner and outer factors: $g = g_i g_o$. Also,

$$|g(j\omega)| = |g_o(j\omega)|.$$

□ *Proof:* Factor g as

$$g = \frac{\prod_{i=1}^{m_u}(s - z_{ui})\prod_{i=m_u+1}^{m}(s - z_{si})}{\prod_{i=1}^{n}(s - p_i)} \qquad \begin{array}{l}\mathrm{Re}(z_{ui}) > 0 \\ \mathrm{Re}(z_{si}) \leq 0.\end{array}$$

Define

$$g_i = \frac{\prod_{i=1}^{m_u}(s - z_{ui})}{\prod_{i=1}^{m_u}(s + \bar{z}_{ui})}, \qquad g_o = \frac{\prod_{i=1}^{m_u}(s + \bar{z}_{ui})\prod_{i=m_u+1}^{m}(s - z_{si})}{\prod_{i=1}^{n}(s - p_i)}.$$

Clearly $g = g_i g_o$ and g_o has no zeros with Re(z) > 0 so g_o is outer. For any z_{ui}, write $z_{ui} = x + jy$. We have

$$\left|\frac{j\omega - z_{ui}}{j\omega + \bar{z}_{ui}}\right| = \left|\frac{-x + j(\omega - y)}{x + j(\omega - y)}\right| = 1.$$

Since this is true for each factor of g_i, g_i is inner. The statement $|g(j\omega)| = |g_o(j\omega)|$ is now obvious. ■

The preceding theorem is essentially true for all $g \in \mathcal{H}_\infty$, but the proof is considerably more difficult.

We can use inner/outer factorizations to analyze the effect of the presence of r.h.p. zeros and poles on achievable robust performance. Factor S and T:

$$S = S_i S_o, \qquad T = T_i T_o.$$

At any r.h.p. zero, z, of P, $T(z) = 0$, $S(z) = 1$ and hence,

$$S_o(z) = \frac{1}{S_i(z)}.$$

Also, since T_o has no r.h.p. zeros, $T_i(z) = 0$. Similarly, at any r.h.p. pole,

p, of P, $S_i(p) = 0$, $T(p) = 1$ and

$$T_o(p) = \frac{1}{T_i(p)}.$$

Thus, for any weights $W_1, W_2 \in RH_\infty$,

$$\begin{aligned} \|W_1 S\|_\infty &= \sup_\omega |W_1(j\omega)S(j\omega)| \\ &= \sup_\omega |W_1(j\omega)S_o(j\omega)| \\ &= \|W_1 S_o\|_\infty \\ &\geq |W_1(z)S_o(z)| \\ &\geq |W_1(z)| \frac{1}{|S_i(z)|}. \end{aligned}$$

Now, $S_i(p) = 0$ at any pole p. If there is a r.h.p. pole close to a r.h.p. zero, then $1/|S_i(z)|$ will be large. Thus, the effect of a r.h.p. zero on sensitivity is increased if there is a r.h.p. pole nearby.

Similarly,

$$\begin{aligned} \|W_2 T\|_\infty &= \|W_2 T_o\|_\infty \\ &\geq |W_2(p)T_o(p)| \\ &\geq |W_2(p)| \frac{1}{|T_i(p)|}. \end{aligned}$$

The effect of a r.h.p. pole on robust stability is increased if there is a r.h.p. zero nearby.

In other words, plants with r.h.p. poles and zeros that are close to each other are difficult to control.

4.6.3 Waterbed Effect

We will now use Cauchy's Integral Formula (Theorem 2.41) to obtain a bound on the achievable sensitivity reduction for systems with right half-plane zeros.

Consider some plant P and controller H, and let S be the corresponding sensitivity, with an inner/outer factorization

$$S = S_i S_o.$$

Assume that neither P nor H have imaginary axis poles so that S_o has no imaginary axis zeros. Also assume that $\lim_{s\to\infty} |S(s)| > 0$, as is usually the case. We have $\ln S_o \in H_\infty$. Set $f = \ln S_o$ in Theorem 2.42, and take real parts of both sides. For any r.h.p. point $s_0 = \sigma_0 + j\omega_0$ (with $\sigma_0 > 0$) we have

$$\mathrm{Re}(\ln S_o)(s_0) = \frac{1}{\pi}\int_{-\infty}^{\infty} \mathrm{Re}(\ln S_o(j\omega))\frac{\sigma_0}{\sigma_0^2 + (\omega - \omega_0)^2}d\omega.$$

Now, $\mathrm{Re}(\ln S_o) = \ln|S_o|$ (definition of ln function for complex numbers). Thus,

$$\ln|S_o(s_0)| = \frac{1}{\pi}\int_{-\infty}^{\infty} \ln|S(j\omega)|\frac{\sigma_0}{\sigma_0^2 + (\omega - \omega_0)^2}d\omega, \tag{4.8}$$

since $|S_o(j\omega)| = |S(j\omega)|$ on the imaginary axis.

Now let s_0 be a r.h.p. zero of P. Since $S(s_0) = 1$,

$$|S_o(s_0)| = \left|\frac{1}{S_i(s_0)}\right|.$$

Also, because $|S_i(j\omega)| = 1$, $|S_i(s_0)| \leq 1$ and $S_o(s_0)| \geq 1$. Therefore,

$$\ln|S_o(z)| \geq 0.$$

Substitute this into (4.8). Let $[\omega_1, \omega_2]$ be any frequency interval and separate the integral into two sections:

$$I_1 : [-\omega_2, -\omega_1] \cup [\omega_1, \omega_2]$$

and

$$I_2 = (-\infty, \infty)\backslash I_1.$$

Define

$$M_1 = \max_{I_1} |S(j\omega)|.$$

Also define

$$M_2 = \|S\|_\infty.$$

Using these bounds in (4.8), and also

$$\ln|S_o(z)| \geq 0,$$

$$0 \leq \ln M_1 \underbrace{\left(\frac{1}{\pi}\int_{I_1} \frac{\sigma_0}{\sigma_0^2 + (\omega - \omega_0)^2} d\omega\right)}_{c_1}$$

$$+ \ln M_2 \underbrace{\left(\frac{1}{\pi}\int_{I_2} \frac{\sigma_0}{\sigma_0^2 + (\omega - \omega_0)^2} d\omega\right)}_{c_2}.$$

The values of the integrals, c_1 and c_2, are independent of S and hence independent of the controller H.

If the sensitivity is made small over some frequency range so $\ln M_1 < 0$, then $\ln\|S\|_\infty$ must be large. This is the "waterbed effect": As $|S|$ is pushed down in some frequency range, it pops up somewhere else.

The foregoing result can be modified to apply to cases where S has imaginary axis zeros (i.e., P or H have imaginary axis poles). However, the waterbed effect applies only to plants with zeros in the open right half-plane.

4.6.4 Area Formula

The area formula is a formula for the area bounded by the graph of the sensitivity, $|S(j\omega)|$. This formula is independent of the controller chosen. It is a similar result to the waterbed effect, but is valid for all plants that satisfy a relative degree assumption, regardless of whether the plant has any right half-plane zeros.

Recall that the relative degree of a transfer function L is

$$\text{deg denom.}(L) - \text{deg numer.}(L)$$

when L is written as the ratio of polynomials. If L is of relative degree k, then for large ω,

$$|L(j\omega)| \approx \frac{c}{\omega^k}.$$

THEOREM 4.14 (*Area Formula*): Assume that the relative degree of PH is at least 2 and that there are no imaginary axis poles. Then

$$\int_0^\infty \log|S(j\omega)|d\omega = \pi(\log e)(\Sigma \text{Re}(p_i))$$

where p_i = r.h.p. poles of $P(s)H(s)$.

□ *Proof:* Set $f = \ln(|S_o(s)|)$ and $s_0 = \sigma$ where σ is real in Theorem 2.42 to obtain

$$\ln|S_o(\sigma)| = \frac{1}{\pi}\int_{-\infty}^{\infty} \ln(|S_o(j\omega)|)\frac{\sigma}{\sigma^2+\omega^2}d\omega.$$

Or

$$\int_0^{\infty} \ln(|S(j\omega)|)\frac{\sigma}{\sigma^2+\omega^2}d\omega = \frac{\pi}{2}\ln|S_o(\sigma)|.$$

Multiply by σ:

$$\int_0^{\infty} \ln|S(j\omega)|\frac{\sigma^2}{\sigma^2+\omega^2}d\omega = \frac{\pi\sigma}{2}\ln|S_o(\sigma)|.$$

The proof will be done in two steps. The first step is to show that

$$\lim_{\sigma\to\infty} L.H.S. = \int_0^{\infty} \ln|S(j\omega)|dw. \tag{4.9}$$

The second step is to show that

$$\lim_{\sigma\to\infty} RHS = \pi(\Sigma\mathrm{Re}(p_i)). \tag{4.10}$$

(a) Proof of (4.9): Because of the assumption that the relative degree is at least 2, for large ω,

$$|S(j\omega)| \approx \frac{1}{1+\frac{c}{\omega^k}} \approx \frac{\omega^k}{\omega^k + c}$$

where $k \geq 2$. Also,

$$\ln f = (f-1) - \frac{1}{2}(f-1)^2 + \cdots,$$

and setting $f = |S(j\omega)|$ in the above, we have

$$|\ln(|S(j\omega)|)| \approx \frac{c}{\omega^k + c} < \frac{c}{\omega^2 + c}.$$

Thus,

$$\begin{aligned}\int_{\omega_1}^{\infty} |\ln(|S(j\omega)|)| \cdot \frac{\sigma^2}{\sigma^2+\omega^2}d\omega &\approx \int_{\omega_1}^{\infty} \frac{c}{\omega^2+c} \cdot \frac{\sigma^2}{\sigma^2+\omega^2}d\omega \\ &< \int_{\omega_1}^{\infty} \frac{c}{\omega^2+c}d\omega \\ &< \epsilon\end{aligned}$$

for ω_1 sufficiently large, independent of σ. Now, choose σ large enough so that for $0 < \omega < \omega_1$,

$$\left| \frac{\sigma^2}{\sigma^2 + \omega^2} - 1 \right| < \frac{\epsilon}{| \int_0^{\omega_1} \ln|S(j\omega)| d\omega |}.$$

Then,

$$\left| \int_0^{\omega_1} \ln|S(j\omega)| d\omega - \int_0^{\omega_1} \ln|S(j\omega)| \cdot \frac{\sigma^2}{\sigma^2 + \omega^2} d\omega \right| < \epsilon.$$

This proves (4.9).

(b) We will now prove (4.10).

$$\begin{aligned}
|\sigma \ln|S(\sigma)|| &\leq \left| \sigma \ln \left(\frac{1}{1 + |L(\sigma)|} \right) \right| \\
&\leq |\sigma| \ln(1 + |L(\sigma)|) \\
&\approx \sigma \ln\left(1 + \frac{c}{\sigma^k}\right) \\
&= \sigma \left(\frac{c}{\sigma^k} - \frac{1}{2} \left(\frac{c}{\sigma_k} \right)^2 \right) + \cdots .
\end{aligned}$$

This function clearly converges to zero as $\sigma \to \infty$.

Now,

$$\ln(S_o(\sigma)) = \ln|S(\sigma)| - \ln|S_i(\sigma)|$$

and so

$$\lim_{\sigma \to \infty} \frac{\sigma}{2} \ln|S_o(\sigma)| = \lim_{\sigma \to \infty} \frac{\sigma}{2} \ln(|S_i(\sigma)|^{-1}).$$

Also,

$$\ln[|S_i(\sigma)|^{-1}] = \ln \left(\prod_{i=1}^{n_u} \left(\left| \frac{\sigma + \bar{p}_i}{\sigma - p_i} \right| \right) \right) = \sum_{i=1}^{n_u} \ln \left| \frac{\sigma + \bar{p}_i}{\sigma - p_i} \right|,$$

and

$$\begin{aligned}
\frac{\sigma}{2} \ln \left(\left| \frac{\sigma + \bar{p}_i}{\sigma - p_i} \right| \right) &= \frac{\sigma}{2} \ln \left| \frac{1 + \bar{p}_i/\sigma}{1 - p_i/\sigma} \right|, \quad p_i = x + jy \\
&= \frac{\sigma}{4} \ln \left(\frac{(1 + x/\sigma)^2 + (y/\sigma)^2}{(1 - x/\sigma)^2 + (y/\sigma)^2} \right) \\
&= \frac{\sigma}{4} [\ln(NUM) - \ln(DEN)].
\end{aligned}$$

Using the series expansion again,

$$\ln(NUM) = \left(2x/\sigma + \text{higher order terms in } \frac{x}{\sigma}, \frac{y}{\sigma}\right)$$

$$\ln(DEN) = \left(-\frac{2x}{\sigma} + \text{higher order terms in } \frac{x}{\sigma}, \frac{y}{\sigma}\right).$$

Hence,

$$\frac{\sigma}{2}\ln\left(\left|\frac{\sigma + \bar{p}_i}{\sigma - p_i}\right|\right) = \frac{\sigma}{4}(4x/\sigma) + \mathcal{O}\left(\left(\frac{x}{\sigma}\right)^2, \left(\frac{y}{\sigma}\right)^2\right)$$

$$= x + \mathcal{O}\left(\frac{x^2}{\sigma}, \frac{y^2}{\sigma}\right)$$

$$\lim_{\sigma\to\infty} \frac{\sigma}{2}\ln\left(\frac{\sigma + p_i}{\sigma - p_i}\right) = x = \text{Re}(p_i).$$

Therefore,

$$\lim_{\sigma\to\infty} \frac{\sigma}{2}\ln|S_o(\sigma)| = \sum_{i=1}^{n_u} \text{Re}(p_i).$$

This proves (4.10), completing the proof. ■

Example 4.15 (*Idle Speed Control, Continued*): If we neglect the delay, our plant is

$$P(s) = \frac{1}{(1 + s/1.1)(1 + s/1.3)}.$$

The plant is second order, and so for any proper controller, the loop has relative degree at least 2. For any controller,

$$\int_0^\infty \log|S(j\omega)|d\omega = \pi \log e \Sigma \text{Re}(p_i) = 0$$

if the controller stabilizes P. Thus, any sensitivity reduction is accompanied by an "equal" increase in sensitivity in some other frequency range.

The waterbed effect applies to nonminimum phase systems. The area effect applies to all systems satisfying the relative degree assumption. It does not imply a peaking phenomenon.

However, a result similar to the waterbed effect does hold when a bandwidth constraint is introduced.

THEOREM 4.16: Suppose that for some constants M and ω_1

$$|P(j\omega)H(j\omega)| < \frac{M^2}{\omega^2}, \quad |\omega| \geq \omega_1 \tag{4.11}$$

where

$$\frac{M^2}{\omega_1^2} < 1. \tag{4.12}$$

Then

$$\int_0^{\omega_1} \log|S(j\omega)|dw > \pi \log e \Sigma_i \mathrm{Re}(p_i) - M(\omega_1).$$

□ *Proof:* If (4.11–4.12) are satisfied, then for $\omega \geq \omega_1$,

$$|S| \leq \frac{1}{1-|PH|} \leq \frac{1}{1-M^2\omega^{-2}} \leq \frac{\omega^2}{\omega^2 - M}.$$

Hence, by the area formula,

$$\begin{aligned} I(\omega_1) = \int_{\omega_1}^{\infty} \log|S(j\omega)|d\omega &\leq \int_{\omega_1}^{\infty} \log \frac{\omega^2}{\omega^2 - M^2} d\omega \\ &= \log e \int_{\omega_1}^{\infty} \ln\left(\frac{\omega^2}{\omega^2 - M^2}\right) d\omega \\ &= \log e \int_{\omega_1}^{\infty} \ln\left(\frac{1}{1-\frac{M^2}{\omega^2}}\right) d\omega. \end{aligned}$$

Define $z = M/\omega$. Then

$$dz = -\frac{M}{\omega^2}d\omega = \frac{z^2}{-M}d\omega,$$

and

$$\int_{\omega_1}^{\infty} \log|S(j\omega)|d\omega = M \log e \int_0^{M/\omega_1} \frac{1}{z^2}\ln(1-z^2)dz.$$

The Taylor series of $\ln(1-z^2)$ around 0 is

$$\ln(1-z^2) = z^2 + \frac{1}{2}z^4 + \frac{1}{3}z^6 + \cdots$$

and so $\ln(1-z^2)/z^2$ is bounded in $[0, M/\omega_1]$. It follows that $I(\omega_1) < M(\omega_1) < \infty$ where $M(\omega_1)$ is independent of the controller.

From the area formula we have that

$$\int_0^{\omega_1} \log |S(j\omega)| dw > \pi \log e(\Sigma \mathrm{Re}(p_i)) - M(\omega_1). \qquad \blacksquare$$

A restriction of the form (4.11), (4.12) occurs when the controller signal is restricted (as is usual) to a specified bandwidth and gain. In this case, if $|S|$ is made smaller and smaller over some subinterval of $[0, \omega_1]$, then $|S|$ must become larger and larger in some other subinterval. The bandwidth constraint (4.11) limits the achievable sensitivity reduction in $[0, \omega_1]$.

4.7 *OTHER PERFORMANCE SPECIFICATIONS

We have studied primarily the reduction and shaping of the closed-loop sensitivity in this chapter. Other performance specifications arise from different choices of signals. Input disturbances occur when there are external disturbances affecting the plant. The gain from input disturbance d to plant output y is

$$\frac{y}{d} = (1 + PH)^{-1} P.$$

For example, in active noise control, it is generally only necessary to reduce the system response to the input disturbance in the audible frequency range. Let W_1 be a function that is large in this range, and small elsewhere. The performance criterion is to minimize

$$\|W_1(1 + PH)^{-1} P\|_\infty.$$

A list of various functions along with possible interpretations and approximations is given in Table 4.2.

TABLE 4.2 Various performance measures

Function	Interpretation	Approximation	
		$\lvert L\rvert \gg 1$	$\lvert L\rvert \ll 1$
$(I + PH)^{-1}$	• Output disturbance to plant output • Reference signal to tracking error	$1/\lvert L\rvert$	
$H(I + PH)^{-1}$	• Output disturbance to controller input • Robust stab. add. plant uncertainty		$\lvert H\rvert$
$(I + PH)^{-1}P$	• Input disturbance to plant output • Rob. stab. add. controller uncertainty	$1/\lvert H\rvert$	
$PH(1 + PH)^{-1}$	• Controller input disturbance to plant output • Robust stab. mult. plant uncertainty		$\lvert L\rvert$

NOTES AND REFERENCES

The basic loopshaping procedures using PID and phase lead/lag controllers are explained in more detail in many introductory engineering books on controller design. See, for instance, [17, 40, 47]. Material on $\mathcal{H}_\infty$-design constraints and restrictions can be found in [13] and in more detail in [53].

EXERCISES

1. Consider the inverting op-amp configuration shown in Fig. 4.30.
   - **(a)** Calculate the transfer function from V_i to V_o.
   - **(b)** Assume that the opamp gain is sufficiently large that $\frac{1}{A} \approx 0$ and so $V_a \approx 0$. (We say that there is a *virtual short circuit* at the op-amp input terminals.) What is the transfer function in this case?
   - **(c)** Calculate the sensitivity of the transfer from V_i to V_o to changes in A. Use the transfer function from (a). What happens to the sensitivity as $A \to \infty$? Compare with the sensitivity using (b).
2. Let $P = \frac{1}{s}$.
   - **(a)** Find a stabilizing controller so that the closed loop tracks a step input $r = 1$.
   - **(b)** Find a stabilizing controller so that the closed loop tracks a periodic input r with frequency 2 rad/s.
   - **(c)** Find a stabilizing controller so that the final value of y is 0 when the plant input disturbance d is periodic with frequency 2 rad/s. (Here r is zero.)

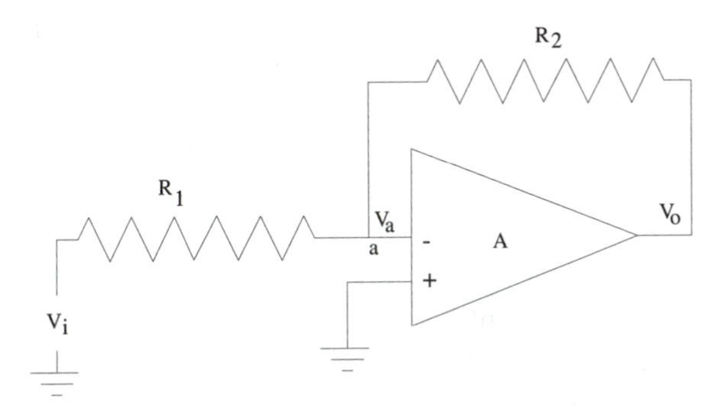

FIGURE 4.30 Inverting operational amplifier.

3. Prove the following statement: For perfect asymptotic tracking of a reference signal (i.e., $\lim_{t\to\infty} e(t) = 0$), the loop $L = PH$ must have poles at the closed r.h.p. poles of the reference input. This is known as the *Internal Model Principle*: In effect, the loop must contain a "model" of the unstable part of the tracked signal.

4. Prove that if $|(PH)(0)| < M$ and the closed loop is stable, then the closed loop tracks a step reference input with steady-state error less than $1/(1 + M)$.

5. **(a)** Consider a plant

$$P(s) = \frac{1 + s/z}{1 + s/p}$$

where $p, z > 0$. Show that arbitrary levels of sensitivity reduction can be achieved. (Hint: Refer to Example 4.8.)

(b) Consider a plant of the form

$$P(s) = \frac{1 - s/z}{1 + s/p}$$

where $p, z > 0$. Prove that for any controller, $S(z) = 1$. Compare with (a).

6. Consider a plant $P(s) = 1/(s + 0.01)$.

(a) Find a controller so the steady-state response to a step input has zero error.

(b) Find a controller that tracks a step (as in (a)) and has setting time of less than 10 minutes.

7. Consider the idle speed control plant with a phase lead controller, that is, let

$$P(s) = \frac{1.1 \times 1.3}{(s + 1.1)(s + 1.3)(1 + 0.1s)}$$

and

$$H(s) = K\frac{1 + \alpha Ts}{1 + Ts}, \quad \alpha > 1.$$

The problem is to asymptotically track a unit step to within a steady-state error of 5%. It is also desirable to improve the gain and phase margins as well as the settling time of the closed-loop response. One phase lead design is done as an example in this chapter.

(a) Choose $\alpha = 4$. Choose the other parameters T and K to obtain a steady-state error less than 5% as well as satisfactory gain and phase margins.

(b) Discuss and compare the results in terms of the closed-loop stability, the settling times, the stability margins, and magnitude of the gain K. Which is the "best" controller design? Explain your choice. Include all relevant plots.

8. For

$$P(s) = \frac{s+1}{(s+2)(s+5)}$$

compute a controller H so that the feedback system is stable and the tracking error e goes to 0 when r is a unit ramp and $d = n = 0$.

9. Consider the plant

$$P(s) = \frac{1}{s^2 + 11s + 10}$$

and a phase lag controller

$$H(s) = K\frac{1+Ts}{1+\alpha Ts}, \qquad \alpha > 1.$$

(a) Obtain a Bode plot of the plant.

(b) Determine the closed-loop steady-state error in response to a unit step. Choose the appropriate controller parameter so that the steady-state error is less than 3%.

(c) Suppose instead that we have

$$H(s) = K.$$

What possible advantages/disadvantages does the phase lag controller have compared to this simpler controller?

10. Consider the plant

$$P(s) = \frac{80}{s^3 + 70s^2 + 1000s + 1}$$

and a phase lag controller

$$H(s) = K\frac{1+Ts}{1+\alpha Ts}, \qquad \alpha > 1.$$

(a) Obtain a Bode plot of the plant.

(b) Determine the closed-loop steady-state error in response to a unit step. Choose the appropriate controller parameter so that the steady-state error is less than 2%.

(c) Complete the controller design so that the closed loop has gain margin at least 25 dB and phase margin at least 45°.

11. Design a controller for

$$G(s) = \frac{80}{s(1 + 0.02s)(1 + 0.05s)}$$

so that the closed loop rejects sinusoidal disturbances at 1 rad/s with steady-state error less than 2%, a phase margin at least 20°, and gain margin at least 8 dB.

12. Let $P(s) = K/s$, where $1 \leq K \leq 10$.

(a) Model the plant uncertainty using multiplicative uncertainty.

(b) Show that a proportional controller provides robust stability for this set.

13. **(a)** Prove that H yields robust performance $\|W_1 S\|_\infty < 1$ for all plants in $\mathcal{A}(P, W_2)$ if and only if

$$\| |W_1 S| + |W_2 H S| \|_\infty < 1.$$

(b) Derive a set of inequalities similar to (4.7) to be used to estimate $|H|$.

14. Consider a plant with nominal model

$$P(s) = \frac{1}{s(s + 0.5)}.$$

A time delay of 0.08 s and some high-order dynamics are neglected by this model. Assume the actual plant is included in the set $\mathcal{M}(P, W_2)$ where

$$W_2(s) = \frac{0.1s}{0.03s + 1}.$$

A controller so that the closed loop system tracks a step and is also insensitive to disturbances in the frequency range [0, 1] is wanted.

(a) Plot the magnitude of W_2 and $P/\tilde{P} - 1$ where $\tilde{P}$ is the nominal plant with the delay included.

(b) Design a controller that tracks a step input and achieves robust stability. (Hint: Try a simple proportional controller.)

(c) Plot

$$W_1(s) = \frac{10}{s^3 + 2s^2 + 2s + 1}.$$

(This is an example of a *Butterworth filter*.) Why is W_1 appropriate as a weight in a performance measure for this problem?

(d) Examine the performance of your controller with the nominal plant P. Plot $W_1 S$ and also the step response.

(e) Examine the performance of your controller with the "actual" plant

$$\frac{e^{-.08s}}{s(s + 0.5)(1 + s/10)}.$$

Plot $W_1 S$ and also the step response.

15. Let the model for a given plant be

$$P(s) = \frac{s - \alpha}{s + 1}$$

where $|\alpha| \leq 1$.

(a) Model the plant uncertainty using additive perturbations.

(b) Find a simple proportional controller that provides robust stability for this additive uncertainty set.

(c) Plot the nominal performance S with your controller from (b), and the performance for several other elements in the set.

16. Design a controller as in Example 4.10, except use a multiplicative uncertainty model. The weight on the uncertainty should be analytic in the right half-plane. In this example it will not be proper.

17. Suppose that a plant has a pole at 0. Derive an inequality, similar to the waterbed effect, that is valid for plants with r.h.p. zero(s) and a pole at 0.

18. Consider

$$P(s) = 2\frac{(s - 1)}{(s + 1)^2}$$

and suppose H is a stabilizing controller with the property that

$$\|S\|_\infty \leq \frac{7}{5}.$$

Give a positive lower bound for

$$\max_{0 \le \omega \le 1} |S(j\omega)|.$$

19. Suppose that P has exactly one pole p and one zero z in the (closed) right half-plane. (It may have other poles and zeros in the left half-plane.) Let H be a stabilizing controller that has no closed right half-plane poles or zeros.

(a) Find the inner factor of the complementary sensitivity function T.

(b) Show that for any $W_2 \in R\mathcal{H}_\infty$,

$$\|W_2 T\|_\infty \ge |W_2(p)| \left| \frac{p+z}{p-z} \right|.$$

(c) What does the inequality in (b) imply about robust stability with respect to multiplicative uncertainty when there are r.h.p. poles near r.h.p. zeros?

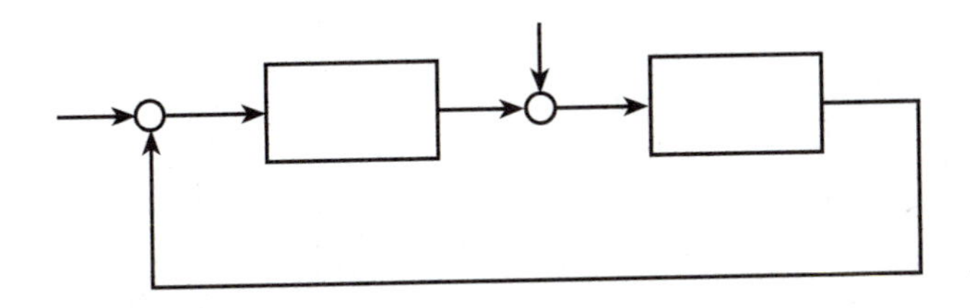

V

BASIC STATE FEEDBACK AND ESTIMATION

An obvious control strategy is to find a constant matrix K and set the plant input to

$$u = d - Kx \tag{5.1}$$

where d is the external input and x is the system state.

Even though the control is restricted to a constant feedback, a great deal of freedom in choosing closed-loop behavior is present. In particular, the closed-loop poles can be located arbitrarily.

THEOREM 5.1: Assume a system (A, B) of order n is controllable. Let $S = \{p_1, \ldots, p_n\}$ be any set of complex numbers with the property that if $p \in S$, then $\bar{p} \in S$. Then there exists a matrix K so $A - BK$ has eigenvalues $p_1, \ldots, p_n$.

□ *Proof:* For simplicity of exposition, we only give the proof for the case where there is one input. In this situation, K will be $1 \times n$.

First, put the system into controllable canonical form (Section 2.4):

$$\dot{z}(t) = \bar{A}z(t) + \bar{B}u(t) \tag{5.2}$$

where

$$\bar{A} = \begin{bmatrix} 0 & 1 & \cdots & 0 \\ 0 & 0 & 1 & \vdots \\ & & & 1 \\ -b_0 & -b_1 & \cdots & -b_n \end{bmatrix}, \bar{B} = \begin{bmatrix} 0 \\ 0 \\ \vdots \\ \vdots \\ 1 \end{bmatrix}. \tag{5.3}$$

The characteristic polynomial

$$\det(sI - A) = \det(sI - \bar{A}) = b_0 + b_1 s + \cdots b_{n-1}s^{n-1} + s^n. \tag{5.4}$$

Write an arbitrary state feedback

$$\bar{K} = [k_0 \quad k_2 \quad \cdots \quad k_{n-1}]. \tag{5.5}$$

Then

$$\bar{A} - \bar{B}\bar{K} = \begin{bmatrix} 0 & 1 & 0 & \cdots & 0 \\ 0 & 0 & 1 & & \vdots \\ -b_0 - k_0 & -b_1 - k_1 & \cdots & \cdots & -b_{n-1} - k_{n-1} \end{bmatrix} \tag{5.6}$$

and the characteristic polynomial

$$\det(sI - \bar{A} + \bar{B}\bar{K}) = (b_0 + k_0) + (b_1 + k_1)s + \cdots (b_{n-1} + k_{n-1})s^{n-1} + s^n. \tag{5.7}$$

By choosing $k_0, \ldots k_{n-1}$ appropriately we can obtain a characteristic polynomial with zeros $p_1, \ldots, p_n$ as required. ■

EXAMPLE 5.2:

$$\dot{x}(t) = \begin{bmatrix} 3 & -2 \\ 0 & 4 \end{bmatrix} x(t) + \begin{bmatrix} 0 \\ 1 \end{bmatrix} u(t). \tag{5.8}$$

We wish a state feedback K so the closed loop has a double pole at -3. We first check that the system is controllable. The controllability matrix

$$\mathcal{C} = \begin{bmatrix} 0 & -2 \\ 1 & 4 \end{bmatrix} \tag{5.9}$$

has rank 2. The system is controllable. Writing $K = [k_0 \;\; k_1]$,

$$A - BK = \begin{bmatrix} 3 & -2 \\ -k_0 & 4 - k_1 \end{bmatrix}. \tag{5.10}$$

The characteristic polynomial is

$$\det(sI - A + BK) = s^2 + s(k_1 - 7) + (12 - 3k_1 - 2k_0). \tag{5.11}$$

For a double eigenvalue at -3,

$$-\frac{(k_1 - 7)}{2} = -3 \tag{5.12}$$

and $k_1 = 13$. We also require that

$$(k_1 - 7)^2 = 4(12 - 3k_1 - 2k_0). \tag{5.13}$$

Substituting in $k_1 = 13$, we obtain that $k_0 = -18$.

There are formal algorithms to calculate a state feedback that place the closed-loop eigenvalues at specified locations. These work well for systems of order up to about 10. One is used by the Matlab routine *"place."*

5.1 FINITE-TIME LINEAR QUADRATIC CONTROL

Consider a system whose dynamics are modeled by

$$\begin{aligned} \dot{x}(t) &= Ax(t) + Bu(t) \\ x(0) &= x_o \end{aligned} \tag{5.14}$$

where A is $n \times n$, B is $n \times m$, $x(t)$ is $n \times 1$, and $u(t)$ is $m \times 1$.

DEFINITION 5.3: A symmetric matrix $G \in R^{n \times n}$ is *positive semidefinite*, or *nonnegative*, if

$$x^* G x \geq 0$$

for all $x \in R^n$.

For an arbitrary initial condition x_o, and symmetric nonnegative matrix G, we wish to choose u so that the quadratic performance criterion

$$J(x_o, u; t_f) = x(t_f)^* G x(t_f) + \int_0^{t_f} \|Cx(t)\|^2 + \|Eu(t)\|^2 dt \tag{5.15}$$

is minimized over the set of admissible functions $u \in L_2(0, t_f, R^m)$. We write this optimal control problem concisely as

$$\inf_{u \in L_2(0, t_f; R^m)} J(x_o, u; t_f) \tag{5.16}$$

subject to $x(t)$ solves (5.14). The minimizing control for the problem is known as the *optimal control.*

The object of the control is to minimize a weighted sum of the L_2 norm of the state $x(t)$ and the control $u(t)$. The choice of C and E dictates the relative weighting between the state and the cost of control, and whether some states will be weighted more than others. It will be convenient to define $Q = C^*C$ and $R = E^*E$. With this notation, the cost functional can be written

$$J(x_o, u; t_f) = x(t_f)^* G x(t_f) + \int_0^{t_f} x(t)^* Q x(t) + u(t)^* R u(t) dt. \tag{5.17}$$

The matrix Q is positive semidefinite, G is assumed symmetric positive semidefinite, and R is symmetric and assumed positive definite.

Before establishing that this problem has a unique solution, and calculating the corresponding optimal control, a technical lemma is needed.

Lemma 5.4: For any $u \in L_2(0, t_f; R^m)$ there is $z_u \in L_2(0, t_f; R^m)$ such that for any $h \in L_2(0, t_f; R^m)$,

$$\begin{aligned} J(x_o, u + h; t_f) - J(x_o, u; t_f) &= \int_0^{t_f} z_u(s)^* h(s) ds + (\Lambda(t_f)h)^* G (\Lambda(t_f)h) \\ &\quad + \int_0^{t_f} (\Lambda(s)h)^* Q (\Lambda(s)h) + h(s)^* R h(s) ds \end{aligned}$$

where

$$\Lambda(t)u = \int_0^t e^{A(t-s)} B u(s) ds.$$

□ *Proof:* Let u, h be any elements of $L_2(0, t_f; R^m)$. Let $x_u(t)$ indicate the state of the system with control u and note that

$$x_u(t) = e^{At} x_0 + \Lambda(t) u. \tag{5.18}$$

This can be substituted into (5.17) to obtain an unconstrained functional of u. Also,

$$x_{u+h}(t) = x_u(t) + \Lambda(t)h.$$

We obtain

$$\begin{aligned}
&J(x_{o,}u + h; t_f) - J(x_o, u; t_f) \\
&\quad = x_{u+h}(t_f)^* G x_{u+h}(t_f) - x_u(t_f)^* G x_u(t_f) \\
&\qquad + \int_0^{t_f} \{x_{u+h}(s)^* Q x_{u+h}(s) - x_u(s)^* Q x_u(s) \\
&\qquad + (u+h)(s)^* R(u+h)(s) - u(s)^* R u(s)\} ds \\
&\quad = \nabla J_{u,t_f}(h) + (\Lambda(t_f)h)^* G(\Lambda(t_f)h) \\
&\qquad + \int_0^{t_f} (\Lambda(s)h)^* Q(\Lambda(s)h) + h(s)^* R h(s) ds
\end{aligned}$$

where

$$\begin{aligned}
\nabla J_{u,t_f}(h) &= 2x_u(t_f)^* G\Lambda(t_f)(h) \\
&\quad + 2\int_0^{t_f} x_u(s)^* Q(\Lambda(s)h) + 2u(s)^* R h(s) ds.
\end{aligned}$$

We will calculate a z_u such that

$$\nabla J_{u,t_f}(h) = \int_0^{t_f} z_u(s)^* h(s) ds.$$

This will complete the proof.

We now compute $\Gamma(s)$ so that for all $x \in L_2(0, t_f; R^n)$, $v \in L_2(0, t_f; R^m)$,

$$\int_0^{t_f} (\Gamma(s)x)^*[v(s)]ds = \int_0^{t_f} x(s)^*[\Lambda(s)v]ds.$$

Now,

$$\begin{aligned}
\int_0^{t_f} x(s)^*[\Lambda(s)v]ds &= \int_0^{t_f} x(s)^* \left[\int_0^s e^{A(s-r)} B v(r) dr\right] ds \\
&= \int_0^{t_f} \int_0^s x(s)^* e^{A(s-r)} B v(r) dr ds \\
&= \int_0^{t_f} \int_r^{t_f} x(s)^* e^{A(s-r)} B ds v(r) dr
\end{aligned}$$

$$= \int_0^{t_f} \int_r^{t_f} \left(B^* e^{A^*(s-r)} x(s)\right)^* ds\, [v(r)] dr$$

$$= \int_0^{t_f} \Bigg[\underbrace{B^* \int_r^{t_f} e^{A^*(s-r)} x(s) ds}_{\Gamma(r)x} \Bigg]^* [v(r)] dr.$$

Thus,

$$\frac{\nabla J_{u,t_f}(h)}{2} = (Gx_u(t_f))^* \int_0^{t_f} e^{A(t_f-s)} B h(s) ds$$

$$+ \int_0^{t_f} [\Gamma(s)(Qx_u)]^* h(s) + [Ru(s)]^* h(s) ds$$

$$= \int_0^{t_f} \left[B^* e^{A^*(t_f-s)} G x_u(t_f) + \Gamma(s)(Qx_u) + Ru(s)\right]^* h(s) ds.$$

Setting

$$z_u(s) = 2\left[B^* e^{A^*(t_f-s)} G x_u(t_f) + \Gamma(s)(Qx_u) + Ru(s)\right],$$

the result follows. ∎

THEOREM 5.5: For every initial condition x_o there is a unique control u_{opt} such that

$$J(x_o, u_{opt}; t_f) = \inf_{u \in L_2(0,t_f;R^m)} J(x_o, u; t_f).$$

□ *Proof:* We will first show that there is a u_{opt} that minimizes the cost function. We will then show that u_{opt} is unique. First, note that

$$0 \le \inf_{u \in L_2(0,t_f;R^m)} J(x_o, u; t_f)$$

and so there is a finite number J_o

$$J_o = \inf_{u \in L_2(0,t_f;R^m)} J(x_o, u; t_f).$$

Now choose a sequence of controls $u_n \in L_2(0, t_f; R^m)$ such that

$$\lim_{n\to\infty} J(x_o, u_n; t_f) = J_o.$$

Let $\lambda_{\min}(R)$ indicate the smallest eigenvalue of R. Since R is positive definite, $\lambda_{\min}(R) > 0$. Thus,

$$\|u_n\|^2_{L_2(0,t_f;R^m)} \le \frac{1}{\lambda_{\min}(R)} J(x_o, u_n; t_f).$$

Since $J(x_o, u_n; t_f)$ is a convergent sequence, it is bounded and so for some constant M and all n,

$$\|u_n\|_{L_2(0,t_f;R^m)} \leq M.$$

This implies (Theorem A.11) that there is a subsequence $\{u_{nk}\}$ of $\{u_n\}$ and $u_{opt} \in L_2(0, t_f; R^m)$ such that

$$\lim_{n\to\infty} \int_0^{t_f} v(s)^* u_{nk}(s)ds = \int_0^{t_f} v(s)^* u_{opt}(s)ds, \quad \text{for each } v \in L_2(0, t_f; R^m).$$

Using Lemma 5.4, with $h = u_{nk} - u_{opt}$, $u = u_{opt}$,

$$\begin{aligned} J(x_o, u_{opt}; t_f) = J(x_o, u_{nk}; t_f) &- \int_0^{t_f} z_{u_{opt}}(s)^*[h(s)]ds \\ &- (\Lambda(t_f)h)^* G(\Lambda(t_f)h) - \int_0^{t_f} (\Lambda(s)h)^* Q(\Lambda(s)h) \\ &+ h(s)^* Rh(s)ds. \end{aligned} \tag{5.19}$$

Since

$$\lim_{n\to\infty} \int_0^{t_f} z_{u_{opt}}(s)^* h(s)ds = 0,$$

and since $\{u_{nk}\}$ is a subsequence of u_n,

$$\lim_{n\to\infty} J(x_o, u_{nk}; t_f) = J_o.$$

The last two terms in (5.19) are always nonnegative, and it follows that

$$J(x_o, u_{opt}; t_f) \leq J_o.$$

However, J_o is defined to be the minimum cost, and so

$$J(x_o, u_{opt}; t_f) = J_o,$$

as was to be shown.

For any two nonzero numbers a and b, $a \neq b$,

$$\left(\frac{a+b}{2}\right)^2 < \frac{1}{2}a^2 + \frac{1}{2}b^2.$$

Similarly, after some calculations, we can show that for any two controls u and v,

$$J\left(x_o, \frac{u+v}{2}; t_f\right) < \frac{1}{2}J(x_o, u_1; t_f) + \frac{1}{2}J(x_o, u_2; t_f). \tag{5.20}$$

Thus, if there are two different controls u_1 and u_2, each minimizing the cost, then

$$J\left(x_o, \frac{u_1+u_2}{2}; t_f\right) < J_o,$$

which contradicts the fact that J_o is defined to be the minimum cost. Hence, the optimal control is unique. ■

LEMMA 5.6: The control u that satisfies

$$u(t) = -R^{-1}B^*\left[e^{A^*(t_f-t)}Gx_u(t_f) + \int_t^{t_f} e^{A^*(s-t)}Qx_u(s)ds\right] \tag{5.21}$$

where

$$x_u(t) = e^{At}x_o + \int_0^t e^{A(t-s)}Bu(s)ds$$

solves (5.16).

□ *Proof:* If the control u is chosen so that $z_u(s) = 0$ then

$$\begin{aligned} J(x_o, u+h; t_f) - J(x_o, u; t_f) &\geq \int_0^{t_f} h(s)^* Rh(s)ds \\ &\geq \lambda_{\min}(R) \int_0^{t_f} h(s)^* h(s)ds \\ &= \lambda_{\min}(R)||h||^2_{L_2(0,t_f;R^m)} \end{aligned}$$

where $\lambda_{\min}(R)$ is the smallest eigenvalue of R. Since R is positive definite, $\lambda_{\min}(R) > 0$. So $J(x_o, u+h; t_f) = J(x_o, u; t_f)$ only if $h = 0$. The control u is the unique optimal control if and only if $z_u = 0$.

Therefore, from definition of z_u, we require that

$$u(s) = -R^{-1}B^*\left[e^{A^*(t_f-s)}Gx_u(t_f) + \int_s^{t_f} e^{A^*(r-s)}Qx_u(r)dr\right]$$

as was to be shown. ■

The expression (5.21) for the optimal control u is unfortunately in terms of future values of the state x_u. Since the state x_u depends on u this makes calculation of the optimal control difficult. The expression for the optimal control is now rewritten so that the optimal control is in terms of the current state. In other words, the optimal control is a feedback control.

THEOREM 5.7: The minimum cost for the problem (5.16) is

$$J(u_{opt}, t_f) = x_o^* P(0, t_f)x_o \tag{5.22}$$

where $P(t, t_f)$ solves the *differential Riccati equation*

$$\dot{P}(t, t_f) + A^* P(t, t_f) + P(t, t_f)A - P(t, t_f)BR^{-1}B^* P(t, t_f) + Q = 0,$$
$$P(t_f, t_f) = G. \tag{5.23}$$

The corresponding optimal control is

$$u_{opt}(t) = -R^{-1}B^* P(t, t_f)x_u(t)$$

where $x_u(t)$ is the system state with control u_{opt} and initial condition x_o.

Furthermore, $0 \leq P(0, t_f) < \infty$ for all times t_f and final conditions G.

□ *Proof:* Defining

$$P(t, t_f)x_u(t) = e^{A^*(t_f - t)}Gx_u(t_f) + \int_t^{t_f} e^{A^*(s-t)}Qx_u(s)ds,$$

we can write the optimal control

$$u(t) = -R^{-1}B^* P(t, t_f)x_u(t).$$

Note that $P(t_f, t_f)x_u(t_f) = Gx_u(t_f)$ and differentiate with respect to t:

$$\dot{P}(t, t_f)x_u(t) + P(t, t_f)\dot{x}_u(t) = -Qx_u(t) - A^*e^{A^*(t_f - t)}Gx_u(t_f)$$
$$- A^* \int_t^{t_f} e^{A^*(s-t)}Qx_u(s)ds.$$

If we note that

$$\dot{x}_u(t) = Ax_u(t) - BR^{-1}B^* P(t, t_f)x_u(t),$$

it follows that

$$\dot{P}(t, t_f)x_u(t) + P(t, t_f)Ax_u(t) - P(t, t_f)BR^{-1}B^* P(t, t_f)x_u(t)$$
$$= -Qx_u(t) - A^* P(t, t_f)x_u(t).$$

Hence, $P(t, t_f)$ must satisfy

$$\dot{P}(t, t_f) + A^* P(t, t_f) + P(t, t_f)A - P(t, t_f)BR^{-1}B^* P(t, t_f) + Q = 0,$$
$$P(t_f, t_f) = G.$$

Thus, solving the differential Riccati equation for $P(t, t_f)$ yields the optimal control

$$u_{opt}(t) = -R^{-1}B^1 P(t, t_f)x_u(t).$$

We have for any x_o, and $u \in L_2(0, t_f; R^m)$,

$$J(x_o, u; t_f) = x_o^* P(0, t_f) x_o$$
$$+ \int_0^{t_f} x(t)^* Q x(t) + u(t)^* R u(t) + \frac{d}{dt}(x(t)^* P(t, t_f) x(t)) dt.$$

Dropping the explicit indication of dependence on t inside the integral, and using the fact that $P(t, t_f)$ solves (5.23),

$$J(x_o, u; t_f) = x_o^* P(0, t_f) x_o + \int_0^{t_f} x^* Q x + u^* R u + (Ax + Bu)^* P x$$
$$+ x^* \frac{dP}{dt} x + x^* P(Ax + Bu) dt$$
$$= x_o^* P(0, t_f) x_o + \int_0^{t_f} u^* R u + (Bu)^* P x + x^* P B R^{-1} B^* P x$$
$$+ x^* P(Bu) dt$$
$$= x_o^* P(0, t_f) x_o + \int_0^{t_f} (u + R^{-1} B^* P x)^* R (u + R^{-1} B^* P x) dt.$$

The optimal controller is

$$u_{opt}(t) = -R^{-1} B^* P(t, t_f) x(t),$$

and the corresponding optimal cost is $x_o^* P(0, t_f) x_o$.

Since $0 \leq J(x_o, u; t_f) \leq \infty$ for all u, it follows immediately that $P(0, t_f)$ is positive semidefinite and that $P(0, t_f) < \infty$ for all t_f. ■

The optimal control $u(t)$ depends on the state $x_u(t)$ and we have a feedback law. It is interesting that despite the fact that we only assumed that the control $u \in L_2(0, t_f; R^m)$, the optimal control takes this nice form.

Example 5.8: $\dot{x}(t) = x(t) + u(t), \quad x(0) = x_o.$

$$J(x_o, t_f; u) = \int_0^{t_f} 3x^2(s) + u^2(s) ds.$$

Here, $G = 0$, $Q = 3$, $R = 1$. The differential Riccati equation is

$$\dot{p}(t) + 2p(t) - p^2(t) + 3 = 0, \quad p(t_f) = 0,$$

which has solution

$$p(t) = \frac{3(1 - e^{-4(t_f - t)})}{(1 + 3e^{-4(t_f - t)})}$$

with optimal control $u(t) = -p(t)x(t)$.

5.2 LINEAR QUADRATIC REGULATORS

In this section we will develop a well-known technique for designing state-feedback controllers. We still assume that the state is available: The system is described by

$$\dot{x}(t) = Ax(t) + Bu(t), \qquad x(0) = x_o. \tag{5.24}$$

The cost associated with a particular control is now over an infinite time interval:

$$J(x_o, u) = \int_0^\infty \|Cx(t)\|^2 + \|Eu(t)\|^2 dt.$$

For simplicity of notation we will often write $Q = C^*C$ and $R = E^*E$. We will assume that R is positive definite. Consider the infinite-time optimization problem

$$\inf_{u \in L_2(0,\infty;R^m)} J(x_o, u) \tag{5.25}$$

subject to $x(t)$ solves (5.14).

Since now the cost is calculated over an infinite time interval, the optimal cost will not be finite in all cases. We must make a fundamental assumption to ensure that the cost is finite.

Definition 5.9: The system (A, B) is open loop stabilizable if for every initial condition x_o there is a control $u \in L_2(0, \infty; R^m)$ so that $x \in L_2(0, \infty; R^n)$.

Several preliminary results are needed before establishing the main result.

Lemma 5.10: Consider the finite-time control problem (5.16). The differential Riccati equation

$$\frac{d\Pi(t)}{dt} = A^*\Pi(t) + \Pi(t)A - \Pi(t)BR^{-1}B^*\Pi(t) + Q \tag{5.26}$$

$$\Pi(0) = G \tag{5.27}$$

has a unique continuous symmetric nonnegative solution. Also,

$$x_o^*\Pi(t_f)x_o = \min_{u \in L_2(0,t_f,R^m)} J(x_o, t_f; u).$$

If $G = 0$, then the matrix-valued sequence $\Pi(t)$ satisfies

$$x_o^*\Pi(t_1)x_o \leq x_o^*\Pi(t_2)x_o$$

for all $x_o \in R^n$ and all times $0 \leq t_1 \leq t_2$.

□ *Proof:* Define

$$\Pi_{t_f}(t) = P(t_f - t, t_f)$$

where P solves the differential Riccati equation (5.23) with $P(t_f, t_f) = G$. Clearly $\Pi_{t_f}(0) = G$ and $\Pi_{t_f}(t)$ satisfies the differential equation (5.26). Uniqueness of the solution follows from standard results in differential equations. The cost (5.27) also follows from Theorem 5.7.

We now show that $\Pi_{t_f}(t)$ is independent of the final time t_f. Choose any t_1, t_2 with $t_1 \leq t_2$. The matrix-valued functions $\Pi_{t_1}(t)$ and $\Pi_{t_2}(t)$ are each symmetric and continuous on t_1 and t_2, respectively. They also each satisfy (5.26) on $[0, t_1]$. Since the solution to (5.26) is unique,

$$\Pi_{t_1}(t) = \Pi_{t_2}(t), \qquad 0 \leq t \leq t_1.$$

Since t_1 and t_2 were arbitrary, we obtain that $\Pi_{t_f}(t)$ is independent of t_f, and we may write simply $\Pi(t)$.

It remains only to prove monotonicity of $\Pi(t)$ with t. For any x_o, and $t_1 \leq t_2$,

$$\begin{aligned} x_o^* \Pi(t_2) x_o &= \min_{u \in L_2(0,t_2;R^m)} J(x_o, t_2; u) \\ &\geq \min_{u \in L_2(0,t_2;R^m)} J(x_o, t_1; u) \\ &= x_o^* \Pi(t_1) x_o. \end{aligned}$$

Thus, for any x_o and any $t_1 \leq t_2$,

$$x_o^* \Pi(t_1) x_o \leq x_o^* \Pi(t_2) x_o$$

as was to be proven. ■

The proof of the following lemma is identical to the second part of the proof of Theorem 5.7 and is left as an exercise.

LEMMA 5.11: Suppose that Π is a symmetric solution to the algebraic Riccati equation (ARE)

$$A^*P + PA - PBR^{-1}B^*P + Q = 0.$$

Then for every $u \in L_2(0, \infty; R^m)$, $x_o \in R^n$ and $t \geq 0$,

$$\begin{aligned} J(x_o, u; t) = x_o^* \Pi x_o - x(t)^* \Pi x(t) + \int_0^t (u(s) + R^{-1}B^*\Pi x(s))^* R(u(s) \\ + R^{-1}B^*\Pi x(s)) ds. \end{aligned} \tag{5.28}$$

THEOREM 5.12: Assume that the system is open loop stabilizable. Then the infinite-time problem (5.25) has a minimum for each initial condition x_o. There exists a symmetric, nonnegative matrix Π such that

$$\inf_{u\in L_2(0,\infty;R^m)} J(x_o,u) = J(x_o,u_{opt}) = x_o^*\Pi x_o$$

where the optimal control is

$$u_{opt}(t) = -R^{-1}B^*\Pi x(t).$$

The matrix

$$\Pi = \lim_{t\to\infty} \Pi(t)$$

where $\Pi(t)$ is the solution to (5.26) with $\Pi(0) = 0$. Equivalently, Π is the minimal nonnegative solution to the ARE

$$A^*P + PA - PBR^{-1}B^*P + Q = 0. \tag{5.29}$$

That is, any other (symmetric) solution Π_1 to the ARE has

$$x_o^*\Pi x_o \le x_o^*\Pi_1 x_o$$

for all $x_o \in R^n$.

□ *Proof:* Since the problem is open-loop stabilizable by some $\tilde{u} \in L_2(0,\infty;R^m)$,

$$x_o^*\Pi(t)x_o = \min_{u\in L_2(0,t;R^m)} J(x_o,u;t) \le J(x_o,\tilde{u}).$$

Thus, the sequence $\Pi(t)$ is bounded. Since it is monotonic, it has a limit, Π. (Think of the limit of each entry in the matrix $\Pi(t)$.) It is clear that Π is nonnegative and symmetric. Since $\Pi(t)$ satisfies the differential Riccati equation (5.26), the right-hand side of (5.26) converges to

$$A^*\Pi + \Pi A - \Pi BR^{-1}B^*\Pi + Q.$$

Since the right-hand sides converge, so do the left-hand sides. Since Π is independent of t, this limit must be 0, and we obtain that Π satisfies the ARE (5.29).

We now prove that Π yields the optimal cost and that u_{opt} is the optimal control. First, for any $t \ge 0$

$$\begin{aligned}\inf_{u\in L_2(0,\infty;R^m)} J(x_o,u) &\ge \inf_{u\in L_2(0,\infty;R^m)} J(x_o,u;t)\\ &= \inf_{u\in L_2(0,t;R^m)} J(x_o,u;t)\\ &= x_o^*\Pi(t)x_o.\end{aligned}$$

Since $\Pi = \lim_{t\to\infty} \Pi(t)$,

$$\inf_{u\in L_2(0,\infty;R^m)} J(x_o, u) \geq x_o^* \Pi x_o. \tag{5.30}$$

Now, from Lemma 5.11, since Π is nonnegative,

$$J(x_o, u; t) \leq x_o^* \Pi x_o + \int_0^t [u(s) + R^{-1} B^* \Pi x(s)]^* R[u(s) + R^{-1} B^* \Pi x(s)] ds. \tag{5.31}$$

If $u(s) = u_{opt}(s)$ where u_{opt} is the feedback control $-R^{-1}B^*\Pi x(s)$, then for all $t \geq 0$,

$$J(x_o, u_{opt}; t) \leq x_o^* \Pi x_o.$$

It follows that

$$J(x_o, u_{opt}) \leq x_o^* \Pi x_o. \tag{5.32}$$

It remains to show that $u_{opt} \in L_2(0, \infty; R^m)$. Recall that R is assumed positive definite and so, letting $r > 0$ indicate the smallest eigenvalue of R, for any $v \in R^m$,

$$v^* v \leq \frac{1}{r} v^* R v.$$

Thus,

$$\begin{aligned} \|u_{opt}\|^2_{L_2(0,\infty;R^m)} &\leq \frac{1}{r} \int_0^\infty u_{opt}^*(t) R u_{opt}(t) dt \\ &\leq \frac{1}{r} J(x_o, u_{opt}) \\ &= \frac{1}{r} x_o^* \Pi x_o \\ &< \infty. \end{aligned}$$

Thus, $u_{opt} \in L_2(0, \infty; R^m)$. Putting (5.30) together with (5.32),

$$x_o^* \Pi x_o \leq \inf_{u\in L_2(0,\infty;R^m)} J(x_o; u) \leq x_o^* \Pi x_o$$

and so,

$$\inf_{u\in L_2(0,\infty;R^m)} J(x_o; u) = J(x_o, u_{opt}) = x_o^* \Pi x_o.$$

Any nonnegative solution to the ARE satisfies (5.31), and so since Π yields the optimal cost, it must be the minimal solution. ■

The fairly weak assumption that some control $u \in L_2(0, \infty; R^m)$ exists that renders the cost finite implies that the optimal control is a feedback control. Furthermore, it is a constant state-feedback control. However, this optimal control is not always stabilizing.

EXAMPLE 5.13: Consider the system

$$\dot{x}(t) = x(t) + u(t), \qquad x(0) = x_o,$$

with cost functional

$$J = \int_0^\infty u(t)^2 dt.$$

Clearly the optimal control is $u(t) \equiv 0$. However, this does not lead to a system state $x(t) \in L_2(0, \infty)$. The problem is that the system is not internally stable, and this instability does not affect the cost.

It turns out that detectability of (A, Q) (or (A, C) where $C^*C = Q$) is sufficient to ensure that the controlled system is internally stable. Several lemmas are needed before establishing this result.

LEMMA 5.14: The function $\exp(At)x_o \in L_2(0, \infty; R^n)$ for every $x_o \in R^n$ if and only if A is Hurwitz.

□ *Proof:* If A is Hurwitz, then $\|\exp(At)x_o\| \leq M \exp(-\alpha t)\|x_o\|$ where $\alpha > 0$ (Lemma 3.5). The conclusion that $\exp(At)x_o \in L_2(0, \infty; R^n)$ follows.

If A is not Hurwitz, then there is an eigenvalue λ with $\mathrm{Re}(\lambda) \geq 0$ and corresponding eigenvector x_λ. From the definition of the matrix exponential

$$\begin{aligned} \|\exp(At)x_\lambda\| &\geq |\exp(\lambda t)| \|x_\lambda\| \\ &\geq \|x_\lambda\| \end{aligned}$$

and so $\exp(At)x_\lambda \notin L_2(0, \infty; R^n)$. ■

LEMMA 5.15: Assume that (A, C) is detectable and define

$$z(t) = \begin{bmatrix} Cx(t) \\ Eu(t) \end{bmatrix}.$$

If $u(t) \in L_2(0, \infty; R^m)$ and $z(t) \in L_2(0, \infty; R^{n+m})$ then $x(t) \in L_2(0, \infty; R^n)$.

□ *Proof:* Choose any F such that $A - FC$ is Hurwitz. Define the system

$$\frac{d\hat{x}(t)}{dt} = (A - FC)\hat{x}(t) + Bu(t) + [F \ \ 0]\, z(t).$$

This can be regarded as a system with inputs u and z, output $\hat{x}$, and state-space realization $(A - FC, [B \quad F], I, 0)$. Since $A - FC$ is Hurwitz, the system is externally stable. Since the inputs u, z are in $L_2(0, \infty; R^m)$ and $L_2(0, \infty; R^{n+m})$, respectively, it follows that $\hat{x}(t) \in L_2(0, \infty; R^n)$. Defining

$$e(t) = \hat{x}(t) - x(t),$$

it follows from $\dot{x}(t) = Ax(t) + Bu(t)$ that

$$\dot{e}(t) = (A - FC)e(t).$$

Since $A - FC$ is Hurwitz, $e(t) \in L_2(0, \infty; R^n)$. Since $x(t) = \hat{x}(t) - e(t)$, $x(t) \in L_2(0, \infty; R^n)$ as was to be proven. ■

Theorem 5.16: Suppose that (A, Q) is detectable and that the problem (5.25) is open-loop stabilizable. Then

1. The optimal feedback $K = R^{-1}B^*\Pi$ is stabilizing, i.e., $A - BK$ is Hurwitz, and
2. The algebraic Riccati equation (ARE) has a unique nonnegative solution Π.

□ *Proof:* We first note that (A, Q) is detectable, if and only if for any factor C such that $Q = C^*C$, we have that (A, C) is detectable. (1) Let Π indicate the minimal nonnegative solution of the ARE. From Theorem 5.12 it follows that for any initial condition x_o,

$$\begin{aligned} x_o^*\Pi x_o &= \int_0^\infty \|Cx(t)\|^2 + \|Eu_{opt}(t)\|^2 dt \\ &= \|z(t)\|^2_{L_2(0,\infty;R^{n+m})} \end{aligned}$$

where

$$z(t) = \begin{bmatrix} Cx(t) \\ Eu_{opt}(t) \end{bmatrix}.$$

Also, since $R = E^*E$ is positive definite, $u_{opt} \in L_2(0, \infty; R^m)$. Since (A, C) is detectable, it follows from Lemma 5.15 that $x(t) \in L_2(0, \infty; R^n)$. Now, $u_{opt}(t) = -Kx(t)$ and so

$$\dot{x}(t) = Ax(t) + B(-Kx(t)), \quad x(0) = x_o.$$

This implies that

$$x(t) = \exp((A - BK)t)x_o.$$

Since x_o is arbitrary, Lemma 5.14 implies that $A - BK$ is Hurwitz.

(2) We will show that any other nonnegative solution $\tilde{\Pi}$ to the ARE satisfies

$$x_o^* \tilde{\Pi} x_o \leq x_o^* \Pi x_o$$

for all $x_o \in R^n$. That is, Π is the maximal nonnegative symmetric solution to the ARE. Since a previous theorem (Theorem 5.12) showed that it is the minimal nonnegative solution, this will prove that Π is the unique nonnegative symmetric solution.

For each x_o, define the following subset of $L_2(0, \infty; R^m)$:

$$\mathcal{U}_{x_o} = \{u \in L_2(0, \infty; R^m) | \lim_{t\to\infty} x(t) = 0, x(t) = \exp(At)x_o + \int_0^t \exp(A(t-r)Bu(r)dr \in L_2(0, \infty; R^n)\}.$$

Since the problem is open-loop stabilizable, $\mathcal{U}_{x_o}$ is nonempty for each x_o. Let $\tilde{\Pi}$ be any symmetric nonnegative solution to the ARE. From Lemma 5.11, for any x_o, u,

$$J(x_o, u; t) = x_o^* \tilde{\Pi} x_o - x(t)^* \tilde{\Pi} x(t) + \int_0^t (u(s) + R^{-1}B^*\tilde{\Pi}x(s))^* R(u(s) + R^{-1}B^*\tilde{\Pi}x(s))ds.$$

If $u \in \mathcal{U}_{x_o}$,

$$J(x_o, u) = \lim_{t\to\infty} J(x_o, u; t) = x_o^* \tilde{\Pi} x_o + \int_0^\infty (u(s)+R^{-1}B^*\tilde{\Pi}x(s))^* R(u(s)+R^{-1}B^*\tilde{\Pi}x(s))ds.$$

Thus, for all $u \in \mathcal{U}_{x_o}$,

$$J(x_o, u) \geq x_o^* \tilde{\Pi} x_o.$$

The optimal control $u_{opt} = -R^{-1}B^*\Pi x(s) \in \mathcal{U}_{x_o}$ and so

$$x_o^* \Pi x_o = J(x_o, u_{opt}) \geq x_o^* \tilde{\Pi} x_o.$$

Since x_o was arbitrary, Π is the maximal symmetric solution to the ARE. This completes the proof. ∎

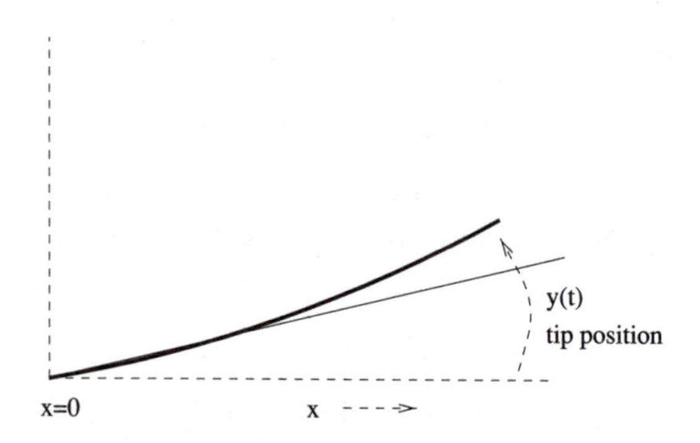

FIGURE 5.1 Flexible beam.

Example 5.17 (*Control of a Flexible Beam*): A flexible slewing beam occurs in such diverse applications as a link of a flexible robot arm, a flexible antenna, a helicopter blade, and also as part of a flexible space structure.

The simplest case of a robotic or structural link with flexibility is a single beam. (See Fig. 5.1.) The beam is moved by a torque applied at the hub ($x = 0$) through a motor. The link rotates about the hub. This system is modeled by a partial differential equation. We can truncate the Fourier series expansion to obtain a system of ordinary differential equations. Let the control u be the applied torque, the measured output y the tip position, and x the state vector. For a particular thin beam of length 1 m we obtain the state space equations

$$\frac{dx}{dt}(t) = Ax(t) + Bu(t)$$

where the state matrix A is

$$\begin{bmatrix} 0 & 1 & & & & & & & & \\ 0 & 0 & & & & & & & & \\ & & 0 & 1 & & & & & & \\ & & -w_1^2 & -2\xi w_1 & & & & & & \\ & & & & 0 & 1 & & & & \\ & & & & -w_2^2 & -2\xi w_2 & & & & \\ & & & & & & 0 & 1 & & \\ & & & & & & -w_3^2 & -2\xi w_3 & & \\ & & & & & & & & 0 & 1 \\ & & & & & & & & -w_4^2 & -2\xi w_4 \end{bmatrix}$$

and

$$B = \frac{1}{0.0829}\begin{bmatrix} 0 \\ 1 \\ 0 \\ 2.886 \\ 0 \\ -2.345 \\ 0 \\ -0.910 \\ 0 \\ -0.454 \end{bmatrix}.$$

The first four natural frequencies are

$$w = 55.89, 131.74, 313.81, 603.67 \quad (\text{rad/s}).$$

We set the damping ratio $\xi = 0.002$ for each frequency. (All unidentified entries in A are 0.) The tip position is

$$y(t) = [1 \ \ 0 \ \ -0.931 \ \ 0 \ \ -1.027 \ \ 0 \ \ 1.169 \ \ 0 \ \ -1.187 \ \ 0]x(t).$$

The matrix A has eigenvalues at 0 and the open loop is not stable. With an initial condition that is zero in the first two states, the system will settle to 0 but the settling time is very long (Fig. 5.2). Figure 5.3 shows the Bode plot. The poor phase and gain margins indicate that simple proportional feedback will not be satisfactory. Furthermore, the multiple peaks in the frequency response make loopshaping difficult. The performance objective is to stabilize the system with settling time less than 5 seconds. A linear quadratic regulator with a state weight of $Q = 10C^*C$ and $R = 1$ was designed by solving an ARE. The response of the controlled system $(A - BK, B, C, O)$ is shown, with the same initial condition as before, in Fig. 5.4. The closed loop is stable, and the settling time is now under 10 s.

5.3 ROBUSTNESS OF LINEAR QUADRATIC REGULATORS

In this section we will show that linear quadratic regulators possess good phase and gain margins. We will assume throughout this section that the system is single input. This means that B is a column vector and the state feedback control K is a row vector. We will also assume that A is Hurwitz. (These assumptions are made solely to simplify the proofs. The results still

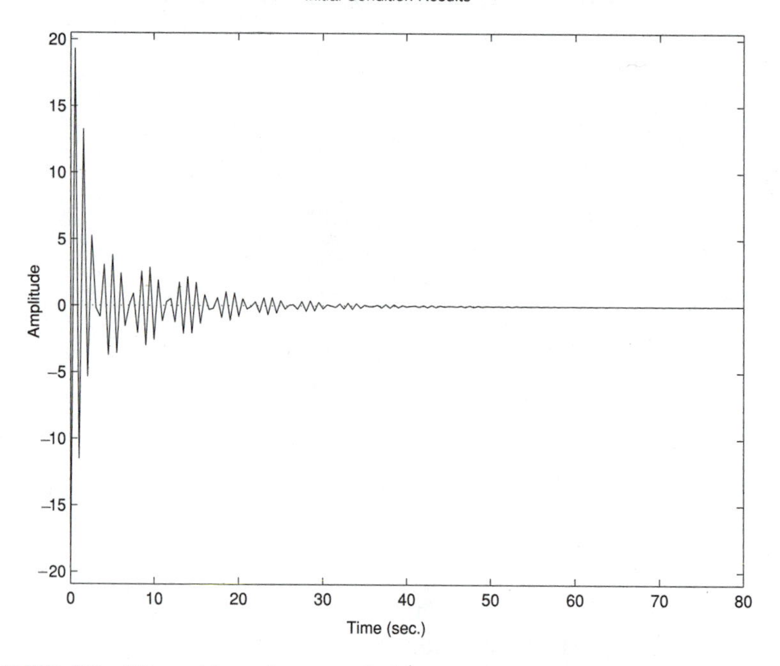

FIGURE 5.2 Tip position of uncontrolled beam with initial condition $x_0 = [0, 0, 1, 1, \ldots, 1]$.

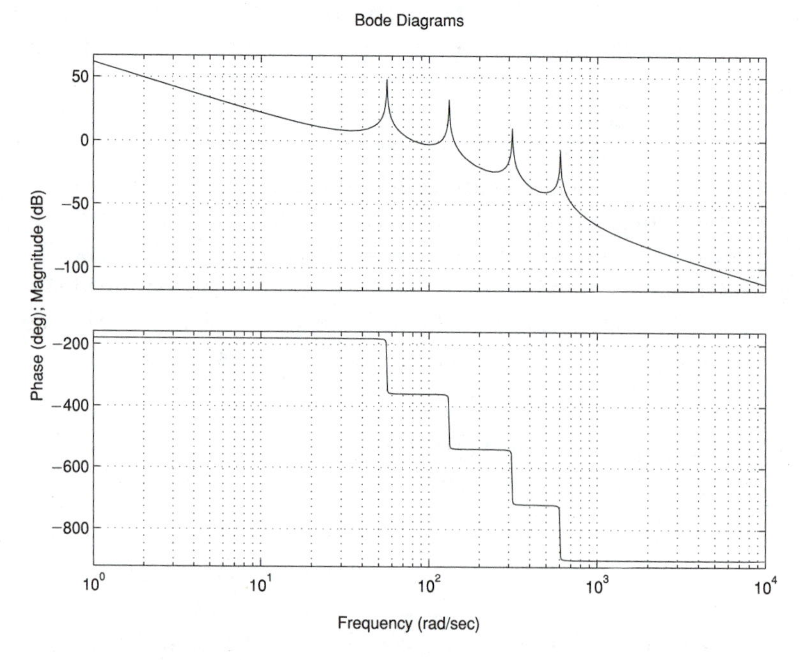

FIGURE 5.3 Bode plot of uncontrolled beam.

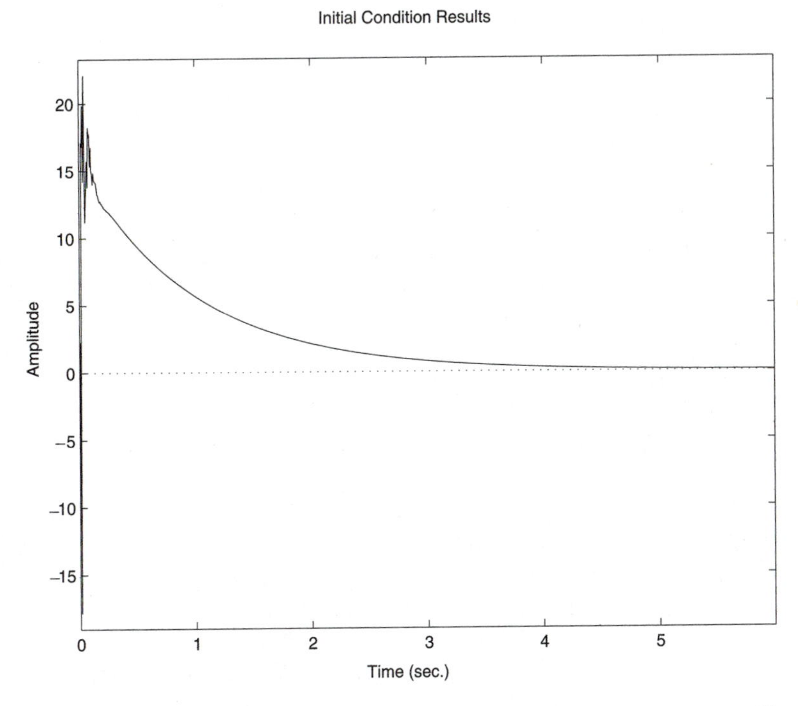

FIGURE 5.4 Beam controlled by linear quadratic regulator, $x(0) = [1, 1 \ldots 1]$.

hold if these two assumptions are removed.) Some preliminary results must first be established.

LEMMA 5.18: Assume that (A, B) is stabilizable. The matrix $A - BK$ is Hurwitz if and only if K externally stabilizes $(sI - A)^{-1}B$ in the sense of Definition 3.3.

□ *Proof:* Thinking of the output operator as the identity operator, it is trivial that (A, I) is detectable. Thus, (A, B, I) is stabilizable and detectable. It follows that the closed-loop system is internally stable if and only if it is externally stable. The result follows. ■

THEOREM 5.19: Assume that A is Hurwitz. The vector K stabilizes (A, B) (that is, $A - BK$ is Hurwitz,) if and only if $[I + K(sI - A)^{-1}B]^{-1} \in M(R\mathcal{H}_\infty)$.

□ *Proof:* We will use the previous lemma to show that the closed-loop matrix for the control system $((sI - A)^{-1}B, K)$ is in $M(R\mathcal{H}_\infty)$ if and only if $H(s)^{-1} \in M(R\mathcal{H}_\infty)$ where $H(s) = I + K(sI - A)^{-1}B$. Write the closed-loop matrix, using (3.16) as

$$\Delta((sI - A)^{-1}B, K) = \begin{bmatrix} I - (sI - A)^{-1}BH(s)^{-1}K & -(sI - A)^{-1}BH(s)^{-1} \\ H(s)^{-1}K & H(s)^{-1} \end{bmatrix}.$$

By assumption, $(sI - A)^{-1}B \in M(R\mathcal{H}_\infty)$. It follows that the closed loop is stable if and only if $H(s)^{-1} \in M(R\mathcal{H}_\infty)$. Since the system $(A, B, I, 0)$ is (trivially) stabilizable and detectable, $A - BK$ is Hurwitz if and only if $[I + K(sI - A)^{-1}B]^{-1} \in M(R\mathcal{H}_\infty)$. ■

This result implies that the stability of a single-input system with constant state feedback can be established by ensuring that the polynomial $1 + K(sI - A)^{-1}B$ has no closed right half-plane zeros.

Lemma 5.20: Let (A, B) be a stabilizable system and Π the unique non-negative solution to the algebraic Riccati equation for some weights $Q \geq 0$, $R > 0$. Defining $K = R^{-1}B^*\Pi$, the optimal state feedback is $u = -Kx(t)$. The following identity holds:

$$\begin{aligned}&[I + B^*(-j\omega I - A^*)^{-1}K^*]R[I + K(j\omega I - A)^{-1}B] \\ &\quad = R + B^*(-j\omega I - A^*)^{-1}Q(j\omega I - A)^{-1}B. \end{aligned} \tag{5.33}$$

□ *Proof:* The algebraic Riccati equation is

$$\Pi A + A^*\Pi - \Pi BR^{-1}B^*\Pi + Q = 0.$$

Or,

$$-\Pi A - A^*\Pi + \Pi BR^{-1}B^*\Pi = Q.$$

Using $K = R^{-1}B^*\Pi$, and adding and subtracting $\Pi j\omega$, we can rewrite this as

$$\Pi(j\omega I - A) + (-j\omega I - A^*)\Pi + KRK = Q.$$

Multiply on the left by $B^*(-j\omega I - A^*)^{-1}$and on the right by $(j\omega I - A)^{-1}B$. We obtain

$$\begin{aligned}&B^*(-j\omega I - A^*)^{-1}\Pi B + B^*\Pi(j\omega I - A)^{-1}B \\ &\qquad + B^*(-j\omega I - A^*)^{-1}KRK(j\omega I - A)^{-1}B \ldots \\ &\quad = B^*(-j\omega I - A^*)^{-1}Q(j\omega I - A)^{-1}B.\end{aligned}$$

Using the fact that $B^*\Pi = RK$, and adding R to both sides leads to

$$\begin{aligned} R + B^*(-j\omega I - A^*)^{-1}K^*R + RK(j\omega I - A)^{-1}B \\ + B^*(-j\omega I - A^*)^{-1}KRK(j\omega I - A) \\ = R + B^*(-j\omega I - A^*)^{-1}Q(j\omega I - A)^{-1}B. \end{aligned}$$

The left-hand side of the equation is now a perfect square and we can rewrite it as $[I + B^*(-j\omega I - A^*)^{-1}K^*]R[I + K(j\omega I - A)^{-1}B] = R + B^*(-j\omega I - A^*)^{-1}Q(j\omega I - A)^{-1}B$. ■

COROLLARY 5.21: Let (A, B) and K be as in the previous lemma. If the system is single input then

$$|1 + K(j\omega I - A)^{-1}B|^2 \geq 1.$$

□ *Proof:* Since the system is single input, R is a scalar, and we can divide by R in Eq. (5.33) to obtain

$$|1 + K(j\omega I - A)^{-1}B|^2 = 1 + R^{-1}B^*(-j\omega I - A^*)^{-1}Q(j\omega I - A)^{-1}B.$$

The matrix Q is positive semidefinite, so writing $p = (j\omega I - A)^{-1}B$, we have

$$\begin{aligned} |1 + K(j\omega I - A)^{-1}B|^2 &= 1 + R^{-1}p^*Qp \\ &\geq 1. \end{aligned}$$

■

The preceding inequality is identical to

$$|-1 - K(j\omega I - A)^{-1}B|^2 \geq 1.$$

This result implies that the Nyquist plot of $K(j\omega I - A)^{-1}B$ lies outside of the circle with center -1 and radius 1.

THEOREM 5.22: Assume that (A, B) is single input and that A is Hurwitz. Then any linear quadratic regulator K stabilizes the system with an infinite gain margin and a phase margin of at least 60°.

□ *Proof:* Theorem 5.19 implies that the closed loop will be stable if and only if the Nyquist plot of $K(j\omega I - A)^{-1}B$ does not encircle the -1 point. It was shown earlier that K does stabilize the system. This fact, together with the previous result, implies that the Nyquist plot of $K(j\omega I - A)^{-1}B$ does not encircle -1 and also lies outside the circle with center -1 and radius 1. Thus, the gain can be multiplied by any number $M > 1$, and stability is maintained.

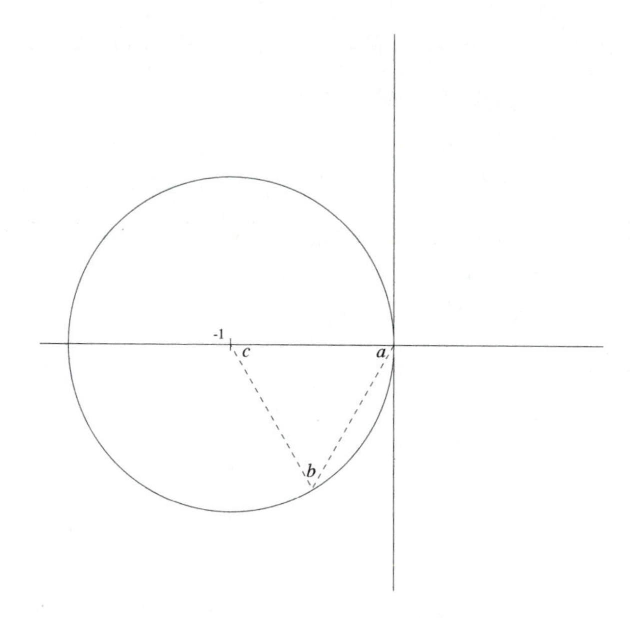

FIGURE 5.5 Figure for Theorem 5.22.

To show that the phase margin is at least $60°$, consider Fig. 5.5. Since the Nyquist plot of $K(j\omega I - A)^{-1}B$ lies outside the circle with center -1 and radius 1, the phase margin is greater than the angle that line ab makes with the negative real axis. The triangle with vertices a,b,c is equilateral and so this angle is $60°$. ∎

EXAMPLE 5.23 (*Flexible Beam Example 5.17 Continued*): Here we examine the robustness of control of the flexible beam using a linear quadratic regulator. An enlarged Nyquist plot (modified around the imaginary axis pole) is shown in Fig. 5.6. The phase and gain margins are generous, as was predicted by the results in this section.

5.4 ESTIMATION

There is a major difficulty in using state feedback. For most systems, the full set of states is not measured. Consider an arbitrary control system (with $E = 0$ for simplicity):

$$\begin{aligned} \frac{dx}{dt} &= Ax(t) + Bu(t), \quad x(0) = x_0 \\ y(t) &= Cx(t). \end{aligned} \tag{5.34}$$

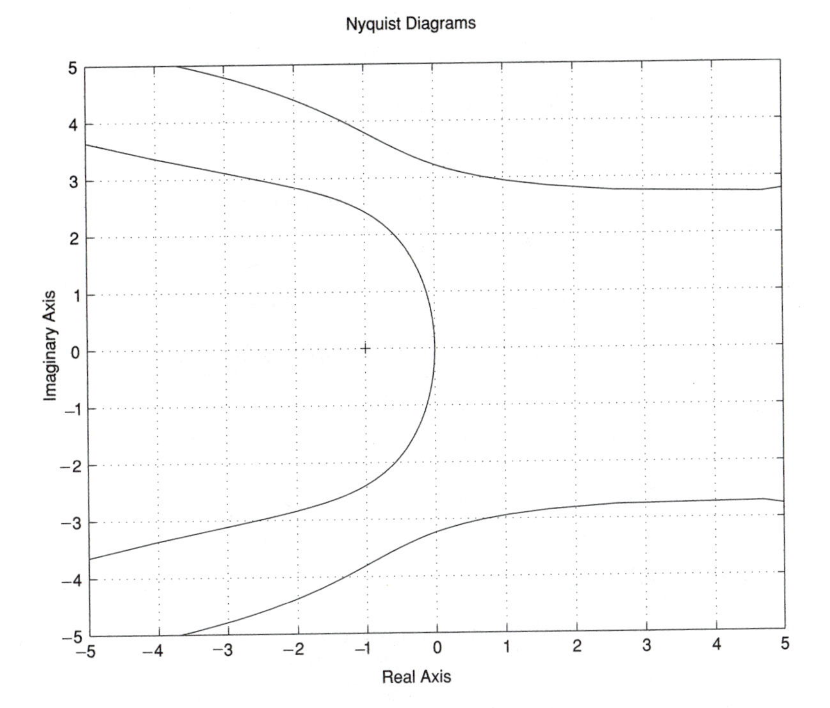

FIGURE 5.6 Nyquist diagram of beam controlled by linear quadratic regulator.

We would like to construct an *estimate* $\hat{x}(t)$ of the state $x(t)$, based only on y and u and our model (5.34).

DEFINITION 5.24: We say $\hat{x}(t)$ *estimates* $x(t)$ if

$$\lim_{t\to\infty} \hat{x}(t) = x(t) \tag{5.35}$$

for any choice of initial condition $\hat{x}(0)$.

The estimate $\hat{x}(t)$ could then be combined with a satisfactory state feedback K to obtain the control

$$u = -K\hat{x}(t). \tag{5.36}$$

An obvious choice of estimate is

$$\frac{d\hat{x}}{dt} = A\hat{x}(t) + Bu(t), \quad \hat{x}(0) = x_0. \tag{5.37}$$

However, even if $x(0)$ can be measured, it will not be perfectly known. An error in the initial condition may cause the estimation error

$$e(t) = \hat{x}(t) - x(t) \tag{5.38}$$

to increase with time. Notice also that this scheme does not use the information from the output $y(t)$. It makes sense to feed back the error in the estimated output to the estimator:

$$\begin{aligned}\frac{d\hat{x}(t)}{dt} &= A\hat{x}(t) + Bu(t) + F(y(t) - Cx(t)) \\ &= (A - FC)\hat{x}(t) + Bu(t) + Fy(t).\end{aligned} \tag{5.39}$$

The following theorem shows that if (A, C) is detectable, then F can be chosen so that the system (5.39) yields an estimate of $x(t)$.

THEOREM 5.25: The system (5.39) is an estimator for $x(t)$ in (5.34) if $A - FC$ is Hurwitz.

□ *Proof:* The error

$$e(t) = x(t) - \hat{x}(t) \tag{5.40}$$

and

$$\frac{de(t)}{dt} = \frac{dx(t)}{dt} - \frac{d\hat{x}(t)}{dt}. \tag{5.41}$$

Substituting in (5.34) and (5.39) for $\frac{dx}{dt}$, $\frac{d\hat{x}}{dt}$,

$$\dot{e}(t) = (A - FC)e(t). \tag{5.42}$$

The estimation error will converge to zero, regardless of whether $x(0) - \hat{x}(0) = 0$, if F is chosen so that $A - FC$ is Hurwitz. ■

THEOREM 5.26: If (A, C) is observable, then the poles of the estimator (5.39) may be chosen arbitrarily.

□ *Proof:* The poles of $A - BC$ are the conjugates of those of $A^* - C^*F^*$. Since (A^*, C^*) is controllable, the result follows. ■

The best choice of F for an estimator, like the best choice of K for feedback control, involves competing objectives and is not straightforward. If F is chosen so that the eigenvalues of $A - FC$ have very large negative real parts, then the error $e(t) \to 0$ quickly. However, if the eigenvalues of $A - FC$ are chosen far to the left in the complex plane, the bandwidth of the estimator will be large. In this situation, high-frequency noise in the measurements of u and y lead to noise in the estimate $\hat{x}$. A useful general guideline in pole placement for estimators of stable systems is that the estimator should have a settling time about 2–3 times faster that of the original system. Other factors to consider are complexity of the estimator (simple and low-order versus high-order) and robustness to plant uncertainty.

We now briefly describe a famous class of estimators usually known as *Kalman filters*. Some statistical terms will be used. The definitions of these can be found in any standard introduction to statistics. The result will be quoted without proof.

Suppose the system is described by

$$\begin{aligned} \dot{x}(t) &= Ax(t) + Bu(t) + d(t) \\ y(t) &= Cx(t) + n(t). \end{aligned} \tag{5.43}$$

The signal d can be interpreted as actuator or plant input noise and n can be regarded as sensor noise. Suppose that noise processes d and n are white, Gaussian, of zero mean, and independent with known covariances Q and R, respectively. The matrices Q and R are each symmetric with $Q \geq 0$ and $R > 0$. Also assume that the initial state of the sytem is a Gaussian random variable of known mean x_o and covariance G. The initial state is assumed independent of $d(t)$ and $n(t)$.

Theorem 5.27 (*Kalman Filter*): Consider the problem of obtaining an estimate $\hat{x}(t)$ of the state $x(t)$ in (5.43). This estimate should minimize the error covariance

$$P_e(t_1) = E\,[x(t_1) - \hat{x}(t_1)]\,[x(t_1) - \hat{x}(t_1)]^*$$

for fixed but arbitrary t_1. Assume that (A, Q) is stabilizable. If we take the initial time $t_o \to -\infty$, the minimum covariance P_e is independent of t and may be calculated by solving any of the following equivalent problems:

1. $\dot{P}_e(t) = AP_e(t) + P_e(t)A^* - P_e(t)C^*R^{-1}CP_e(t) + Q$ with $\lim_{t_o \to -\infty} P(t_o) = 0$ and $P_e = P_e(t)$ at any time t, or
2. $\dot{P}_e(t) = AP_e(t) + P_e(t)A^* - P_e(t)C^*R^{-1}CP_e(t) + Q$ with $P_e(0) = 0$ and $P_e = \lim_{t \to \infty} P(t)$, or
3. $AP_e + P_eA^* - P_eC^*R^{-1}CP_e + Q = 0$ where P_e is non-negative definite.

Defining $F = P_eC^*R^{-1}$, the optimal estimator is

$$\frac{d\hat{x}(t)}{dt} = A\hat{x}(t) + Bu(t) - F(C\hat{x}(t) - y(t)).$$

Moreover, $A - FC$ is Hurwitz.

Note that just as controllability and observability are dual concepts, the problems of designing an optimal state feedback and an optimal estimator are dual.

The assumptions on the nature of the noise processes are reasonable in many instances. Unfortunately, in practice, the noise covariances Q and R are rarely known. Two aspects of this result are very useful in estimator design. First, since reliable methods for solution of the ARE for systems of order up to 100 (and sometimes larger) exist, the solution of the ARE can be used to design a nonstatistical estimator. This is known as the linear quadratic estimator (LQE). Second, Q and R play a role similar to the control and weighting parameters in the dual control problem. Although the noise characteristics are rarely known, a desired filter bandwidth is often known. Adjustments in the weights can be used to design an acceptable filter. For instance, suppose that in an initial design, the filter bandwidth is too high. Increasing $\|R\|$ with respect to $\|Q\|$ means that the plant output noise is more significant. The resulting estimator will have a smaller bandwidth in order to reduce the effect of this high-frequency noise.

In the next section several examples of estimator design in conjunction with state feedback are given.

5.5 OUTPUT FEEDBACK

In many cases we do not have the full state available for feedback control:

$$\dot{x}(t) = Ax(t) + Bu(t), \qquad x(0) = x_o \tag{5.44}$$

$$y(t) = Cx(t) \tag{5.45}$$

where C is not invertible. One approach is to design a state feedback, as though the full state was available and then implement this state feedback in series with an estimator.

Example 5.28 (*Control of Loudspeaker Response Continued*): In a previous example on shaping the response of a loudspeaker (Example 4.7) we used a simple constant state feedback to shape the loudspeaker response so that its frequency response was close to constant over the frequency range of interest. This state feedback only used part of the state, the velocity of the loudspeaker cone $\dot{x}$. Unfortunately, it is difficult to measure velocity. However, we can easily measure current i.

Recall that the loudspeaker was modeled by the equations

$$\dot{z}(t) = \begin{bmatrix} 0 & 1 & 0 \\ -\frac{k}{m} & -\frac{c}{m} & \frac{\psi}{m} \\ 0 & -\frac{\psi}{L} & -\frac{R_d}{L} \end{bmatrix} z(t) + \begin{bmatrix} 0 \\ 0 \\ \frac{1}{L} \end{bmatrix} u(t)$$

with state $z = (x, v, i)$ where $v(t) = \dot{x}(t)$ and the control $u(t)$ is voltage. Our output is now

$$y = Cz$$

with $C = [0\ 0\ 1]$.

We need to choose

$$F = \begin{bmatrix} f_1 \\ f_2 \\ f_3 \end{bmatrix}$$

so that $A - FC$ is Hurwitz. Now,

$$A - FC = \begin{bmatrix} 0 & 1 & -f_1 \\ -\frac{k}{m} & -\frac{c}{m} & \frac{\psi}{m} - f_2 \\ 0 & -\frac{\psi}{L} & -\frac{R_d}{L} - f_3 \end{bmatrix}.$$

If we choose $f_1 = 0,\ f_2 = \frac{\psi}{m}$, and $f_3 = 0$ then

$$A - FC = \begin{bmatrix} 0 & 1 & 0 \\ -\frac{k}{m} & -\frac{c}{m} & 0 \\ 0 & -\frac{\psi}{L} & -\frac{R_d}{L} \end{bmatrix}.$$

This matrix is Hurwitz. The full estimator is

$$\dot{\hat{z}}(t) = \begin{bmatrix} 0 & 1 & 0 \\ -\frac{k}{m} & -\frac{c}{m} & 0 \\ 0 & -\frac{\psi}{L} & -\frac{R_d}{L} \end{bmatrix} \hat{z}(t) + \begin{bmatrix} 0 \\ 0 \\ \frac{1}{L} \end{bmatrix} u(t) + \begin{bmatrix} 0 \\ \frac{\psi}{m} \\ 0 \end{bmatrix} i(t).$$

Since our state feedback is of the form $[0\ K_v\ 0]$, we only need an estimate of $\dot{x}$. The advantage to the choice of F used here is that it decouples the dynamics of $x, \dot{x}$ from i. We only need the equations

$$\frac{d}{dt}\begin{bmatrix} \hat{x} \\ \dot{\hat{x}} \end{bmatrix} = \begin{bmatrix} 0 & 1 \\ -\frac{k}{m} & -\frac{c}{m} \end{bmatrix}\begin{bmatrix} \hat{x} \\ \dot{\hat{x}} \end{bmatrix} + \begin{bmatrix} 0 \\ \frac{\psi}{m} \end{bmatrix} i(t).$$

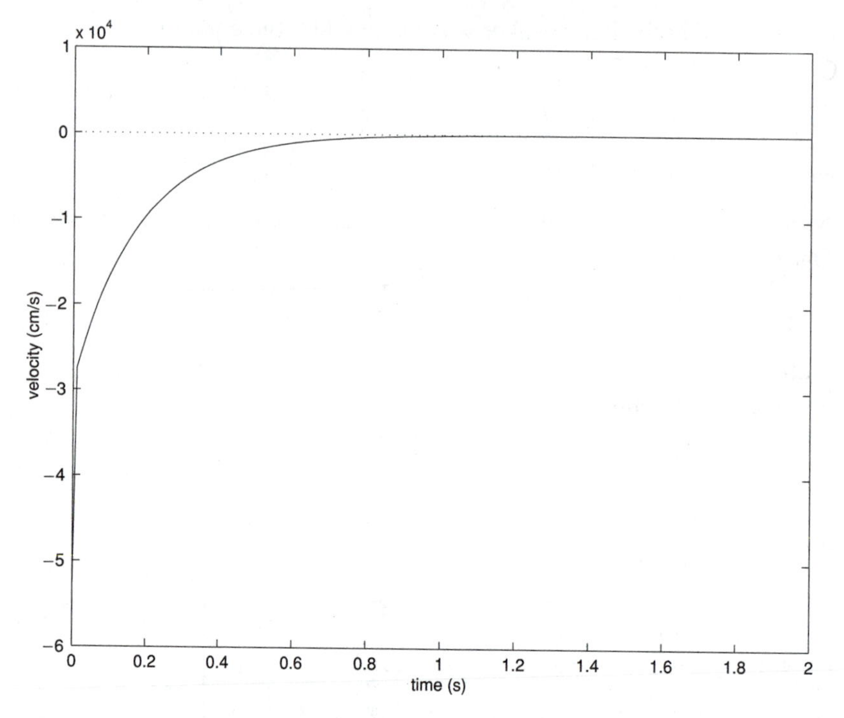

FIGURE 5.7 Actual velocity with initial condition [0.0001 −0.0002 0.00002] (solid) and estimated velocity with zero initial condition (dotted).

This reduced-order estimator is of order one less than the full system. Figure 5.7 compares actual and estimated velocities with different initial conditions.

Figure 5.8 compares the closed-loop transfer function of the speaker with velocity feedback $K_v = 300$ to that of the speaker with the controller

$$\frac{d}{dt}\begin{bmatrix} x_{H_1} \\ x_{H_2} \end{bmatrix} = \begin{bmatrix} 0 & 1 \\ -\frac{k}{m} & -\frac{c}{m} \end{bmatrix}\begin{bmatrix} x_{H_1} \\ x_{H_2} \end{bmatrix} + \begin{bmatrix} 0 \\ \frac{\psi}{m} \end{bmatrix} y(t)$$

$$y_H(t) = [0 \;\; K_v]\begin{bmatrix} x_{H_1} \\ x_{H_2} \end{bmatrix}$$

where $K_v = 300$. The transfer functions are identical. This is to be expected, since with $\hat{z}(0) = 0$, $z(0) = 0$, the estimate of the velocity equals the velocity.

We now describe the general structure of a controller designed by placing state feedback in series with an estimator. Letting $A_F = A - FC$,

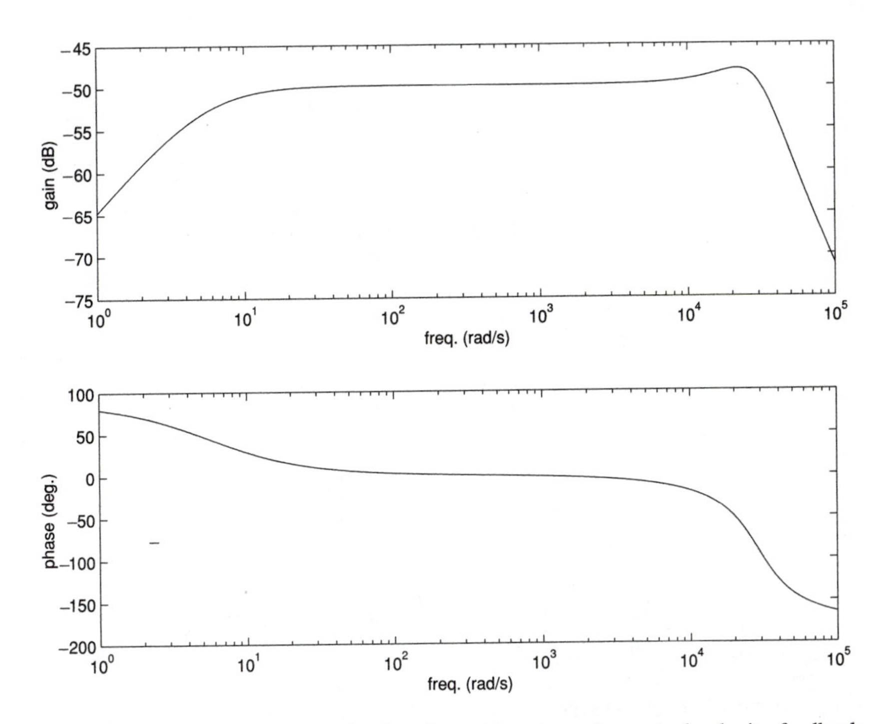

FIGURE 5.8 Closed-loop transfer function with estimated or actual velocity feedback ($k = 300$).

consider the system

$$\dot{x}_H(t) = A_F x_H(t) + Bu + Fy, \qquad x_H(0) = x_{Ho}.$$

In the previous section on estimation it was shown if F is chosen so that A_F is Hurwitz, then x_H is an estimate of the state of the original system. This suggests that to obtain the effect of state feedback Kx, we define the controller output to be

$$y_H(t) = -Kx_H(t)$$

and connect the controller to the original system (5.44 and 5.45) via the feedback connection

$$u = d + y_H.$$

We can put this control system into the standard feedback framework, and include the effect of a second external input as follows. (See Fig. 5.9.)

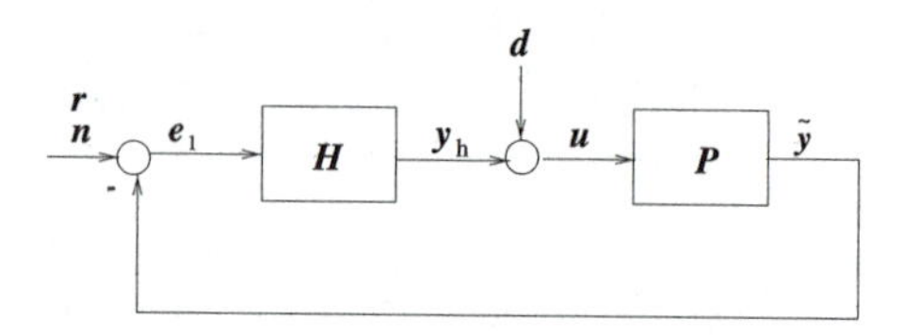

FIGURE 5.9 Standard feedback configuration.

Define an augmented plant output

$$\tilde{y}(t) = \begin{bmatrix} y(t) \\ u(t) \end{bmatrix}$$
$$= \begin{bmatrix} C \\ 0 \end{bmatrix} x(t) + \begin{bmatrix} 0 \\ I \end{bmatrix} u(t). \tag{5.46}$$

The controller is

$$\dot{x}_H(t) = A_F x_H(t) + [-F - B]e_1(t), \qquad x_H(0) = x_{Ho} \tag{5.47}$$
$$y_H(t) = -K x_H(t). \tag{5.48}$$

It is connected to the system via the usual connection:

$$e_1(t) = \begin{bmatrix} r(t) \\ n(t) \end{bmatrix} - \tilde{y}(t) \tag{5.49}$$
$$u(t) = d(t) + y_H(t).$$

The following theorem is often referred to as the *Separation Theorem* since it states that the dynamics of the output feedback controller are those of the estimator plus those of state feedback.

THEOREM 5.29: Consider the closed-loop system (5.44 and 5.45, 5.47–5.49) with external inputs d, n, r, outputs y and y_H, and state $[x(t), x_H(t)]$. The closed-loop eigenvalues are the union of the eigenvalues of A_F and A_K:

$$\lambda(A_{cls}) = \lambda(A_K) \cup \lambda(A_F).$$

Assume that K, F are chosen so that $A - BK$ and $A - FC$ are Hurwitz. Let $M \geq 1, \alpha_F > 0$, and $\alpha_K > 0$ be constants such that $\|\exp((A - BK)t)\| \leq Me^{-\alpha_K t}$ and $\|\exp(A_F t)\| \leq Me^{-\alpha_F t}$.

For any $\alpha < \min(\alpha_F, \alpha_K)$, the closed-loop semigroup is internally stable with decay rate α.

Furthermore, the map from d to x and the map from d to y are identical to those of full state feedback.

□ *Proof:* Elementary manipulations can be used to calculate a state-space realization for the closed loop:

$$\begin{bmatrix} \dot{x}(t) \\ \dot{x}_H(t) \end{bmatrix} = \begin{bmatrix} A & -BK \\ FC & A - FC - BK \end{bmatrix} \begin{bmatrix} x(t) \\ x_H(t) \end{bmatrix} + \begin{bmatrix} B & 0 & 0 \\ B & -F & -B \end{bmatrix} \begin{bmatrix} d(t) \\ r(t) \\ n(t) \end{bmatrix}$$

$$\begin{bmatrix} y(t) \\ u(t) \\ y_H(t) \end{bmatrix} = \begin{bmatrix} C & 0 \\ 0 & -K \\ 0 & -K \end{bmatrix} \begin{bmatrix} x(t) \\ x_H(t) \end{bmatrix} + \begin{bmatrix} 0 & 0 & 0 \\ I & 0 & 0 \\ 0 & 0 & 0 \end{bmatrix} \begin{bmatrix} d(t) \\ r(t) \\ n(t) \end{bmatrix}.$$

Setting the output

$$\tilde{y} = \begin{bmatrix} y \\ u \end{bmatrix}$$

was a device introduced to put the control system into the standard framework. We reduce the output to y, y_H. Also, since the external input n has the same effect as $-d$, write the external input as

$$u_{ext} = [d \quad r].$$

We have

$$\begin{bmatrix} \dot{x}(t) \\ \dot{x}_H(t) \end{bmatrix} = \begin{bmatrix} A & -BK \\ FC & A - FC - BK \end{bmatrix} \begin{bmatrix} x(t) \\ x_H(t) \end{bmatrix} + \begin{bmatrix} B & 0 \\ B & -F \end{bmatrix} u_{ext}(t)$$

$$\begin{bmatrix} y(t) \\ y_H(t) \end{bmatrix} = \begin{bmatrix} C & 0 \\ 0 & -K \end{bmatrix} \begin{bmatrix} x(t) \\ x_H(t) \end{bmatrix}.$$

Indicate this state-space realization by $(A_o, B_o, C_o, 0)$.

To prove that

$$\lambda(A_o) = \lambda(A_K) \cup \lambda(A_F),$$

introduce the state transformation

$$L = \begin{bmatrix} I & 0 \\ I & -I \end{bmatrix}.$$

The new state is

$$\begin{bmatrix} x \\ e \end{bmatrix} = L \begin{bmatrix} x \\ x_H \end{bmatrix}$$

where $e = x - x_H$. Noting that $L^{-1} = L$, the state-space equations become

$$\begin{bmatrix} \dot{x}(t) \\ \dot{e}(t) \end{bmatrix} = \begin{bmatrix} A - BK & BK \\ 0 & A - FC \end{bmatrix} \begin{bmatrix} x(t) \\ e(t) \end{bmatrix} + \begin{bmatrix} B & 0 \\ 0 & F \end{bmatrix} u_{ext}(t)$$

$$\begin{bmatrix} y(t) \\ y_H(t) \end{bmatrix} = \begin{bmatrix} C & 0 \\ -K & K \end{bmatrix} \begin{bmatrix} x(t) \\ e(t) \end{bmatrix}.$$

Now,

$$\lambda\left(\begin{bmatrix} A - BK & BK \\ 0 & A - FC \end{bmatrix}\right) = \lambda(A - BK) \cup \lambda(A - FC),$$

as was to be shown.

We now show that the map from d to $x(t)$ and the map from d to y are those of full state feedback. Since the system is linear, let $x(0) = \hat{x}(0) = 0$ and $r \equiv 0$. Straightforward manipulations show that $x(t) = \int_0^t \exp(A_K(t - s))Bd(s)ds$. Therefore, the input/output maps $d \to x$ and $d \to y$ of the closed loop are exactly those of the system with full state feedback. Thus the input/output maps are identical. ■

A classical method for design of an output feedback controller is to choose the constant state feedback vector to be a linear quadratic regulator and then to use a well-known estimator design, the Kalman filter. Such a controller is known as a linear–quadratic–Gaussian (LQG) controller.

EXAMPLE 5.30 (*Flexible Beam Example 5.17 Continued*): In the previous example a linear quadratic regulator was designed for a flexible beam. In practice, most of the states must be estimated. Suppose that the tip position is measured,

$$y(t) = [1 \;\; 0 \;\; -0.931 \;\; 0 \;\; -1.027 \;\; 0 \;\; 1.169 \;\; 0 \;\; -1.187 \;\; 0]\, x(t),$$

and that the plant input noise covariance is $Q = 10I$ while the sensor noise covariance is $R = 1$. Solution of an algebraic Riccati equation yields the optimal Kalman gain F. This was used, with the previously designed optimal state feedback K, to construct an output feedback controller as explained earlier. Figure 5.10 displays the controlled tip position with output feedback. Even though a large error in the estimated initial condition is used, the response is good.

Unfortunately, the robustness of linear quadratic regulators does not extend to LQG controllers. Arbitrarily small stability margins can result, leading to instability if the system is modeled incorrectly.

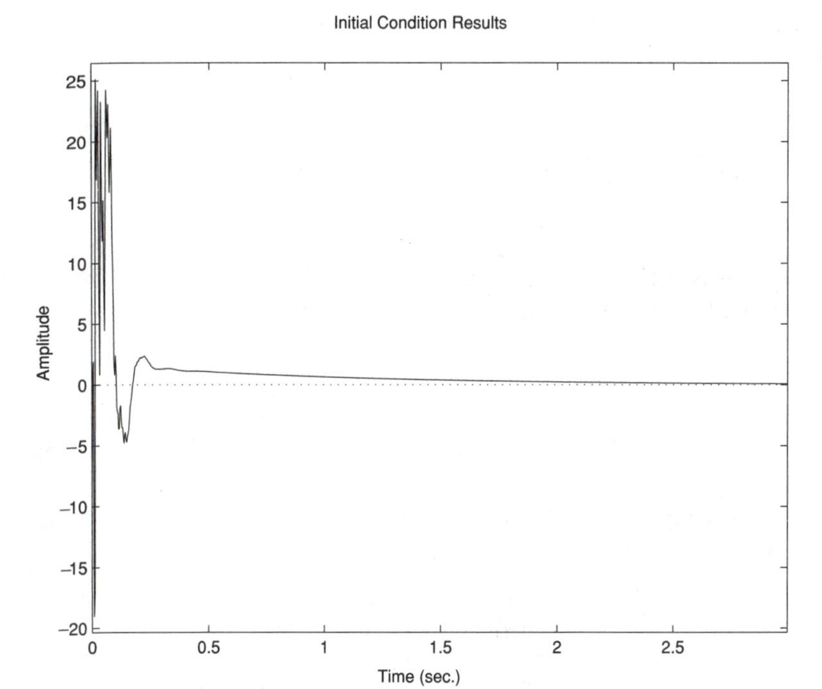

FIGURE 5.10 Comparision of state and output feedback (estimated initial condition is zero and actual initial condition is $x_o = [11 \ldots 1]$).

Example 5.31: Consider

$$\begin{bmatrix} \dot{x}_1(t) \\ \dot{x}_2(t) \end{bmatrix} = \begin{bmatrix} 1 & 1 \\ 0 & 1 \end{bmatrix} \begin{bmatrix} x_1(t) \\ x_2(t) \end{bmatrix} + \begin{bmatrix} 0 \\ 1 \end{bmatrix} u(t) + d(t)$$

$$y(t) = [1 \quad 0] \begin{bmatrix} x_1(t) \\ x_2(t) \end{bmatrix} + n(t).$$

The noise signals $d(t)$ and $n(t)$ are each white, Gaussian with zero mean. They have covariances

$$Q_e = \begin{bmatrix} 12 & 12 \\ 12 & 12 \end{bmatrix}$$

and $R_e = 1$, respectively. This system is both controllable and observable.

Using the Kalman filter to design an estimator, we obtain

$$F = \begin{bmatrix} 6 \\ 6 \end{bmatrix}.$$

Design a linear quadratic regulator for the state feedback, with state weight $Q = Q_e$, $R = 1$. We obtain

$$K = [6 \quad 6].$$

The controller has internal dynamics determined by

$$\lambda(A - FC - BK) = -5 \pm j3.46,$$

and so, as is common in state-space based controller design, the controller is internally stable. It was shown earlier that the closed loop has "A" matrix

$$A_o = \begin{bmatrix} A & -BK \\ FC & A - FC - BK \end{bmatrix}.$$

In this example, A_o has eigenvalues -0.2679, -3.73 (each with multiplicity 2).

Now, suppose that the input matrix B and output matrix C are slightly in error, and that their actual values are

$$\tilde{B} = \begin{bmatrix} 0 \\ 1.05 \end{bmatrix}, \qquad \tilde{C} = [1.05 \quad 0].$$

The matrices $A - \tilde{B}K$ and $A - F\tilde{C}$ are still Hurwitz. However, the closed loop now has "A"-matrix

$$\tilde{A}_o = \begin{bmatrix} A & -\tilde{B}K \\ F\tilde{C} & A - FC - BK \end{bmatrix}.$$

This matrix has eigenvalues -5.07, $-1.54 \pm j1.11, 0.1477$. This small perturbation in the system leads to an unstable closed loop.

An LQG controller design can be very useful. However, an LQG or LQE controller is in general only used when the system is very accurately modeled.

NOTES AND REFERENCES

The loudspeaker example is from [28]. One book that deals with state feedback and estimation in more detail is [1], although there are many more. The technique of loop transfer recovery has been proposed to improve the robustness of LQG controllers and it is discussed in [29, 1], among others. The proof of the linear quadratic regulator given here is along the lines of that in [10]. The *IEEE Transactions on Automatic Control* published a special issue on linear quadratic control in 1971. This issue contains the classic paper on Riccati equations by Willems [46]. The "bad" example of an LQG controller (Example 5.31) is from [16]. Exercise 5 is from [1]. The parameters in Exercise 7 are based on those of the laboratory apparatus sold by Quanser.

EXERCISES

1. Prove inequality (5.20). This can be done by one of two similar ways. The least sophisticated is to substitute $(u+v)/2$ into the cost functional and expand and then simplify.
   Alternatively, note that $Q = C^*C$, $R = E^*E$ and

$$J(x_o, u; t_f) = \left\| \begin{bmatrix} G^{1/2}x(t_f) \\ Cx(\cdot) \\ Eu(\cdot) \end{bmatrix} \right\|_Z$$

   where Z is the normed linear space formed by elements $R \times L_2(0, t_f; R^p) \times L_2(0, t_f; R^m)$ with the obvious inner product and norm. Show that for any inner product space,

$$\|a+b\|^2 = \|a\|^2 + \|b\|^2 + 2\mathrm{Re}(\langle a, b\rangle)$$

   and also

$$\|a-b\|^2 = \|a\|^2 + \|b\|^2 - 2\mathrm{Re}(\langle a, b\rangle).$$

   Use these equations to show that for non-zero a, b if $a \neq b$,

$$\left\|\frac{a+b}{2}\right\|^2 < \frac{1}{2}\|a\|^2 + \frac{1}{2}\|b\|^2.$$

   The inequality (5.20) should then follow easily.

2. Use a "complete the square" argument similar to that used in Theorem 5.7 to prove Lemma 5.11.
3. Prove that if (A, C) is observable and the optimal control problem (5.25) is open-loop stabilizable, then the non-negative solution to the ARE is positive definite.
4. Prove that if (A, B) is open-loop stabilizable then it is feedback stabilizable. That is, there is K so that $A - BK$ is Hurwitz. Thus, show that open-loop stabilizability is equivalent to stabilizability by a constant state feedback.
5. An angular position control system is described by

$$\begin{bmatrix} \dot{x}_1 \\ \dot{x}_2 \end{bmatrix} = \begin{bmatrix} 0 & 1 \\ 0 & -10 \end{bmatrix} \begin{bmatrix} x_1 \\ x_2 \end{bmatrix} + \begin{bmatrix} 0 \\ 1 \end{bmatrix} u(t).$$

(Here x_1 is angular position, x_2 is angular velocity, and u is torque.)

(a) For $q > 0$, choose

$$R = 1, \quad Q = \begin{bmatrix} q & 0 \\ 0 & 0 \end{bmatrix},$$

so that angular position but not velocity is penalized. Show analytically that the optimal control is

$$u_{opt} = -\left[\sqrt{q} \quad \sqrt{100 + 2\sqrt{q}} - 10\right]x.$$

(Hint: Write down the algebraic Riccati equation and examine it to find each entry of the solution Π. Use the facts that Π is symmetric and nonnegative definite to eliminate some solutions.) Show that the optimal feedback K leads to a stable closed loop: $A - BK$ is Hurwitz.

(b) Repeat (a) for the weights

$$R = 1, Q = \begin{bmatrix} 0 & 0 \\ 0 & q \end{bmatrix}.$$

Show that $A - BK$ in this case is not Hurwitz. Why is the optimal feedback not stabilizing in this case?

(c) Calculate the optimal state feedback for each of the following choice of weights: $R = 1$ and

$$Q_1 = \begin{bmatrix} 1 & 0 \\ 0 & 0 \end{bmatrix} \qquad R_1 = 1$$

$$Q_2 = \begin{bmatrix} 10 & 0 \\ 0 & 0 \end{bmatrix} \qquad R_2 = 1$$

$$Q_3 = \begin{bmatrix} 1 & 0 \\ 0 & 0 \end{bmatrix} \qquad R_3 = 0.1$$

$$Q_4 = \begin{bmatrix} 0.1 & 0 \\ 0 & 0 \end{bmatrix} \qquad R_4 = 1$$

$$Q_5 = \begin{bmatrix} 1 & 0 \\ 0 & 1 \end{bmatrix} \qquad R_5 = 1.$$

It is probably easier to use a computer Riccati equation solver (such as "ARE" in MATLAB) than to solve the problems analytically. For each calculated state feedback K_i, plot the closed loop step responses of the position, velocity, and torque.

Compare the different step responses. How does the choice of weights affect the closed loop response?

6. **(a)** Show that if $(A + \alpha I, B)$ is stabilizable, then an optimal control problem can be solved that yields a state feedback K such that the eigenvalues of $A - BK$ all have real parts less than $-\alpha$.

(b) Use this technique to design a state feedback for the beam in Example 5.17 so that the closed-loop settling time is less than 5 s. Include a plot with your solution.

(c) Design an output feedback controller for the beam where only tip position is measured so that the settling time is less than 5 s. Include a plot.

7. A linearized model of an inverted pendulum attached to moving cart is given in Chapter 2, Exercise 10. The model in that exercise is based on a force F applied to the cart. The actual system is controlled by voltage V (in volts)

$$F(t) = aV(t) - b\dot{x}(t),$$

where $a = 2$ and $b = 8$. For this exercise, use the following values for the remaining parameters:

m_p	Mass of rod (kg)	0.2 kg
m_c	Mass of cart (kg)	0.5 kg
I_p	Center of gravity of rod (m)	0.3 m.

(a) Rewrite the state-space equations with $V(t)$ as the control variable.

(b) Use this model to design a state-feedback controller that will stabilize the system about the up position. The closed loop settling time should be under 0.15 s and it should be slightly underdamped: $\xi \simeq 0.7$.

(c) What are the gain and phase margins of the closed loop?

(d) How does your controller from (b) perform with the nonlinear model?

8. Consider again the inverted pendulum of Exercise 7, and the linear quadratic state feedback K. In practice, only the position x and angle θ can be measured.

(a) Design a linear quadratic estimator so that the settling time of the estimator is about three times as fast as that of the state-feedback controlled system.

TABLE 5.1 Spring-mass parameters

Parameter	Value
m_1	36 kg
m_2	240 kg
k_1	160,000 N/m
k_2	16,000 N/m
d_1	100 Ns/m
d_2	980 Ns/m

(b) Use this estimator to construct an output feedback controller.

(c) How well does the controller perform with the linear model? With the non linear model? Compare to the state-feedback controller.

9. Consider the double-spring-mass system of Example 2.2 with parameters in Table 5.1.

(a) Use pole-placement to design a constant feedback controller that uses only the velocities to obtain a closed loop with little overshoot.

(b) Use pole-placement to design a state-feedback controller so that the closed loop has settling time less than 1 s, with little overshoot. (Use an algorithm such as Ackerman's method in the MATLAB routine "place.")

(c) Use optimal linear quadratic control to design a state-feedback controller so that the closed loop has settling time less than 1 s, with little overshoot.

(d) Compare the two state-feedback controllers. Which component of the state is more useful for feedback?

(e) Compare the gain and phase margins of the loop transfer functions $K(sI - A)^{-1}B$.

(f) In general, only the positions can be measured. Design an estimator and implement it with one of the previously designed state-feedback controllers. Let P indicate the transfer function matrix from force to position and let H indicate the transfer function of the controller. What is the gain and phase margin of the loop HP?

10. Consider the flexible beam of Example 5.17 with measurement as in Example 5.30. Design an output feedback controller so the closed loop settling time is less than two seconds with little overshoot.

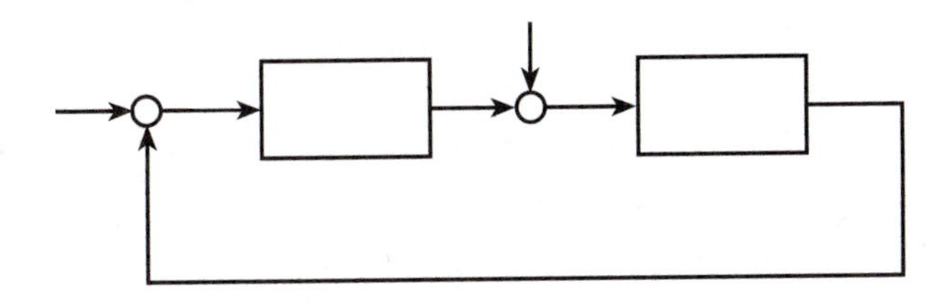

VI
CONTROLLER PARAMETRIZATION

We have expressed performance conditions in such forms as disturbance rejection,

$$\|W_1(1+PH)^{-1}\|_\infty < 1,$$

and robustness (in the case of multiplicative uncertainty) as

$$\|W_2H(1+PH)^{-1}\|_\infty < 1.$$

In all cases, the requirement that the controller stabilize the plant is implicit. In this chapter we derive an explicit description of the family of all controllers that stabilize a given plant.

6.1 STABLE PLANTS

Assume that P is SISO and stable: Since we are only concerned with systems modeled by linear time-invariant ODE's, P will also be rational, with real coefficients. Thus $P \in \mathcal{RH}_\infty$.

Suppose we have a rational H that stabilizes P. In other words, the four maps from (r, d) to (e_1, e_2) are stable:

$$\frac{1}{1+PH}\begin{bmatrix} 1 & -P \\ H & 1 \end{bmatrix} \in M(R\mathcal{H}_\infty).$$

(Recall that $M(R\mathcal{H}_\infty)$ consists of matrices with entries in $R\mathcal{H}_\infty$.) Denote the map from r to e_2 by Q:

$$e_2 = Qr.$$

(This is the map from r to the controller output.) By assumption, $Q \in R\mathcal{H}_\infty$. It follows that

$$\frac{H}{1+PH} = Q$$

and so

$$\begin{aligned} H &= Q(1+PH) \\ &= Q(1-QP)^{-1} \end{aligned}$$

if $1 - QP$ is not identically zero. Indicate the zero function by Θ. Assume $1 - QP \neq \Theta$. We have H in terms of Q and P. Also,

$$\begin{aligned} \frac{1}{1+PH} &= \frac{Q}{H} \\ &= (1-QP) \\ \frac{P}{1+PH} &= P(1-QP). \end{aligned}$$

The closed-loop transfer matrix is

$$\Delta(P, H) = \begin{bmatrix} 1-PQ & -P(1-PQ) \\ Q & 1-PQ \end{bmatrix}. \tag{6.1}$$

All four functions are *affine* in Q; that is, each is of the form $T_i + T_j Q$ for some $T_i, T_j \in R\mathcal{H}_\infty$.

We have shown that if H stabilizes P, H can be written in terms of Q and P, provided that $1 - PQ \neq \Theta$. This condition only fails if $Q = 1/P$. If P is strictly proper, as is usual, this is impossible.

Suppose now that

$$H = Q(1-PQ)^{-1}, \qquad 1 - PQ \neq \Theta,$$

for some arbitrary $Q \in R\mathcal{H}_\infty$. Substituting this into the transfer matrix, we get again (6.1). Clearly, every entry is in $R\mathcal{H}_\infty$ and the closed loop is stable. We have proven the following theorem.

THEOREM 6.1: Assume $P \in R\mathcal{H}_\infty$. The set of all controllers in $R\mathcal{H}_\infty$ that stabilize P is

$$\left\{\frac{Q}{1-PQ}; \qquad Q \in R\mathcal{H}_\infty, 1-PQ \neq \Theta\right\}.$$

6.1.1 Asymptotic Performance

Suppose we want the plant output y to asymptotically track a step. This is equivalent to the error $e_1(t) \to 0$ as $t \to \infty$. From the Final Value Theorem (Section 4.2), we know that the sensitivity S must have a zero at $s = 0$. In terms of Q,

$$S(0) = 1 - P(0)Q(0).$$

Therefore, the problem is solvable if and only if $P(0) \neq 0$. If this holds, the set of all suitable controllers is

$$\left\{H = \frac{Q}{1-PQ}; \quad Q \in R\mathcal{H}_\infty, Q(0) = \frac{1}{P(0)}, \quad 1-PQ \neq \Theta\right\}.$$

EXAMPLE 6.2 (*Idle Speed Control Continued*): If we neglect the time delay, the nominal plant is

$$P = \frac{1.43}{(s+1.1)(s+1.3)}.$$

Any stabilizing controller for P is of the form

$$H = Q(1-PQ)^{-1}, \qquad Q \in R\mathcal{H}_\infty.$$

To asymptotically track a step, it is necessary (and sufficient) to choose a Q so

$$Q(0) = \frac{1}{P(0)} = 1.$$

The simplest such Q is just

$$Q = 1,$$

which yields the controller

$$H = \frac{0.05(10s+11)(10s+13)}{s(5s+12)}.$$

Let us check the robustness of this controller with respect to additive uncertainty:

$$H(1+PH)^{-1} = Q = 1.$$

Therefore, the controller will be stable for all additive perturbations

$$\tilde{P} = P + \Delta$$

such that $\tilde{P}$ is stable and

$$|\Delta(j\omega)| < 1, \qquad \forall \omega.$$

Consider the error due to neglecting the time delay of $0.1s$:

$$\begin{aligned}\tilde{P} &= Pe^{-0.1s} \\ &= P + \underbrace{P(e^{-0.1s} - 1)}_{\Delta}.\end{aligned}$$

A simple numerical calculation indicates that $\|\Delta\|_\infty < 1$ and so H stabilizes $Pe^{-0.1s}$.

Alternatively, if we model the error as a multiplicative uncertainty we obtain

$$\tilde{P} = e^{-Ts}P = (1+\Delta)P$$

where

$$\Delta = e^{-Ts} - 1.$$

We previously showed in Chapter 5 that the perturbation Δ due to neglecting the delay satisfies, for $T \leq 0.1s$,

$$|\Delta(j\omega)| \leq |W_2(j\omega)|$$

where

$$W_2 = \frac{0.21s}{0.1s+1}.$$

The complementary sensitivity is

$$T = PH(1+PH)^{-1} = PQ = P.$$

Therefore, for robust stability we require

$$\|W_2 P\|_\infty < 1.$$

A numerical calculation indicates that this is also satisfied.

EXAMPLE 6.3:

$$P = \frac{1}{(s+1)(s+2)}.$$

It is desired to find a (stabilizing) controller so that y asymptotically tracks a ramp. The sensitivity must have two zeros at $s = 0$. Since this is two constraints, try

$$Q(s) = \frac{as+b}{s+1}.$$

We want to choose a, b so

$$\begin{aligned} S(0) &= 0 \\ S'(0) &= 0. \end{aligned}$$

This leads to

$$\left.\begin{aligned} 0 = S(0) &= \frac{2-b}{2} \\ 0 = S'(0) &= \frac{(5-a)2 - 5(2-b)}{2} \end{aligned}\right\} \Rightarrow \begin{aligned} b &= 2 \\ a &= 5. \end{aligned}$$

Therefore

$$\begin{aligned} Q(s) &= \frac{5s+2}{s+1} \\ H(s) &= \frac{Q}{1-QP} \\ &= \frac{(5s+2)(s+1)(s+2)}{s^2(s+4)}. \end{aligned}$$

By construction, H stabilizes P and leads to a closed loop that tracks a ramp with zero asymptotic error.

In general, to find a stabilizing controller that satisfies asymptotic performance constraints:

1. Select a form of Q with as many variables as there are constraints.
2. Reduce the specifications to interpolation conditions on Q.
3. Solve the constraint equations (if possible) for the variables in Q to obtain Q .
4. Set $H = Q(1 - PQ)^{-1}$.

A practical disadvantage to this design procedure is that the designed controller may be unnecessarily complicated. For instance, for the plant in

Example 6.2, it was shown in Chapter 4 by a simple loopshaping procedure that a simple proportional-integral (PI) controller will stabilize the plant and lead to a closed loop that tracks a step. In the example just given,

$$H(s) = \frac{0.5 + s}{s^2}$$

stabilizes the plant and leads to a closed loop that tracks a ramp.

6.2 GENERAL PLANTS

We have shown that if a scalar system P is stable, then the set of all controllers that stabilize P is

$$\mathcal{S}(P) = \left\{ \frac{Q}{1 - PQ}, \quad Q \in R\mathcal{H}_\infty; \quad 1 - PQ \neq \Theta \right\}.$$

What if P is not stable? Suppose we still write

$$H = \frac{Q}{1 - PQ}$$

where $Q \in R\mathcal{H}_\infty$. The closed-loop transfer matrix is

$$\begin{bmatrix} 1 - PQ & -P(1 - PQ) \\ Q & 1 - PQ \end{bmatrix},$$

which is not always in $M(R\mathcal{H}_\infty)$. Write P as the ratio of two coprime polynomials,

$$P = \frac{n_P}{d_P},$$

and write similarly a controller H:

$$H = \frac{n_H}{d_H}.$$

We know that H stabilizes P if and only if

$$(n_H n_P + d_H d_P)^{-1} \in R\mathcal{H}_\infty.$$

Since n_P, d_P are coprime polynomials, we can obtain polynomials x, y (using, for instance, Euclid's Algorithm) so that

$$x n_P + y d_P = 1.$$

This suggests

$$H = \frac{x}{y}$$

as a stabilizing controller. There are two problems with this, however:

1. H may possibly be improper.
2. This is only one controller: We want a parametrization of all stabilizing controllers.

We now address the first difficulty.

Since we want transfer functions in $R\mathcal{H}_\infty$, regard $R\mathcal{H}_\infty$ as the basic algebra and write all other functions in terms of functions in $R\mathcal{H}_\infty$.

Example 6.4:

$$P = \frac{s}{s-1} = \frac{\left(\frac{s}{s+1}\right)}{\left(\frac{s-1}{s+1}\right)} = \frac{N}{D}.$$

Definition 6.5: Two functions N, D in $R\mathcal{H}_\infty$ are *coprime* in $R\mathcal{H}_\infty$ if (1) they have no common zeros in $\mathrm{Re}(s) \geq 0$ and (2) $\lim_{s\to\infty} |N(s)| + |D(s)| \neq 0$.

Definition 6.6: If $P = N_P/D_P$ where $N_P, D_P \in R\mathcal{H}_\infty$ are coprime, we say that (N_P, D_P) is a *coprime factorization (c.f.)* of P (over $R\mathcal{H}_\infty$).

Computing a coprime factorization of a low-order scalar function is fairly easy: Write P as the ratio of two coprime polynomials and divide numerator and denominator by something like $(s+1)^n$ where n is the degree of the denominator. We discuss the general problem in a later section.

Theorem 6.7: Let (N_H, D_H) be a coprime factorization over $R\mathcal{H}_\infty$ of H, and let (N_P, D_P) similarly be a c.f. of P. The controller H stabilizes P if and only if

$$(N_H N_P + D_H D_P)^{-1} \in R\mathcal{H}_\infty.$$

□ *Proof:* Identical to the earlier stability theorem with polynomials (Theorem 3.30). ■

A parametrization of all stabilizing controllers for a general plant will now be derived.

We define a *degree function* $\delta : R\mathcal{H}_\infty \to I^+$ as follows:

$$\begin{aligned} \delta(N) &= \text{number of zeros of } N \text{ in } C_{+e} \\ &= \text{number of r.h.p. zeros of } N + \text{relative degree of } N. \end{aligned}$$

Notice that $\delta(N) = 0$ if and only if $N^{-1} \in R\mathcal{H}_\infty$.

Example 6.8:

$$N(s) = \frac{(s+1)(s-2)}{(s-1)(s+3)^3},$$
$$\delta(N) = 3.$$

Lemma 6.9: For every $N, D \in R\mathcal{H}_\infty$, with $D \neq \Theta$, and $\delta(N) \geq \delta(D)$ there is $Q \in R\mathcal{H}_\infty$ such that either $R = N - QD = \Theta$ or $\delta(R) < \delta(D)$.

□ *Proof:* If $\delta(D) = 0$ then $Q = ND^{-1}$ yields $R = \Theta$. Suppose then that $\delta(D) = m > 0$. Write

$$D = \frac{n_{D+}n_{D-}}{d_D},$$

where n_{D+}, n_{D-}, and d_D are polynomials, n_{D+} has all its zeros in the closed r.h.p., and n_{D-} has all its zeros in the open r.h.p. Also write N as a ratio of polynomials:

$$N = \frac{n_N}{d_N}.$$

Since $N \in R\mathcal{H}_\infty$, d_N has no zeros in the closed r.h.p. The polynomials d_N and n_{D+} are therefore coprime. By Theorem B.11 in Appendix B there are polynomials a, b such that $\deg(a) < \deg(d_N)$ and

$$an_{D+} + bd_N = (s+1)^{m-1}n_N.$$

If we divide by $(s+1)^{m-1}d_N$,

$$\frac{an_{D+}}{(s+1)^{m-1}d_N} + \frac{b}{(s+1)^{m-1}} = N. \tag{6.2}$$

Defining

$$Q = \frac{ad_D}{(s+1)^{m-1}d_N n_{D-}}$$
$$R = \frac{b}{(s+1)^{m-1}},$$

we can write Eq. (6.2) as

$$N = QD + R.$$

It remains to be shown that $Q, R \in R\mathcal{H}_\infty$ and that $\delta(R) < m$. The function Q has no r.h.p. poles, and writing

$$Q = \frac{(s+1)a}{d_N}\frac{d_D}{(s+1)^m n_{D-}},$$

we see that Q is proper and we can conclude $Q \in R\mathcal{H}_\infty$. The function $R = N - QD$ and so R is also proper. Since it has no r.h.p. poles, $R \in R\mathcal{H}_\infty$. From the definition of R, $\delta(R) \leq m - 1$, completing the proof. ■

THEOREM 6.10: A pair $N, D \in R\mathcal{H}_\infty$ are coprime if and only if there are $X, Y \in R\mathcal{H}_\infty$ such that

$$X(s)N(s) + Y(s)D(s) = 1, \qquad \text{Re}(s) \geq 0. \tag{6.3}$$

□ *Proof:* First, if (6.3) is satisfied for some $X, Y \in R\mathcal{H}_\infty$, then N and D cannot have any common zeros in C_{+e} and they are coprime.

Now, assume that N and D are coprime and prove that (6.3) must be satisfied for some $X, Y \in R\mathcal{H}_\infty$. Define the set $\mathcal{I}$ of all linear combinations of N and D,

$$\mathcal{I} = \{AN + BD; A, B \in R\mathcal{H}_\infty\}.$$

Choose $Z \in \mathcal{I}$ so that $\delta(Z)$ is a minimum over all elements of $\mathcal{I}$.

Let C be an arbitrary element of $\mathcal{I}$. From Lemma 6.9, there exists $Q \in R\mathcal{H}_\infty$ such that either $R = C - ZQ$ is zero, or $\delta(R) < \delta(Z)$. The second alternative contradicts how Z was chosen, and so $C = ZQ$ for some $Q \in R\mathcal{H}_\infty$. Since C was arbitrary, we have proven that

$$\mathcal{I} = \{AZ; A \in R\mathcal{H}_\infty\}.$$

Also, since $N = 1N + 0D$, $N \in \mathcal{I}$ and similarly, $D \in \mathcal{I}$. Thus, N/Z and D/Z are in $R\mathcal{H}_\infty$. Since N and D are coprime, this implies that $\delta(Z) = 0$, or, equivalently, $Z^{-1} \in R\mathcal{H}_\infty$.

Since $Z \in \mathcal{I}$, there must exist $E, F \in R\mathcal{H}_\infty$ such that

$$EN + FD = Z.$$

Dividing through by Z, we obtain

$$XN + YD = 1$$

with $X = E/Z, Y = F/Z$, as required. ■

Equation (6.3) is known as the *Bezout identity.*

THEOREM 6.11: Let (N_P, D_P) be a coprime factorization for a plant P. A controller H stabilizes P if and only if H has a coprime factorization (N_H, D_H) satisfying

$$N_H N_P + D_H D_P = 1.$$

□ *Proof:* Let X, Y be functions satisfying the Bezout identity with respect to (N_P, D_P). It follows immediately from the previous theorem and Theorem 6.7 that

$$H = \frac{X}{Y}$$

stabilizes P. Since X and $Y \in R\mathcal{H}_\infty$, H is always proper, although it may be unstable.

Conversely, if a given H stabilizes P, then write

$$H = \frac{N}{D},$$

where $N, D \in R\mathcal{H}_\infty$ and are coprime. From Theorem 6.7,

$$NN_P + DD_P = U,$$

where $U^{-1} \in R\mathcal{H}_\infty$. Rewriting this equation, we obtain

$$\underbrace{(U^{-1}N)}_{N_H}N_P + \underbrace{(U^{-1}D)}_{D_H}D_P = 1.$$

Defining $N_H = U^{-1}N$, $D_H = U^{-1}D$, we obtain the Bezout identity. Thus, every stabilizing controller can be written N_H/D_H where $N_H N_P + D_H D_P = 1$. ■

Theorem 6.12: Let $(N_P, D_P) \in R\mathcal{H}_\infty$ be a coprime factorization for P and $X, Y \in R\mathcal{H}_\infty$ such that

$$XN_P + YD_P = 1.$$

The set of all stabilizing controllers for P is

$$\mathcal{S}(P) = \left\{ \frac{X + D_P Q}{Y - N_P Q}, \qquad Q \in R\mathcal{H}_\infty, \quad Y - N_P Q \neq \Theta \right\}.$$

□ *Proof:* Once one stabilizing controller with coprime factorization (X, Y) is found, then for any $Q \in R\mathcal{H}_\infty$,

$$(X + QD_P)N_P + (Y - QN_P)D_P = XN_P + YD_P = 1.$$

Thus, every H of the form

$$H = \frac{X + QD_P}{Y - QN_P}$$

stabilizes P.

Suppose we have a stabilizing controller

$$H = \frac{N_H}{D_H}.$$

Without loss of generality, assume

$$N_H N_P + D_H D_P = 1. \tag{6.4}$$

We require a $Q \in R\mathcal{H}_\infty$ that solves

$$X + QD_P = N_H \tag{6.5}$$

$$Y - QN_P = D_H. \tag{6.6}$$

Multiply (6.5) by D_H and (6.6) by N_H. Then subtract the second equation from the first. After using (6.4) to simplify, we obtain

$$Q = N_H Y - X D_H. \qquad \blacksquare$$

Note that the closed-loop transfer matrix is affine in the *Youla parameter* Q:

$$\Delta\left(\frac{N_P}{D_P}, \frac{N_H}{D_H}\right) = \begin{bmatrix} (Y - N_P Q)D_P & -(Y - N_P Q)N_P \\ (X + D_P Q)D_P & (Y - N_P Q)D_P \end{bmatrix}.$$

The controller parametrization

$$H = \frac{X + QD_P}{Y - QN_P}, \qquad Q \in R\mathcal{H}_\infty,$$

where

$$XN + YD = 1,$$

should reduce to the original parametrization we obtained (6.1) if $P \in R\mathcal{H}_\infty$. Assume that $P \in R\mathcal{H}_\infty$. Then $(P, 1)$ is a c.f. and

$$0 \cdot P + 1 \cdot 1 = 1,$$

so $X = 0, Y = 1$ satisfy the Bezout identity for $(P, 1)$. Substituting into the general controller parametrization, we obtain

$$H = \frac{Q}{1 - QP}, \qquad Q \in R\mathcal{H}_\infty,$$

which is what was obtained previously.

EXAMPLE 6.13:

$$P = \frac{1}{s(s + a)}; \qquad a > 0.$$

Define

$$D_P = \frac{s(s+a)}{(s+a)^2} = \frac{s}{s+a}$$

$$N_P = \frac{1}{(s+a)^2}.$$

The pair (N_P, D_P) is a c.f. for P.

The controller $H = 1$ stabilizes P; so

$$U = 1 \cdot N_P + 1 \cdot D_P = \frac{s^2 + as + 1}{s^2 + 2as + a^2}$$

has a stable inverse U^{-1}. Setting

$$X = U^{-1}, \quad Y = U^{-1},$$

we obtain

$$XN_P + YD_P = 1.$$

The set of all stabilizing controllers for P is

$$\mathcal{S}(P) = \{(X + QD_P)(Y - QN_P)^{-1}, \quad Q \in R\mathcal{H}_\infty\}.$$

Using the Youla parametrization, we can write the sensitivity as

$$S = (Y - N_P Q)D_P.$$

In this example, the plant has a pole at 0, and so $D_P(0) = 0$. Therefore, for any choice of parameter Q, $S(0) = 0$ and the closed loop will track a step. That is, any stable closed loop will track a step.

6.3 COMPUTING A COPRIME FACTORIZATION

The remaining difficulty in obtaining a controller parametrization is in computing X, Y in the Bezout identity when no stabilizing controller is known.

For the scalar systems with which we are concerned, a procedure based on Euclid's algorithm can be used. Since this procedure is rather tedious, it should only be used if no stabilizing controller is known. For instance, for a stable plant P we can always choose $N = P$, $D = 1$ and $X = 0$, $Y = 1$. For many other plants, a root locus will yield a stabilizing constant controller.

The key step in this algebraic procedure is to transform $P(s)$ to $\tilde{P}(z)$ using the bilinear transformation $s = (1+z)/(1-z)$. The inverse of this transformation is $z = (s-1)/(s+1)$. The important points about this transformation are that it is analytic in the right half-plane and it transforms the right half-plane to the unit circle. A function analytic and bounded in the right half-plane is transformed to one analytic and bounded in the unit circle. (There are numerous other rational transformations that may be used.) This is useful because polynomials can be used to obtain coprime factors. We use Euclid's algorithm to obtain the factors for the Bezout identity and then transform the calculated factors back to the right-half-plane.

This procedure is best illustrated via example.

Calculation of Coprime Factors

The procedure is explained using

$$P(s) = \frac{2s+1}{(s+1)(s-1)}$$

as an example.

1. Applying the transformation $s = (1+z)/(1-z)$ to the example $P(s) = (2s+1)/[(s+1)(s-1)]$ yields

$$\tilde{P}(z) = \frac{(1-z)(3+z)}{4z}.$$

2. Write $\tilde{P}$ as a ratio of coprime polynomials:

$$\tilde{n}(z) = (1-z)(3+z), \tilde{d}(z) = 4z.$$

(The pair (n, d) where

$$n(s) = \tilde{n}\left(\frac{s-1}{s+1}\right), d(s) = \tilde{d}\left(\frac{s-1}{s+1}\right)$$

is a c.f. for P.)

3. Use Euclid's algorithm to find $\tilde{x}$, $\tilde{y}$ such that

$$\tilde{x}\tilde{n} + \tilde{y}\tilde{d} = 1.$$

Several steps are usually required to find $\tilde{x}$ and $\tilde{y}$:

(a) Divide $\tilde{d}$ into $\tilde{n}$:

$$\tilde{n} = \underbrace{\left(-\frac{z}{4} - \frac{1}{2}\right)}_{q_1}\tilde{d} + \underbrace{(3)}_{r_1}.$$

(The "rem" and "quo" functions in MAPLE are useful for this.)

(b) Divide r_1 into $\tilde{d}$ to get quotient q_2 and remainder r_2:

$$\tilde{d} = q_2 \cdot r_1 + r_2.$$

Continue until the remainder is a constant.
(In this example, r_1 is a constant, so only one division was required.)

4. The functions $\tilde{x}, \tilde{y}$ are now found by back substitution. In this example,

$$\tilde{n} = q_1\tilde{d} + r_1,$$

$$\frac{1}{r_1}\tilde{n} - \frac{q}{r_1}\tilde{d} = 1.$$

Thus, $\tilde{x} = \frac{1}{3}$, $\tilde{y} = \frac{z}{12} + \frac{1}{6}$.

5. Use the inverse transformation $z = (s-1)/(s+1)$ to obtain the functions X, N, Y, D in terms of s:

$$N = \tilde{n}\left(\frac{s-1}{s+1}\right) = \frac{4(2s+1)}{(s+1)^2}, \qquad D = \tilde{d}\left(\frac{s-1}{s+1}\right) = \frac{4(s-1)}{(s+1)}$$

$$X = \tilde{x}\left(\frac{s-1}{s+1}\right) = \frac{1}{3}, \qquad Y = \tilde{y}\left(\frac{s-1}{s+1}\right) = \frac{(3s+1)}{12(s+1)}.$$

6. As a check for mistakes, verify that $P = N/D$ and $XN + YD = 1$.

7. The set of all rational stabilizing controllers for P is

$$\mathcal{S}(P) = \left\{\frac{X + QD}{Y - QN}; \quad Q \in R\mathcal{H}_\infty\right\}.$$

Example 6.14:

$$P(s) = \frac{2s+1}{(s+1)(s-1)}.$$

Follow the same procedure as in the preceding example, except use the transformation

$$s = 3\frac{1+z}{1-z}.$$

(The inverse transformation is $z = (s-3)/(s+3)$.)

1. Write P in terms of z:

$$\tilde{P}(z) = \frac{(1-z)(5z+7)}{4(2z+1)(z+2)}.$$

2. Coprime polynomials for $\tilde{P}$:

$$\tilde{n}(z) = (1-z)(5z+7)$$
$$\tilde{d}(z) = 4(1+2z)(2+z).$$

3. Use Euclid's algorithm until a constant remainder is obtained:

(a) $\tilde{n} = \underbrace{(-\tfrac{5}{8})}_{q_1}\tilde{d} + \underbrace{(\tfrac{21}{2}z+12)}_{r_1}$

(b) $\tilde{d} = \underbrace{(\tfrac{16}{21}z + \tfrac{152}{147})}_{q_2} r_1 + \underbrace{(-\tfrac{216}{49})}_{r_2}$.

4. Solve for $\tilde{x}$ and $\tilde{y}$:

$$\tilde{d} = q_2 r_1 + r_2$$
$$\tilde{d} - q_2(\tilde{n} - q_1\tilde{d}) = r_2$$
$$(1+q_2q_1)\tilde{d} + (-q_2)\tilde{n} = r_2$$
$$\underbrace{\frac{(1+q_2q_1)}{r_2}}_{\tilde{y}}\tilde{d} + \underbrace{\left(-\frac{q_2}{r_2}\right)}_{\tilde{x}}\tilde{n} = 1.$$

Hence,

$$\tilde{x} = \frac{14}{81}z + \frac{19}{81}, \quad \tilde{y} = \frac{35}{324}z - \frac{13}{162}.$$

5. Transform back to the s-plane using $z = (s-3)/(s+3)$ and simplify:

$$X_2 = \frac{(11s+5)}{27(s+3)}, \quad Y_2 = \frac{(3s-61)}{108(s+3)}$$

$$N_2 = \frac{36(2s+1)}{(s+3)^2}, \; D_2 = 36\frac{(s-1)(s+1)}{(s+3)^2}.$$

6. Check for mistakes by verifying that $X_2N_2 + Y_2D_2 = 1$, $P = N_2/D_2$.

7. The set of all rational stabilizing controllers for P is

$$\mathcal{S}(P) = \left\{ \frac{X_2 + QD_2}{Y_2 - QN_2}, \qquad Q \in R\mathcal{H}_\infty \right\}.$$

This is the same set of stabilizing controllers obtained in the previous example. The central controller X_2/Y_2 is different, yielding a different parametrization of the set $\mathcal{S}(P)$.

6.4 MULTI-INPUT/MULTI-OUTPUT SYSTEMS

The theory for multi-input/multi-output systems is identical. The only significant difference is that since matrix multiplication is not commutative, it is necessary to distinguish between right coprime factorizations

$$P = ND^{-1}$$

and left coprime factorizations

$$P = \tilde{D}^{-1}\tilde{N}.$$

Definition 6.15: The pair (N, D) is a *right-coprime factorization* (r.c.f.) for a plant P if $N, D \in M(R\mathcal{H}_\infty)$, $P = ND^{-1}$, and there exists $X, Y \in M(R\mathcal{H}_\infty)$ such that

$$XN + YD = I.$$

Definition 6.16: The pair $(\tilde{N}, \tilde{D})$ is a *left-coprime factorization* (l.c.f.) for a plant P if $\tilde{N}, \tilde{D} \in M(R\mathcal{H}_\infty)$, $P = \tilde{D}^{-1}\tilde{N}$, and there exists $\tilde{X}, \tilde{Y} \in M(R\mathcal{H}_\infty)$ such that

$$\tilde{N}\tilde{X} + \tilde{D}\tilde{Y} = I.$$

We can always choose $X, Y, \tilde{X}, \tilde{Y}$ so that the factorizations are *bi-coprime*:

$$\begin{bmatrix} Y & X \\ -\tilde{N} & \tilde{D} \end{bmatrix}\begin{bmatrix} D & -\tilde{X} \\ N & \tilde{Y} \end{bmatrix} = \begin{bmatrix} I & 0 \\ 0 & I \end{bmatrix}. \tag{6.7}$$

To see this, suppose we have left- and right-coprime factorizations $ND^{-1} = \tilde{D}^{-1}\tilde{N}$ with $X, Y, A, B \in M(R\mathcal{H}_\infty)$ such that

$$XN + YD = I$$
$$\tilde{N}A + \tilde{D}B = I.$$

Then

$$\begin{bmatrix} Y & X \\ -\tilde{N} & \tilde{D} \end{bmatrix}\begin{bmatrix} D & -A \\ N & B \end{bmatrix} = \begin{bmatrix} I & \Delta \\ 0 & I \end{bmatrix}$$

where $\Delta = -YA + XB$. Multiplying each side on the right by

$$\begin{bmatrix} I & -\Delta \\ 0 & I \end{bmatrix},$$

we obtain

$$\begin{bmatrix} Y & X \\ -\tilde{N} & \tilde{D} \end{bmatrix}\begin{bmatrix} D & -(D\Delta + A) \\ N & -N\Delta + B \end{bmatrix} = \begin{bmatrix} I & 0 \\ 0 & I \end{bmatrix}.$$

Define $\tilde{X} = D\Delta + A$, $\tilde{Y} = -N\Delta + B$ to complete the bicoprime factorization.

For multi-input/multi-output systems, two parametrizations of the set of stabilizing controllers can be obtained: one for the controller left-coprime factors in terms of right-coprime factors of the plant, and one for the controller right-coprime factors in terms of the plant left-coprime factors.

THEOREM 6.17: Let $(N, D) \in M(R\mathcal{H}_\infty)$ be a right coprime factorization for P and $X, Y \in M(R\mathcal{H}_\infty)$ such that

$$XN + YD = I.$$

Let $(\tilde{N}, \tilde{D}) \in R\mathcal{H}_\infty$ be a left coprime factorization for P and $\tilde{X}, \tilde{Y} \in R\mathcal{H}_\infty$ such that

$$\tilde{N}\tilde{X} + \tilde{Y}\tilde{D} = I.$$

The set of all stabilizing controllers for P is

$$\mathcal{S}(P) = \{(\tilde{X} + DQ)(\tilde{Y} - NQ)^{-1}, \quad Q \in M(R\mathcal{H}_\infty), \quad \tilde{Y} - NQ \neq \Theta\},$$

or equivalently,

$$\mathcal{S}(P) = \{(Y - R\tilde{N})^{-1}(X + R\tilde{D}), \quad R \in M(R\mathcal{H}_\infty), \quad \tilde{Y} - NR \neq \Theta\}.$$

For multi-input/multi-input systems it is more convenient to calculate coprime factorizations using the state-space formulas described later than the algebraic procedure discussed in the previous section. These state-space formulas can of course also be used for scalar systems.

The following proof uses an estimator-based controller (Section 5.5) to construct a coprime factorization for the plant. The controller factors are then shown to satisfy the Bezout identity for the plant. The following technical lemma is required. The proof is straightforward and left as an exercise.

LEMMA 6.18: If P has state-space realization (A, B, C, E) where E^{-1} exists, then P^{-1} has realization $(A - BE^{-1}C, BE^{-1}, -E^{-1}C, E^{-1})$.

THEOREM 6.19: Let (A, B, C, E) be a stabilizable/detectable realization of a system with transfer function P. Choose K so that $A_K = A - BK$ is Hurwitz and F so that $A - FC$ is Hurwitz.

1. Defining

$$\begin{aligned} N(s) &= (C - EK)(sI - A_K)^{-1}B + E, \\ D(s) &= -K(sI - A_K)^{-1}B + I, \\ P &= ND^{-1}. \end{aligned}$$

2. Defining

$$\begin{aligned}\tilde{N}(s) &= C(sI - A_F)^{-1}(B - FE) + E,\\ \tilde{D}(s) &= -C(sI - A_F)^{-1}F + I,\\ P &= \tilde{D}^{-1}\tilde{N}.\end{aligned}$$

3. Defining

$$\begin{aligned}Y(s) &= K(sI - A_F)^{-1}(B - FE) + I,\\ X(s) &= K(sI - A_F)^{-1}F,\\ \tilde{Y}(s) &= (C - EK)(sI - A_K)^{-1}F + I,\\ \tilde{X}(s) &= K(sI - A_K)^{-1}F.\end{aligned}$$

we have a bicoprime factorization:

$$\begin{bmatrix} Y & X \\ -\tilde{N} & \tilde{D} \end{bmatrix}\begin{bmatrix} D & -\tilde{X} \\ N & \tilde{Y} \end{bmatrix} = \begin{bmatrix} I & 0 \\ 0 & I \end{bmatrix}.$$

□ *Proof:* A proof based on the output feedback controller discussed in Chapter 5 is given here. This theorem can be also be verified via elementary matrix operations and this is outlined in the exercises.

Write the system equations as usual:

$$\begin{aligned}\dot{x}(t) &= Ax(t) + Bu(t)\\ y(t) &= Cx(t) + Eu(t).\end{aligned}$$

Choose K so that $A - BK$ is Hurwitz and define $u = -Kx + r$. Consider the system equations describing u:

$$\begin{aligned}\dot{x}(t) &= (A - BK)x(t) + Br(t)\\ u(t) &= -Kx(t) + r(t).\end{aligned}$$

The transfer function of this system with input r and output u is D as written above. The overall system with input r and output y is

$$\begin{aligned}\dot{x}(t) &= (A - BK)x(t) + Br(t)\\ y(t) &= (C - EK)x(t) + Er(t),\end{aligned}$$

and this system has transfer function N. Since $P = \hat{y}/\hat{u}$, $D = \hat{u}/\hat{r}$, and $N = \hat{y}/\hat{r}$, $N = PD$ and ND^{-1} is a right factorization of P over $M(R\mathcal{H}_\infty)$.

Now consider an observer for the system P. Choose F so $A - FC$ is Hurwitz and define the observer

$$\dot{z}(t) = (A - FC)z(t) + (B - FE)u(t) + Fy(t)$$
$$\eta(t) = y(t) - Cz(t) - Eu(t)$$

The signal η is the error in the observation:

$$\eta(t) = Cx(t) - Cz(t).$$

Since $\dot{x}(t) - \dot{z}(t) = (A - FC)(x(t) - z(t))$, any error in observation is due solely to differences in initial conditions. (See Section 5.4.) Thus, with zero initial conditions, $\eta \equiv 0$. The transfer function from u to η is $-\tilde{N}$ and that from y to η is $\tilde{D}$. Thus,

$$0 = -\tilde{N} + \tilde{D}P,$$

and so $\tilde{D}^{-1}\tilde{N} = P$. Since $\tilde{D}, \tilde{N} \in M(R\mathcal{H}_\infty)$, it remains only to show that (N, D) are right coprime and $(\tilde{N}, \tilde{D})$ are left coprime.

Define an estimator-based controller (as in Section 5.5) with input y and output u:

$$\dot{z}(t) = (A - BK - FC + FEK)z(t) + Fy(t)$$
$$u(t) = Kz(t).$$

Denote the transfer function from y to u by H. As above, we can use state feedback to obtain a right factorization $\tilde{X}\tilde{Y}^{-1}$ of H. Define

$$y = (C - EK)z(t) + v(t)$$

and note that $\tilde{X}$ is the transfer function from v to u and $\tilde{Y}$ is the transfer function from v to y. Clearly, $\tilde{X} = H\tilde{Y}$ and so $H = \tilde{X}\tilde{Y}^{-1}$. Similarly, we can use $-B + FE$ to obtain a left factorization $H = Y^{-1}X$.

The state-space realizations of the transfer function matrix

$$\begin{bmatrix} D & -\tilde{X} \\ N & \tilde{Y} \end{bmatrix} \tag{6.8}$$

can be written in compact form as

$$\left[\begin{array}{c|cc} A - BK & B & F \\ \hline -K & I & 0 \\ C - EK & E & I \end{array}\right].$$

(Here the "A" matrix is $A - BK$, the "B" matrix is $[B \;\; F]$, etc.) The state-space realization of the inverse of the transfer function matrix (6.8) can be calculated using the previous lemma to be

$$\left[\begin{array}{c|cc} A - FC & (B - FE) & \vdots \; F \\ \hline K & I & \vdots \; 0 \\ \cdots\cdots & \cdots\cdots\cdots & \cdots \\ -C & -E & \vdots \; I \end{array}\right].$$

This is exactly the state-space realization of

$$\begin{bmatrix} Y & X \\ -\tilde{N} & \tilde{D} \end{bmatrix},$$

written in compact form. Thus,

$$\begin{bmatrix} Y & X \\ -\tilde{N} & \tilde{D} \end{bmatrix} \begin{bmatrix} D & -\tilde{X} \\ N & \tilde{Y} \end{bmatrix} = \begin{bmatrix} I & 0 \\ 0 & I \end{bmatrix}.$$

■

6.5 *OTHER DEFINITIONS OF STABILITY

Thus far we have been trying to obtain a closed-loop transfer matrix with all its entries in $R\mathcal{H}_\infty$. Stable transfer functions have all their poles in the left half-plane $\text{Re}(s) < 0$. Consider now the problem of obtaining a "stable" closed loop with a stricter definition of "stable." For instance, if settling time is an issue we may demand that all the transfer functions have all their poles to the left of some line $\text{Re}(s) = -\sigma$, $\sigma > 0$. In this situation, our domain of "instability" is

$$\mathcal{D} = \{s; \text{Re}(s) \geq -\sigma\}$$

and our set of acceptable transfer functions is

$$\mathcal{H}_{\infty,\mathcal{D}} = \{P(s) \text{ analytic in } \mathcal{D} \text{ and with } \sup_{s \in \mathcal{D}} |P(s)| < \infty\}.$$

The only difference is that the region $\{\text{Re}(s) \geq 0\}$ has been replaced by $\mathcal{D}$. The results of the previous section depended only on algebraic manipulations of functions in $R\mathcal{H}_\infty$ and they can be extended to problems with a different definition of stability.

Example 6.20: Consider again the plant

$$P = \frac{2s + 1}{(s + 1)(s - 1)},$$

but now let the domain of instability be

$$\mathcal{D} = \{s; \text{Re}(s) \geq -2\}.$$

We want to parametrize all controllers so that the closed loop has no poles in $\mathcal{D}$. Such systems will be stable and have a decay rate bounded by e^{-2t}.

The first step is to find $N, D \in R\mathcal{H}_{\infty,\mathcal{D}}$ such that $P = N/D$ and there are $X, Y \in R\mathcal{H}_{\infty,\mathcal{D}}$ satisfying

$$XN + YD = 1.$$

That is, (N, D) is a c.f. for P over $R\mathcal{H}_{\infty,\mathcal{D}}$. The transformation $(s-3)/(s+3)$ used in Example 6.14 is in $R\mathcal{H}_{\infty,\mathcal{D}}$ and yielded $X_2, N_2, Y_2, D_2 \in R\mathcal{H}_{\infty,\mathcal{D}}$. Therefore, the parametrization is

$$\mathcal{S}_{\mathcal{D}}(P) = \left\{ H = \frac{X_2 + QD_2}{Y_2 - QN_2}; \quad Q \in R\mathcal{H}_{\infty,\mathcal{D}} \right\}.$$

The expressions in $\mathcal{S}_{\mathcal{D}}(P)$ and $\mathcal{S}(P)$ in the previous example are the same. The difference is that the parameter Q ranges over a different set.

6.6 *STRONG AND SIMULTANEOUS STABILIZABILITY

A plant P is *strongly stabilizable* if it can be stabilized by a stable controller. A study of strong stabilizability is important for several reasons. It is desirable, particularly if the plant is stable, to control it with a stable controller. (Recall that even if a plant is stable, not every stable controller will lead to a stable closed loop.) Situations where this might occur are sensor and/or actuator failure, startup and shutdown. Also, stable controllers are easier to implement.

The problem of strong stabilization is related to *simultaneous stabilization*: finding a single controller that stabilizes two plants.

First, consider stable plants. How many stabilizing controllers for a stable plant P are stable, besides $H = 0$? We have

$$H = Q(1 - PQ)^{-1}, \quad Q \in R\mathcal{H}_{\infty}.$$

If $\|PQ\|_\infty < 1$, then $PQ \neq -1$ for all s and $(1 - PQ)^{-1} \in R\mathcal{H}_{\infty}$. The controller is stable. Thus, as long as Q is not "too large," H is stable. Recall that

$$\frac{u}{r} = Q = H(I + PH)^{-1},$$

so this can be interpreted as a constraint on the controller gain. However, other choices of Q may also lead to stable controllers.

We now consider plants that may be unstable. Not every unstable plant can be stabilized by a stable controller.

Example 6.21:

$$P(s) = \frac{s-1}{s(s-2)}.$$

$$P = \frac{N}{D}, \quad N = \frac{s-1}{(s+1)^2}, \quad D = \frac{s(s-2)}{(s+1)^2}$$

and $XN + YD = 1$, where

$$X = \frac{14s-1}{s+1}, \qquad Y = \frac{s-9}{s+1}.$$

Every stabilizing controller for P is of the form

$$H = \frac{X + QD}{Y - QN}, \qquad Q \in R\mathcal{H}_\infty,$$

and

$$(X + QD)N + (Y - QN)D = 1. \tag{6.9}$$

We now show that $Y - QN$ has a zero in Re(s) ≥ 0 for every $Q \in R\mathcal{H}_\infty$. This will imply that every stabilizing controller is unstable. Now,

$$N(1) = 0$$

and

$$\lim_{s\to\infty} N(s) = 0.$$

The Bezout identity (6.9) then implies that for every $Q \in R\mathcal{H}_\infty$,

$$(Y - QN)(1) = \frac{1}{D(1)}$$
$$= -4$$

and

$$\lim_{s\to\infty} (Y - QN)(s) = 1.$$

Since $Y - QN$ is a continuous function of s, it must have a real zero in the interval $[1, \infty)$.

The plant in the next example is very similar to that in the previous example. However, there does exist a stable, stabilizing controller for this plant.

Example 6.22: Find a stable, stabilizing controller for $P = (s-1)/(s-2)^2$. We need to find a $Q \in R\mathcal{H}_\infty$ such that the inverse of $U = Y - NQ$ is in $R\mathcal{H}_\infty$.

Defining

$$X = \frac{27}{s+1}, \qquad Y = \frac{s+7}{s+1}$$
$$N = \frac{s-1}{(s+1)^2}, \quad D = \frac{(s-2)^2}{(s+1)^2},$$

we have that

$$P = \frac{N}{D}$$
$$XN + YD = 1.$$

Thus,

$$Y(1) = \frac{1}{D(1)} = 4, \quad Y(\infty) = \frac{1}{D(\infty)} = 1.$$

Since at these points

$$U(1) = Y(1), \quad U(\infty) = Y(\infty),$$

we need a $U \in R\mathcal{H}_\infty$ with $U^{-1} \in R\mathcal{H}_\infty$ and with the required values at the r.h.p. zeros of N. We start with

$$U(s) = 1,$$

which has the required value at ∞. Any function of the form

$$U(s) = 1 + \frac{as}{(s+1)^2}$$

where a is a constant also has $U(\infty) = 1$. Choosing $a = 12$ yields $U(1) = 4$. Also,

$$U^{-1}(s) = \frac{(s+1)^2}{s^2 + 14s + 1} \in R\mathcal{H}_\infty.$$

In summary, the function

$$U(s) = 1 + \frac{12s}{(s+1)^2}$$

interpolates the required values of Y and also $U, U^{-1} \in R\mathcal{H}_\infty$.

Recall that

$$U = Y - NQ,$$

$$Q = \frac{Y - U}{N}.$$

Since U was chosen to equal Y at the r.h.p. zeros of N, $Q \in R\mathcal{H}_\infty$. Since $U = Y - NQ$ was also chosen so that $U^{-1} \in R\mathcal{H}_\infty$, the corresponding controller

$$H = \frac{X + DQ}{U}$$

is stable.

To complete this example, construct H. The algebraic procedure in the previous section can used to obtain a coprime factorization for P. Solving for Q,

$$\begin{aligned} Q &= \frac{Y - U}{N} \\ &= -6 \end{aligned}$$

and

$$\begin{aligned} H &= \frac{X + DQ}{U} \\ &= -3\frac{2s^2 - 17s - 1}{s^2 + 14s + 1}. \end{aligned}$$

Note that $H \in R\mathcal{H}_\infty$ as required.

The problem with Example 6.20 was that Y changed sign between r.h.p. zeros of N, so that it was impossible to choose $U = Y - QN$ so that it has no r.h.p. zeros. Since for z_i such that $N(z_i) = 0$,

$$Y(z_i)D(z_i) = 1,$$

we can rewrite the interpolation condition without Y as $U(z_i) = 1/D(z_i)$.

Therefore, in order for a plant P with coprime factorization N/D to be strongly stabilizable, it is necessary that D have the same sign at all real r.h.p. zeros of N (or equivalently P). Is this condition sufficient?

Consider a general plant

$$P = \frac{N}{D}$$

where N, D are coprime. Order the r.h.p. real zeros of N:

$$0 \leq z_1 \leq z_2 \cdots\cdots z_\ell \leq \infty.$$

We require a $U \in R\mathcal{H}_\infty$ such that

1. $U^{-1} \in R\mathcal{H}_\infty$ and
2. $U(z_i) = \frac{1}{D(z_i)} \qquad i = 1 \ldots \ell.$

This is necessary for

$$\frac{Y - U}{N} = Q \in R\mathcal{H}_\infty.$$

We also need U to interpolate the other, complex, r.h.p. zeros of N. However, one consequence of the following lemma on interpolation is that the values of D at the complex zeros are irrelevant.

LEMMA 6.23: Let $\{z_1, \ldots, z_\ell\}$ be extended r.h.p. real numbers, $\{z_{\ell+1}, \ldots, z_n\}$ extended r.h.p. complex numbers (nonzero imaginary part), m_i a corresponding set of positive integers, and $r_{ij}, i = 1, \ldots, n, j = 0 \ldots m_i - 1$, a corresponding set of complex numbers. The r_{ij} are ordered so that $r_{ij}, i \leq \ell, j = 0 \ldots m_i - 1$ are real.

There exists $U \in R\mathcal{H}_\infty$ with $U^{-1} \in R\mathcal{H}_\infty$ such that U satisfies the interpolation constraints

$$\frac{d^j}{ds^j} U(z_i) = r_{ij} \qquad j = 0, \ldots m_i - 1, i = 1, \ldots, n$$

if and only if the numbers $\{r_{10}, \ldots r_{\ell 0}\}$ all have the same sign.

Thus, we can construct an invertible U with the required values at all the r.h.p. zeros of N if and only if $D(z_i)$ has the same sign for all real extended r.h.p. zeros z_i of N. In this case, by construction

$$Q = \frac{Y - U}{N} \in R\mathcal{H}_\infty$$

and, also, $U = Y - QN$ is invertible in $R\mathcal{H}_\infty$ so that $H \in R\mathcal{H}_\infty$.

THEOREM 6.24: A plant is strongly stabilizable if and only if it has an even number of real poles between every pair of real zeros in $\mathrm{Re}(s) \geq 0$.

□ *Proof:* Let (N, D) be a coprime factorization for a plant P and $X, Y \in R\mathcal{H}_\infty$ such that

$$XN + YD = 1.$$

Every stabilizing controller H can be written

$$H = \frac{X + QD}{Y - QN}$$

for some $Q \in R\mathcal{H}_\infty$. Also, $(X + QD, Y - QN)$ is a coprime factorization for H. This means the number of closed right half-plane poles of H is equal to the number of closed right half-plane zeros of

$$U = Y - QN.$$

Define the degree function δ as in Section 6.2 and consider

$$\inf_{Q \in R\mathcal{H}_\infty} \delta(Y - QN).$$

The plant is strongly stabilizable if and only if the foregoing infinum is 0.

The real closed right half-plane zeros z_i, $i = 1 \ldots \ell$, of P are identical to the real closed right half-plane zeros of N. Thus,

$$U(z_i) = (Y - QN)(z_i) = Y(z_i).$$

Also, since $XN + YD = 1$,

$$Y(z_i) = \frac{1}{D(z_i)}$$

and

$$U(z_i) = \frac{1}{D(z_i)}.$$

Now suppose that P does not have an even number of poles between real r.h.p. zeros. That is, D does not have an even number of zeros between each real r.h.p. zero z_i. Since D is continuous in the right half-plane (actually it is analytic), this implies that it changes sign between successive zeros, say, z_i and z_{i+1}. Since U is also a continuous function of s, if $D(z_i) > 0$ and $D(z_{i+1}) < 0$ for some i, then U must have a real root in the interval $[z_i, z_{i+1}]$. The same conclusion holds if $D(z_i) < 0$ and $D(z_{i+1}) > 0$. This argument is independent of the choice of Q, and hence, if D changes sign between real r.h.p. zeros of P, then any stabilizing controller H will have real right half-plane poles and will be unstable.

Now suppose that D has the same sign at all z_i, $i = 1 \ldots \ell$. Let z_i, $i = \ell + 1 \ldots n$ be the nonreal r.h.p. zeros of N and let m_i, $i = 1 \ldots n$ be the multiplicity of each zero. At each z_i we require for $i = 1 \ldots n$

$$\begin{aligned} U(z_i) &= Y(z_i) \\ &= \frac{1}{D(z_i)}, \end{aligned}$$

and for zeros with multiple multiplicity ($m_i > 1$)

$$\frac{d^j}{ds^j}U(z_i) = \frac{d^j}{ds^j}Y(z_i).$$

The previous lemma implies the existence of $U \in R\mathcal{H}_\infty$ with $U^{-1} \in R\mathcal{H}_\infty$ such that U has the above required values at all r.h.p zeros z_i of N. Now define

$$Q = \frac{Y - U}{N}$$

and

$$H = \frac{X + QD}{Y - NQ} = \frac{X + QD}{U}.$$

By construction, $Q \in R\mathcal{H}_\infty$ and $H \in R\mathcal{H}_\infty$ as required. ■

A plant that satisfies the assumptions of the foregoing theorem is said to have the *parity interlacing property.*

EXAMPLE 6.25:

$$P(s) = \frac{s}{(s+1)(s-1)}.$$

$$\left.\begin{array}{l}\text{r.h.p. real zeros: } 0, \infty \\ \text{r.h.p. poles: } 1\end{array}\right\} \begin{array}{l}\text{number of poles} \\ \text{between two zeros is odd.}\end{array}$$

This plant does not have the parity interlacing property and therefore is not strongly stabilizable.

EXAMPLE 6.26:

$$\frac{s(s-.4)}{(s+1)(s-1)(s-2)}$$

$$\begin{array}{rl}\text{r.h.p real zeros:} & 0, .4, \infty \\ \text{r.h.p. poles:} & 1, 2\end{array}$$

$$\left.\begin{array}{l}0 \text{ poles between } [0, 0.4] \\ 2 \text{ poles between } [0.4, \infty]\end{array}\right\}.$$

This plant does have the parity interlacing property and it is strongly stabilizable.

Theorem 6.24 can be used to prove a result on simultaneous stabilization.

Theorem 6.27: Let (N_1, D_1) and (N_2, D_2) be coprime factorizations for P_1, P_2, respectively. Choose $X_1, Y_1 \in \mathcal{RH}_\infty$ such that

$$X_1N_1 + Y_1D_1 = 1$$

and similarly $X_2, Y_2 \in \mathcal{RH}_\infty$ such that $X_2N_2 + Y_2D_2 = 1$. The two plants P_1 and P_2 are stabilizable by the same controller if and only if N/D is strongly stabilizable, where

$$N = N_2D_1 - N_1D_2,\ D = N_2X_1 + D_2Y_1.$$

□ *Proof:* For simultaneous stabilization, we require $Q_1 \in \mathcal{RH}_\infty$, $Q_2 \in \mathcal{RH}_\infty$ such that

$$\frac{X_1 + D_1Q_1}{Y_1 - N_1Q_1} = \frac{X_2 + D_2Q_2}{Y_2 - N_2Q_2}.$$

Since $(X_1 + D_1Q_1, Y_1 - N_1Q_1)$ is coprime, we require $U \in \mathcal{RH}_\infty$ with $U^{-1} \in \mathcal{RH}_\infty$ such that

$$\begin{aligned} X_1 + D_1Q_1 &= U(X_2 + D_2Q_2) \\ (Y_1 - N_1Q_1) &= U(Y_2 - N_2Q_2). \end{aligned}$$

Rearranging,

$$\begin{aligned} D_1Q_1 - UD_2Q_2 &= -X_1 + UX_2 \\ -N_1Q_1 + UN_2Q_2 &= -Y_1 + UY_2. \end{aligned} \tag{6.10}$$

Multiply the first equation by N_2 and add it to the second equation multiplied by D_2:

$$(N_2D_1 - N_1D_2)Q_1 = -N_2X_1 - Y_1D_2 + U.$$

Multiply the first equation in (6.10) by Y_2 and subtract the second equation multiplied by X_2:

$$(D_1Y_2 + N_1X_2)Q_1 - UQ_2 = -X_1Y_2 + Y_1X_2.$$

Define $D = N_2X_1 + D_2Y_1$, $N = N_2D_1 - N_1D_2$, $X = X_1Y_2 - Y_1X_2$, $Y = D_1Y_2 + N_1X_2$. We rewrite the previous two equations as

$$\begin{aligned} D + NQ_1 &= U \\ X + YQ_1 &= UQ_2. \end{aligned}$$

If we can find a $Q_1 \in \mathcal{RH}_\infty$ to satisfy the first equation for some U, with $U^{-1} \in \mathcal{RH}_\infty$ then Q_2 can be found from the second equation. Thus, P_1 and P_2 are simultaneously stabilizable if and only if N/D is stabilizable by some stable controller, Q_1. ■

COROLLARY 6.28: Two stable plants P_1, P_2 are stabilizable by the same controller if and only if $P_2 - P_1$ is strongly stabilizable.

□ *Proof:* Define N_1, D_1 etc. as in the previous theorem. Since P_1 and P_2 are stable, $N/D = P_2 - P_1$. Then apply the theorem. ■

6.7 PERFORMANCE LIMITATIONS (CONTINUED)

Some performance limitations were obtained in Chapter 4. The Youla parametrization can be used to obtain further bounds on achievable controller performance.

Using the Youla parametrization for a plant:

$$P = \frac{N}{D},$$

$$XN + YD = 1,$$

any stabilizing controller can be written, for some $Q \in R\mathcal{H}_\infty$, as

$$H = (Y - QN)^{-1}(X + QD).$$

The weighted sensitivity can be written, for any $W_1 \in R\mathcal{H}_\infty$, as

$$W_1 S = W_1 D(Y - NQ).$$

As discussed in Chapter 4, reducing the weighted sensitivity is a common performance objective.

THEOREM 6.29: Let z_i, $\text{Re}(z_i) \geq 0$, be the r.h.p. zeros of a plant P.

$$\inf_{H \in \mathcal{S}(P)} \|W_1 S\|_\infty \geq \max_{z_i} |W_1(z_i)|. \tag{6.11}$$

If P has no r.h.p. zeros and is nonstrictly proper, then the optimal weighted sensitivity is zero, for any choice of weight W_1.

□ *Proof:* At the r.h.p. zeros of P,

$$Y(z_i)D(z_i) = 1$$

and

$$|W_1 S(z_i)| = |W_1(z_i)|.$$

The lower bound (6.11) follows.

Now consider the case of no r.h.p. zeros:

$$W_1 S = \underbrace{W_1 D Y}_{T_1} - \underbrace{W_1 D N}_{T_2} Q.$$

The assumption of no r.h.p. zeros implies that N is an outer (or minimum-phase) function. Do an inner/outer factorization of T_2:

$$T_{2i} = W_{1i} D_i,$$

$$T_{2o} = W_{1o} D_o N$$

and since $|W_{1i}(jw)| = 1$,

$$\begin{aligned}\|W_1 S\|_\infty &= \|W_{1i}(W_{1o} D_o Y - W_{1o} D_o N Q)\|_\infty \\ &= \|\underbrace{W_{1o} D_o Y}_{R} - \underbrace{W_{1o} D_o N Q}_{X}\|_\infty.\end{aligned}$$

Since R is stable, choose $X = R$.

If P is nonstrictly proper, then $N^{-1} \in R\mathcal{H}_\infty$. The choice

$$\tilde{Q} = N^{-1} Y$$

yields

$$\|W_1 S\|_\infty = 0. \qquad \blacksquare$$

Now consider the common situation where P has no r.h.p. zeros but is strictly proper. The optimal

$$\tilde{Q} = N^{-1} Y$$

is improper. Let ℓ be the relative degree of P and choose $m > 0$. The function

$$Q_m = \frac{Y}{N(1 + s/m)^\ell}$$

is proper, and with this sequence Q_m (i.e., this controller sequence),

$$\begin{aligned}\lim_{m\to\infty} \|W_1 S_m\|_\infty &= \lim_{m\to\infty} \left\| W_1 D Y \left(1 - \frac{1}{(1 + s/m)^2}\right)\right\|_\infty \\ &= \lim_{s\to\infty} |W_1(s)|.\end{aligned}$$

We can achieve performance arbitrarily close to optimal, as long as W_1 is strictly proper. (This is analogous to the previous result, if we consider that P being strictly proper is equivalent to zeros "at infinity.")

The corresponding controller is

$$\begin{aligned} H_m &= \frac{X + DQ_m}{Y - Q_m N} \\ &= \frac{\frac{1}{N}\left(XN + YD\left(\frac{1}{1+s/m}\right)^{\ell}\right)}{Y\left(1 - \left(\frac{1}{1+s/m}\right)^{\ell}\right)} \\ &= \frac{1 + DY\left[\left(\frac{1}{1+s/m}\right)^{\ell} - 1\right]}{NY\left(1 - \left(\frac{1}{1+s/m}\right)^{\ell}\right)}. \end{aligned}$$

As $m \to \infty$, the numerator of $H_m \to 1$ and the denominator of $H_m \to 0$. The gain $\|H_m\|_\infty$ of the controller becomes increasingly large.

Consider also the complementary sensitivity

$$T = \frac{PH}{1 + PH}.$$

Recall from Chapter 3 that multiplicative uncertainty is described as follows: For some $W_3 \in R\mathcal{H}_\infty$, the family of plants is

$$\mathcal{M}(W_3, P) = \left\{ \tilde{P};\ \left| \frac{\tilde{P}(jw)}{P(jw)} - 1 \right| \le |W_3(jw)| \quad \text{and} \right.$$
$$\left. \tilde{P} \text{ has the same number of r.h.p. poles as} P \right\}.$$

Recall that a controller H stabilizes every element of $\mathcal{M}(W_3, P)$ if and only if $\|W_3 T\|_\infty < 1$. Equivalently, there exists a controller stabilizing every plant in $\mathcal{M}(W_3, P)$ if and only if

$$\inf_{H \in S(P)} \|W_3 T\|_\infty < 1.$$

Now,

$$\begin{aligned} T &= 1 - D(Y - NQ_m) \\ &= 1 - D\left(Y - \frac{Y}{(1 + s/m)^{\ell}}\right) \\ &= 1 - DY\left(1 - \frac{1}{(1 + s/m)^{\ell}}\right). \end{aligned}$$

Thus, as m is increased, and $(1 - \frac{1}{(1+s/m)^{\ell}})$ becomes closer to 1, the gain

$\|W_3\|_\infty$ so that

$$\|W_3 T\|_\infty < 1$$

becomes smaller. The price of optimal performance is not only large controller gain but less robust stability.

We have established robust stability conditions for a number of uncertainty models in a previous chapter. Now consider the problem of finding a controller that is optimally robust with respect to multiplicative uncertainty:

$$\gamma = \inf_{H \in \mathcal{S}(P)} \|W_3 T\|_\infty.$$

The corresponding optimal controller H_{opt} will stabilize every plant such that

$$\tilde{P} = (1 + \Delta)P,$$

where $\tilde{P}$ has the same number of unstable poles as P and

$$|\Delta(jw)| < \frac{1}{\gamma}|W_3(jw)|.$$

Let (N, D) be a coprime factorization for P and choose $X, Y \in R\mathcal{H}_\infty$ such that $XN + YD = 1$. Then

$$W_3 T = W_3(NX + NDQ).$$

If P is stable, then the zero controller is a stabilizing controller ($X + DQ = 0$). The zero controller yields an infinite stability margin. However, no improvement in performance is achieved. This is a "dual" result to that for optimal sensitivity: The controller that yields optimal sensitivity for a plant with no r.h.p zeros has "infinite" gain and a 0 stability margin.

If the plant is unstable, with r.h.p. poles p_i,

$$\|W_3 T\|_\infty \geq \max_{p_i} |W_3(p_i)|.$$

We have the same themes as before: r.h.p. zeros limit performance while r.h.p. poles limit robustness.

For a stable plant with no r.h.p zeros we can achieve arbitrarily good performance with some controller, but the controller gain will be large and the stability margin small. Similarly, for a plant with no r.h.p. poles, infinite stability margin can be achieved with a controller that has small gain, but performance will be poor. Controller synthesis is discussed in more detail in subsequent chapters.

NOTES AND REFERENCES

The state-space procedure for calculating coprime factorizations is in [25, 35]. The material in this chapter is covered in more detail in [44]. Procedures that use the Youla parametrization to calculate controllers that optimize weighted sensitivity or robust stability are discussed in Chapter 9.

EXERCISES

1. For $P = 1/(s+1)$ find the set of all stabilizing controllers for which the closed loop tracks a step.
2. $P = 1/[s(s+1)]$
   (a) Describe the set of all stabilizing controllers for which the closed loop tracks a step.
   (b) Describe the set of all stabilizing controllers for which the closed loop tracks a ramp.
3. Find the set of all stabilizing controllers for
$$P = \frac{s+1}{s(s+2)}.$$
4. Let $P(s) = 1/s$. Use the Youla parametrization to find a proper stabilizing controller H so that
   (a) The closed loop asymptotically tracks a unit step.
   (b) The final value of y is zero when d is a sinusoid of frequency 2 rad/s and $r = 0$.
   (c) How does your solution compare to that obtained to Exercise 2 in Chapter 4?
5. Suppose that
$$P(s) = \frac{1}{s}, \quad H(s) = \frac{Q}{1 - PQ}$$
where Q is a real-rational function. Find a condition(s) on Q so that the feedback system is stable.
6. (a) Plot the step response for
$$P(s) = \frac{3s}{s^2 + 0.01s + 4}.$$

(b) Find all controllers for $P(s)$ such that the closed loop is in $\mathcal{S}_{\mathcal{D}}$ where

$$\mathcal{D} = \{s; \mathrm{Re}(s) < -3, |\mathrm{Im}(s)| \leq |\mathrm{Re}(s)|\}.$$

(c) Calculate one controller in $\mathcal{S}_{\mathcal{D}}(P)$. Plot the closed-loop step response.

(d) What does this definition of stability imply about the time domain behavior of systems that are stable in this sense?

7. Find a stable, stabilizing controller for

$$P(s) = \frac{s(s-1)}{(s^2+1)(s-2)(s+3)}.$$

8. Consider the linear model for the inverted pendulum and cart (Exercise 2.10).

(a) Calculate the transfer function from force F to position d. Show that this system is not strongly stabilizable.

(b) Calculate the transfer function from F to θ. Show that this is strongly stabilizable. Will the closed loop be internally stable?

(c) Repeat (b) for the observation $d + I_p\theta$.

9. Show that if P has no r.h.p. zeros then over any range $[\omega_1, \omega_2]$ sensitivity $(I + PH)^{-1}$ as close to 0 as desired over this interval can be achieved.

10. This question is concerned with verifying the state-space realizations for coprime factorizations (Section 6.4) through matrix manipulations.

(a) Verify that if P has a state-space realization (A, B, C, E) where E^{-1} exists, then P^{-1} has state-space realization $(A - BE^{-1}C, BE^{-1}, -E^{-1}C, E^{-1})$.

(b) Show that $PD = N$.

(c) Show that $\tilde{D}P = \tilde{N}$.

(d) Verify that $XN + YD = I$.

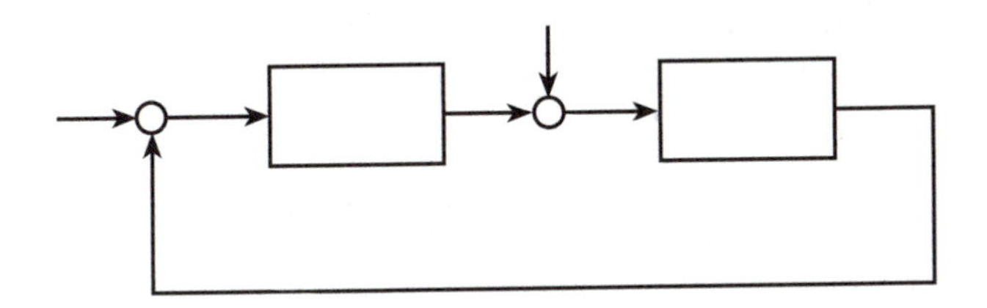

VII
GENERALIZED PLANTS

Recall that a controller H for a SISO plant P in the standard feedback configuration achieves robust performance with respect to multiplicative uncertainty if and only if

$$\| |W_1 S| + |W_3 T| \|_\infty < 1$$

where $S = (1 + PH)^{-1}, T = PH(I + PH)^{-1}$. Robust performance with respect to additive uncertainty is achieved if and only if

$$\| |W_1 S| + |W_2 H S| \|_\infty < 1.$$

(For MIMO plants, replace absolute value by the matrix norm.) This problem is computationally intensive to solve, even approximately. We will solve a related problem:

$$\|(|W_1 S|^2 + |W_3 T|^2)^{1/2}\|_\infty = \left\| \begin{bmatrix} W_1 S \\ W_3 T \end{bmatrix} \right\|_\infty .$$

Constructive methods are available to find controllers that satisfy bounds on the preceding quantity. The inequality

$$\left\| \begin{bmatrix} W_1 S \\ W_3 T \end{bmatrix} \right\|_\infty = \| [|W_1 S|^2 + |W_3 T|^2]^{1/2} \|_\infty < \frac{1}{\sqrt{2}}$$

implies

$$\| |W_1 S| + |W_3 T| \|_\infty < 1.$$

Similarly,

$$\left\| \begin{bmatrix} W_1 S \\ W_3 T \end{bmatrix} \right\|_\infty = \| [W_1 S]^2 + |W_3 T|^2]^{1/2} \|_\infty < 1$$

implies

$$\| |W_1 S| + |W_3 T| \|_\infty < \sqrt{2}.$$

(See Fig. 7.1.) Thus, the problem

$$\left\| \begin{bmatrix} W_1 S \\ W_3 T \end{bmatrix} \right\|_\infty \tag{7.1}$$

is an approximation to the robust performance problem. We will solve this problem, and related problems such as

$$\left\| \begin{bmatrix} W_1 S \\ W_2 H S \\ W_3 T \end{bmatrix} \right\|_\infty ,$$

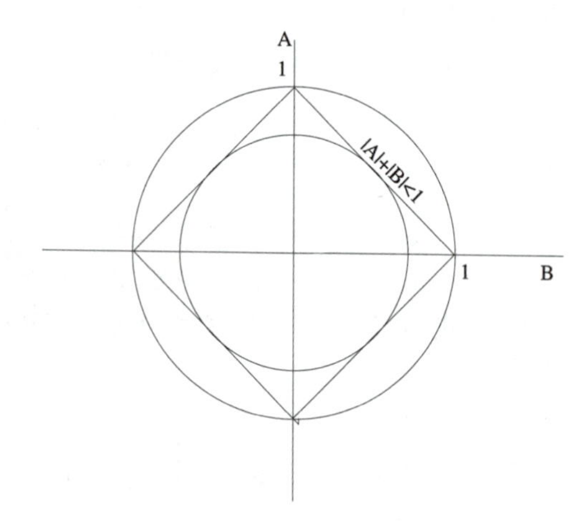

FIGURE 7.1 Relationship between different norms on R^2.

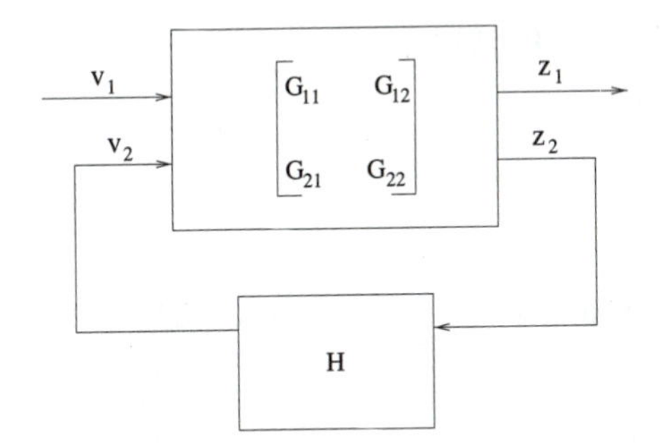

FIGURE 7.2 General feedback configuration.

in both the SISO and MIMO contexts. In this chapter we put the problem (7.1) and other problems into the very general control configuration shown in Fig. 7.2. In this configuration, z_1 is the measured output, and z_2 is controller input. The variables v_1 are uncontrolled inputs (such as disturbances) and v_2 is the controller output.

We have

$$\begin{aligned} z_2 &= G_{21}v_1 + G_{22}v_2 \\ &= G_{21}v_1 + G_{22}z_2 \\ v_2 &= Hz_2. \end{aligned}$$

We define

$$G = \begin{bmatrix} G_{11} & G_{12} \\ G_{21} & G_{22} \end{bmatrix}.$$

This is known as an *augmented plant* or the *generalized plant*. If we use the third equation to eliminate v_2,

$$\begin{aligned} z_2 &= (I - G_{22}H)^{-1}G_{21}v_1 \\ z_1 &= G_{11}v_1 + G_{12}(Hz_2) \\ &= [G_{11} + G_{12}H(I - G_{22}H)^{-1}G_{21}]v_1. \end{aligned}$$

The controller design problem is to find H,

$$\|z_1\|_2 < 1 \quad \text{for all} \quad v_1 \text{ with } \|v_1\|_2 \leq 1.$$

(Here $\| \|_2$ indicates the L_2 norm in frequency-domain, that is, the L_2 norm in the time domain.) In other words, find H so that G_{22} is stabilized and

$$\|G_{11} + G_{12}H(I - G_{22}H)^{-1}G_{21}\|_\infty < 1.$$

We use the concise notation

$$G = \left[\begin{array}{c|c} A & B \\ \hline C & E \end{array}\right]$$

to indicate a state-space realization (A, B, C, E) of a system with transfer function G. If

$$\left[\begin{array}{c|cc} A & B_1 & B_2 \\ \hline C_1 & E_{11} & E_{12} \\ C_2 & E_{21} & E_{22} \end{array}\right],$$

then each subsystem G_{ij} has the state-space realization (A, B_j, C_i, E_{ij}).

The framework is best illustrated via a few examples.

Example 7.1: In the standard mixed sensitivity problem (7.1), the objective is to find H to stabilize P and so that

$$\left\| \begin{matrix} W_1 S \\ W_3 T \end{matrix} \right\|_\infty < 1.$$

In this case the measured output is

$$z_1 = \begin{bmatrix} W_1 S \\ W_2 T \end{bmatrix}.$$

Referring to the standard feedback control diagram (Fig. 7.3), $v_1 = r$ (reference input or output disturbance) and the controller output $v_2 = u$. Rewrite z_1 (performance measure) and z_2 (controller input) in terms of v_1

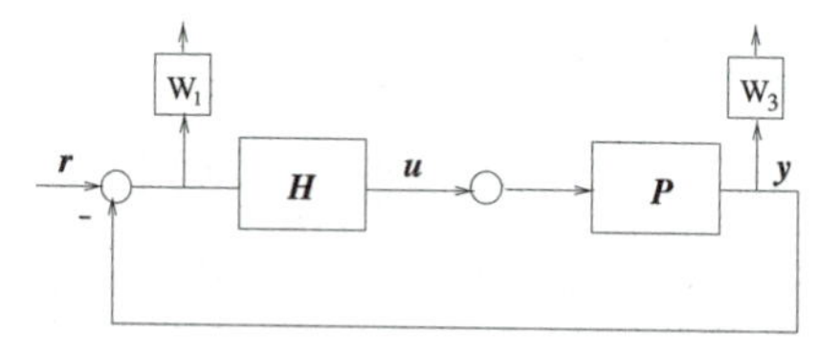

FIGURE 7.3 Robust performance (multiplicative uncertainty) feedback configuration.

and v_2:

$$
\begin{aligned}
z_2 &= r - Pu \\
&= v_1 - Pv_2 \\
z_1 &= \begin{bmatrix} W_1 Sr \\ W_3 Tr \end{bmatrix} \\
&= \begin{bmatrix} W_1 e_1 \\ W_3 y \end{bmatrix} \\
&= \begin{bmatrix} W_1(v_1 - Pv_2) \\ W_3 P v_2 \end{bmatrix} \\
&= \begin{bmatrix} W_1 \\ 0 \end{bmatrix} v_1 + \begin{bmatrix} -W_1 P \\ W_3 P \end{bmatrix} v_2.
\end{aligned}
$$

Define

$$
\begin{aligned}
G_{11} &= \begin{bmatrix} W_1 \\ 0 \end{bmatrix}, & G_{12} &= \begin{bmatrix} -W_1 P \\ W_3 P \end{bmatrix} \\
G_{21} &= I, & G_{22} &= [-P].
\end{aligned}
$$

Suppose the plant P and weights W_1, W_3 have state-space realizations as follows:

$$
P = \left[\begin{array}{c|c} A_P & B_P \\ \hline C_P & E_P \end{array}\right], \quad W_1 = \left[\begin{array}{c|c} A_{w1} & B_{w1} \\ \hline C_{w1} & E_{w1} \end{array}\right], \quad W_3 = \left[\begin{array}{c|c} A_{w3} & B_{w3} \\ \hline C_{w3} & E_{w3} \end{array}\right].
$$

A realization of $W_1 P$ is (Theorem C.1)

$$
\left[\begin{array}{cc|c} A_{w1} & 0 & B_P \\ B_{w1} C_P & A_P & B_{w1} E_P \\ \hline E_{w1} C_P & C_{w1} & E_{w1} E_P \end{array}\right]. \tag{7.2}
$$

The realization for $W_3 P$ is similar. It is straightforward then to show that the generalized plant G has the realization

$$A = \begin{bmatrix} A_P & 0 & 0 \\ B_{w1}C_P & A_{w1} & 0 \\ B_{w3}C_P & 0 & A_{w3} \end{bmatrix}, \qquad B = \begin{bmatrix} 0 & \vdots & B_P \\ -B_{w1} & \vdots & B_{w1}E_P \\ 0 & \vdots & B_{w3}E_P \end{bmatrix}$$

$$C = \begin{bmatrix} -E_{w1}C_P & -C_{w1} & 0 \\ E_{w3}C_P & 0 & C_{w3} \\ \cdots & \cdots & \cdots \\ -C_P & 0 & 0 \end{bmatrix}, \quad E = \begin{bmatrix} E_{w1} & \vdots & -E_{w1}E_P \\ 0 & \vdots & E_{w3}E_P \\ \cdots & \cdots & \cdots \\ I & \vdots & -E_P \end{bmatrix},$$

Example 7.2: Consider a different mixed sensitivity problem that arises when additive uncertainty (or a bound on the controller output) is considered. The problem now is to find a stabilizing controller H for P so that

$$\left\| \begin{bmatrix} W_1(I+PH)^{-1} \\ W_2H(I+PH)^{-1} \end{bmatrix} \right\|_\infty < 1. \tag{7.3}$$

Referring to Fig. 7.4,

$$\begin{aligned} v_1 &= r \\ v_2 &= u \\ z_2 &= v_1 - Pv_2 \\ z_1 &= \begin{bmatrix} W_1 \\ 0 \end{bmatrix} v_1 + \begin{bmatrix} -W_1P \\ W_2 \end{bmatrix} v_2. \end{aligned}$$

Notice that we have of course $v_2 = Hz_2$. Thus, the generalized plant is

$$G = \begin{bmatrix} W_1 & \vdots & -W_1P \\ 0 & \vdots & W_2 \\ \cdots & \cdots & \cdots \\ I & \vdots & -P \end{bmatrix}. \tag{7.4}$$

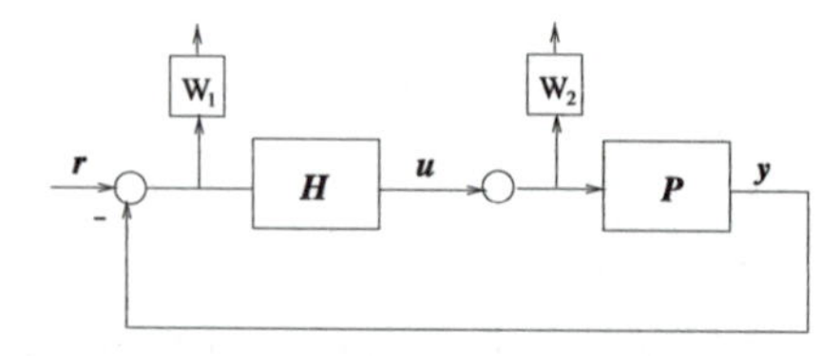

FIGURE 7.4 Robust performance (additive uncertainty) feedback configuration.

Assume that P has realization (A_P, B_P, C_P, E_P), W_1 has realization $(A_{w1}, B_{w1}, C_{w1}, E_{w1})$, and W_2 has realization $(A_{w2}, B_{w2}, C_{w2}, E_{w2})$. Using Theorem C.1, the "A" matrix of the generalized plant is

$$A = \begin{bmatrix} A_{w1} & B_{w1}C_P & 0 \\ 0 & A_P & 0 \\ 0 & 0 & A_{w2} \end{bmatrix}. \tag{7.5}$$

A realization for the generalized plant is

$$\left[\begin{array}{ccc|c:c} & & & B_{w1} & -B_{w1}E_P \\ & A & & 0 & -B_P \\ & & & 0 & B_{w2} \\ \hline C_{w1} & E_{w1}C_P & 0 & E_{w1} & -E_{w1}E_P \\ 0 & 0 & C_{w2} & 0 & E_{w2} \\ \hdashline 0 & C_P & 0 & I & -E_P \end{array}\right]. \tag{7.6}$$

EXAMPLE 7.3: In active noise control, the aim is to reduce the acoustic noise level at a point or number of points. We consider here the problem of reducing the noise level at just a single point in a duct. Figure 7.5 shows the possible controller configurations for reduction of noise level in a duct. In order to determine the best controlled pressure, a measurement of the pressure at the sensing point h is made and used as input to a controller H. The controller calculates a pressure $-u$ that is applied at the control point x_a. There are essentially two configurations: the so-called feedforward $(x_a > h)$ and feedback $(x_a < h)$ arrangements. The uncontrolled input is

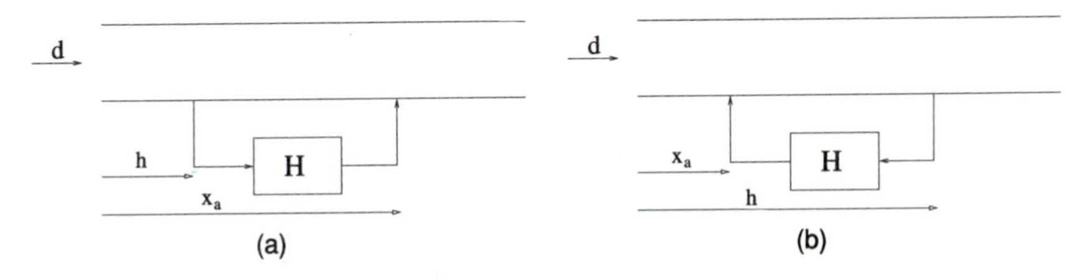

FIGURE 7.5 Active noise control in a duct. (a) Feedforward: $x_a > h$. (b) Feedback: $x_a < h$.

$v_1 = d$, the controller output is $v_2 = u$, and the controller input is $z_2 = T(h)$. The signal to be reduced is $z_1 = T(x)$, where $T(x)$ indicates the acoustic pressure at a point x.

Let $P(x, x_s)$ indicate the transfer function from a source at $x = x_s$ (either disturbance or control) to the measured acoustic pressure at x. In either configuration, the pressure at x due to a disturbance pressure d at $x = 0$ and the control pressure u at $x = x_a$ will be

$$T(x) = P(x, 0)d + P(x, x_a)u.$$

Thus,

$$G_{11} = P(x, 0), \quad G_{12} = P(x, x_a)$$
$$G_{21} = P(h, 0), \quad G_{22} = P(h, x_a).$$

If the disturbance $v_1 = d$ is a single random noise signal with power spectrum W_1, then the problem of minimizing the energy of the resulting output z_1 at x can be written

$$\inf_{Q \in \mathcal{H}_\infty} \|(G_{11} + G_{12}H(I - G_{22}H)^{-1}G_{21})W_1\|_2$$

where

$$\|z_1\|_2 = \left[\frac{1}{2\pi}\int_{-\infty}^{\infty} \|z_1(j\omega)\|^2 d\omega\right]^{1/2}.$$

This leads to a generalized plant with $\tilde{G}_{11} = G_{11}W_1$, etc.

There is often more than one disturbance to be considered, although disturbances are generally concentrated in a known frequency range. Alternatively, we are only concerned with the system response in a frequency range. In either case, we wish to reduce the system's response at x to all signals in a given frequency range. We need a stabilizing controller so that

$$\left\| W_1 \frac{T(x)}{d} \right\|_\infty < \gamma$$

or

$$\|W_1(G_{11} + G_{12}H(I - G_{22}H)^{-1}G_{21})\|_\infty < \gamma$$

where W_1 is a function that is large in the frequency range of interest and small outside of that range and γ is an acceptable noise level.

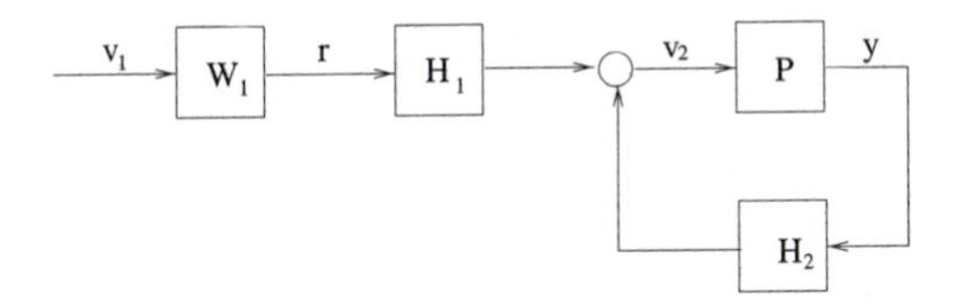

FIGURE 7.6 Two controller design.

EXAMPLE 7.4: Here we consider a quite different control problem. (See Fig. 7.6.)

It is desired that the plant output y should track a reference signal r where

$$r = W_1 v_1, \quad \|v_1\|_2 \leq 1. \tag{7.7}$$

There is extra freedom in this controller design problem, because there are two controllers: H_1 in a feedforward open loop path and H_2 in the feedback closed loop path.

The uncontrolled input v_1 and the controller output v_2 are shown in Fig. 7.6. Here

$$v_2 = [H_1 \quad H_2] \begin{bmatrix} r \\ y \end{bmatrix}, \tag{7.8}$$

and so we define

$$z_2 = \begin{bmatrix} r \\ y \end{bmatrix}. \tag{7.9}$$

The controller input is

$$\begin{aligned} z_2 = \begin{bmatrix} r \\ y \end{bmatrix} &= \begin{bmatrix} W_1 v_1 \\ P v_2 \end{bmatrix} \\ &= \begin{bmatrix} W_1 \\ 0 \end{bmatrix} v_1 + \begin{bmatrix} 0 \\ P \end{bmatrix} v_2. \end{aligned} \tag{7.10}$$

The tracking error is $r - y$, or $W_1 v_1 - P v_2$. An obvious choice of design objective is to minimize

$$\|r - y\|_2^2. \tag{7.11}$$

However, this does not place any constraint on the control effort. Thus, we choose to minimize, for some $\delta > 0$,

$$
\begin{aligned}
z_1 &= \begin{bmatrix} r - y \\ \delta v_2 \end{bmatrix} \\
&= \begin{bmatrix} W_1 \\ 0 \end{bmatrix} v_1 + \begin{bmatrix} -P \\ \delta I \end{bmatrix} v_2. \qquad (7.12)
\end{aligned}
$$

The generalized plant is

$$
\left[\begin{array}{c:c} W_1 & -P \\ 0 & \delta I \\ \hdashline W_1 & 0 \\ 0 & P \end{array}\right]. \qquad (7.13)
$$

7.1 LINEAR FRACTIONAL TRANSFORMATIONS

The generalized plants described in the previous section are special cases of *linear fractional transformations* (LFTs).

Definition 7.5: Let

$$
G = \begin{bmatrix} G_{11} & G_{12} \\ G_{21} & G_{22} \end{bmatrix} \qquad (7.14)
$$

be a complex matrix where $G_{11} \in \mathbb{C}^{p_1 \times m_1}$, $G_{12} \in \mathbb{C}^{p_1 \times m_2}$, $G_{21} \in \mathbb{C}^{p_2 \times m_1}$, and $G_{22} \in \mathbb{C}^{p_2 \times m_2}$. Let Δ be any $m_2 \times p_2$ matrix. If $\lim_{s \to \infty} \det(I - G_{22}(s)\Delta(s)) \neq 0$, we say that $\mathcal{F}_L(G, \Delta)$ is *well-posed.* If this condition holds, we define the *lower linear fractional transformation*

$$
\mathcal{F}_L(G, \Delta) = G_{11} + G_{12}\Delta(I - G_{22}\Delta)^{-1}G_{21}. \qquad (7.15)
$$

The generalized feedback diagram in the previous section is a diagram of a lower LFT: $z_1 = \mathcal{F}_L(G, H)v_1$.

As for the usual feedback configuration (Section 3.4), if G_{22} has state-space realization (A, B, C, E) and H has realization (A_H, B_H, C_H, E_H), then well-posedness is equivalent to the invertibility of $I - EE_H$.

We can similarly define an *upper linear fractional transformation*

$$
F_U(G, \Delta) = G_{22} + G_{21}\Delta(I - G_{11}\Delta)^{-1}G_{12} \qquad (7.16)
$$

for any $\Delta \in \mathbb{C}^{m_1 \times p_1}$ such that $\lim_{s \to \infty} \det(I - G_{11}(s)\Delta(s)) \neq 0$.

An upper LFT is shown in Fig. 7.7.

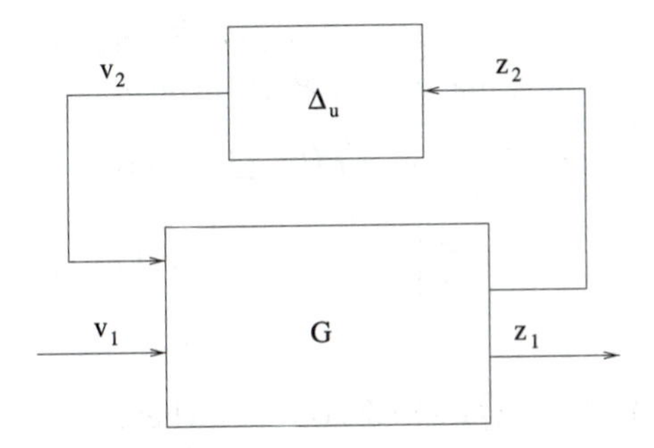

FIGURE 7.7 Upper linear fractional transformation.

EXAMPLE 7.6: As a simple example of an upper LFT, consider

$$\begin{aligned} E + C(sI - A)^{-1}B &= E + C\frac{1}{s}\left(I - A\frac{1}{s}\right)^{-1} B \\ &= \mathcal{F}_U\left(\left[\begin{array}{c|c} A & B \\ \hline C & E \end{array}\right], \frac{1}{s}I\right). \end{aligned} \tag{7.17}$$

Thus, a transfer function can be obtained as the upper LFT of its state-space realization.

The term linear fractional transformation is commonly applied to fractions of the form

$$\frac{s + qt}{p + qr}$$

where here we take all the parameters s, etc., to be scalars. However, if p is invertible, then we can define $a = s/p$, $b = t/p$, and $c = r/p$, to obtain

$$\begin{aligned} \frac{s+qt}{p+qr} &= \frac{a+bq}{1+cq} \\ &= \frac{a(1+cq)+q(b-ac)}{1+cq} \\ &= a + q(b-ac)(1+cq)^{-1} \\ &= \mathcal{F}_L(G, q) \end{aligned}$$

where

$$G = \begin{bmatrix} a & b - ac \\ 1 & -c \end{bmatrix}.$$

This is a lower linear fractional transformation. This equivalence extends to matrices.

THEOREM 7.7: Let $P, Q, R, S, \tilde{P}, \tilde{Q}, \tilde{R}$ and $\tilde{S}$ be matrices with complex entries. Assume that P and $\tilde{P}$ are invertible. Then

1. $(S+TQ)(P+RQ)^{-1} = \mathcal{F}_L(G, Q)$ where

$$G = \begin{bmatrix} SP^{-1} & T - SP^{-1}R \\ P^{-1} & -P^{-1}R \end{bmatrix}.$$

Similarly,

2. $(\tilde{P}+\tilde{Q}\tilde{R})^{-1}(\tilde{S}+\tilde{Q}\tilde{T}) = \mathcal{F}_L(\tilde{G}, \tilde{Q})$ where

$$\tilde{G} = \begin{bmatrix} \tilde{P}^{-1}\tilde{S} & \tilde{P}^{-1} \\ \tilde{T} - \tilde{R}\tilde{P}^{-1}\tilde{S} & -\tilde{R}\tilde{P}^{-1} \end{bmatrix}.$$

□ *Proof:* We show item (1). The proof of (2) is identical and left as an exercise.

$$\begin{aligned} \mathcal{F}_L(G,Q)(P+RQ) &= [SP^{-1} + (T - SP^{-1}R)Q(I + P^{-1}RQ)^{-1}P^{-1}] \\ &\quad \times (P+RQ) \\ &= S + SP^{-1}RQ + (T - SP^{-1}R)Q(I + P^{-1}RQ)^{-1} \\ &\quad \times (I + P^{-1}RQ) \\ &= S + SP^{-1}RQ + TQ - SP^{-1}RQ \\ &= S + TQ. \end{aligned}$$

■

EXAMPLE 7.8: The Youla parametrization of all stabilizing controllers obtained in Chapter 6 can be written as a linear fractional transformation. For some plant P, obtain a bicoprime factorization $P = ND^{-1} = \tilde{D}^{-1}\tilde{N}$ with $XN + YD = I$, $\tilde{N}\tilde{X} + \tilde{D}\tilde{Y} = I$. Any stabilizing controller H for P can be written

$$H = (Y - \tilde{Q}\tilde{N})^{-1}(X + \tilde{Q}\tilde{D})$$

for some $\tilde{Q} \in M(R\mathcal{H}_\infty)$. Notice that since Y is nonstrictly proper, it is invertible for all complex s. Using Theorem 7.7, we can rewrite this immediately as (see Fig. 7.8) $H = \mathcal{F}_L(H_o, \tilde{Q})$ where

$$H_o = \begin{bmatrix} Y^{-1}X & Y^{-1} \\ \tilde{D} - \tilde{N}Y^{-1}X & -\tilde{N}Y^{-1} \end{bmatrix}.$$

One advantage to using LFTs is that quite complicated block diagrams can be put into a standard, compact form. The unknown (or to be designed) parts of the system are isolated while the generalized plant G is entirely determined by the problem statement. In the next example, a model with structured uncertainty is written as a linear fractional transformation. Note that this is not a controller design problem.

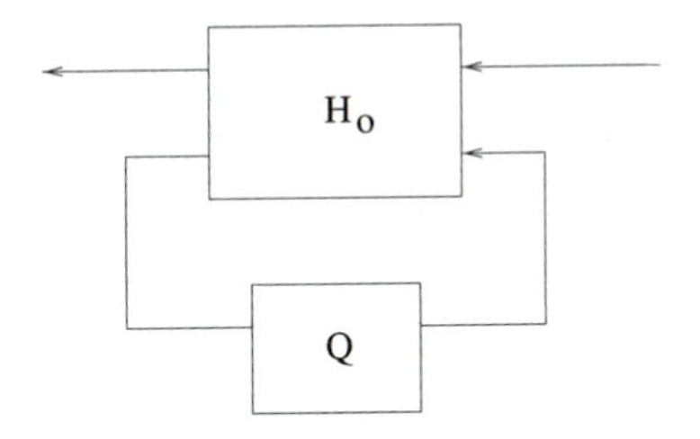

FIGURE 7.8 Family of all stabilizing controllers as a lower LFT.

Example 7.9: Consider a second-order system (such as a mass–spring–damper):

$$\ddot{x}(t) + d\dot{x}(t) + kx(t) = F(t). \tag{7.18}$$

The damping is known only to within 30% of a nominal value d_0,

$$d = (1 + 0.3\Delta_d)d_0, \tag{7.19}$$

and the stiffness k is known to within 10% of a nominal value:

$$k = (1 + 0.1\Delta_k)k_0. \tag{7.20}$$

The block diagram is drawn in Fig. 7.9. Notice that the uncertainties Δ_k, Δ_d are in separate blocks. We will rewrite this model as a linear fractional

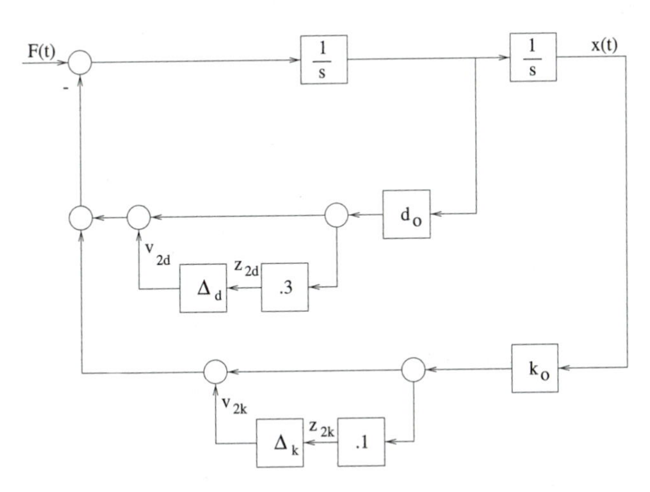

FIGURE 7.9 Second-order system with uncertain stiffness k and damping d.

transformation. Define

$$\Delta = \begin{bmatrix} \Delta_k & 0 \\ 0 & \Delta_d \end{bmatrix}. \tag{7.21}$$

The state variables are defined to be

$$\begin{aligned} x_1 &= x \\ x_2 &= \dot{x}. \end{aligned}$$

We then define the uncertainty "input"

$$z_2 = \begin{bmatrix} z_{2k} \\ z_{2d} \end{bmatrix} \tag{7.22}$$

and the uncertainty "output"

$$v_2 = \begin{bmatrix} v_{2k} \\ v_{2d} \end{bmatrix} \tag{7.23}$$

as shown on the diagram. We select the measured output to be

$$z_1 = \begin{bmatrix} \dot{x} \\ \ddot{x} \end{bmatrix} \tag{7.24}$$

and the "inputs"

$$v_1 = \begin{bmatrix} x \\ \dot{x} \\ F \end{bmatrix} \tag{7.25}$$

with

$$v_2 = \begin{bmatrix} \Delta_k & 0 \\ 0 & \Delta_d \end{bmatrix} z_2. \tag{7.26}$$

The generalized plant is

$$\left[\begin{array}{ccc:cc} 0 & 1 & 0 & 0 & 0 \\ -k_0 & -d_0 & 1 & -1 & -1 \\ \hdashline .1k_0 & 0 & 0 & 0 & 0 \\ 0 & .3d_0 & 0 & 0 & 0 \end{array}\right]. \tag{7.27}$$

Then

$$z_1 = G_{11} + G_{12}\Delta(I - G_{22}H)^{-1}G_{21}.$$

Alternatively,

$$z_1 = \mathcal{F}_L(G, \Delta)v_1 \tag{7.28}$$

where

$$G = \left[\begin{array}{ccc:cc} 0 & 1 & 0 & 0 & 0 \\ -k_0 & -d_0 & 1 & -1 & -1 \\ \hdashline 0.1k_0 & 0 & 0 & 0 & 0 \\ 0 & 0.3d_0 & 0 & 0 & 0 \end{array}\right]. \tag{7.29}$$

7.2 STABILITY

We now have a very general framework:

$$z_1 = G_{11}v_1 + G_{12}v_2 \tag{7.30}$$

$$z_2 = G_{21}v_1 + G_{21}v_2 \tag{7.31}$$

$$v_2 = Hz_2. \tag{7.32}$$

Assuming that the closed-loop is well-posed, we can eliminate z_2, v_2 to obtain

$$z_1 = [G_{11} + G_{12}H(I - G_{22}H)^{-1}G_{21}]v_1, \tag{7.33}$$

or, more concisely,

$$z_1 = \mathcal{F}_L(G, H)v_1.$$

Suppose G and H have the state-space realizations

$$G = \left[\begin{array}{c|cc} A & B_1 & B_2 \\ \hline C_1 & E_{11} & E_{12} \\ C_2 & E_{21} & E_{22} \end{array}\right] \tag{7.34}$$

$$H = \left[\begin{array}{c|c} A_H & B_H \\ \hline C_H & E_H \end{array}\right]. \tag{7.35}$$

Define

$$L_1 = (I - E_{22}E_H)^{-1}, \quad L_2 = (I - E_H E_{22})^{-1}. \tag{7.36}$$

A state-space realization for the closed loop is

$$\left[\begin{array}{cc|c} A + B_2 E_H L_1 C_2 & B_2 L_2 C_H & B_1 + B_2 E_H L_1 E_{21} \\ B_H L_1 C_2 & A_H + B_H L_1 E_{22} C_H & B_H L_1 E_{21} \\ \hline C_1 + E_{12} L_2 E_H C_2 & E_{12} L_2 C_H & E_{11} + E_{12} E_H L_1 E_{21} \end{array}\right]. \tag{7.37}$$

A fundamental objective in choosing H is that, as always, the closed loop must be internally stable.

Definition 7.10: The closed-loop system in Fig. 7.2 is *internally stable* if and only if the "A" matrix of the closed-loop system (7.37) is Hurwitz.

The generalized feedback configuration introduced at the beginning of this chapter differs from the standard feedback system used in previous chapters (Fig. 3.2) in several respects. One is the structure. Another is that the feedback connection is positive: $v_2 = Hz_2$, not $H(-z_2)$ as previously.

Theorem 7.11: A controller H internally stabilizes a generalized plant G if and only if it internally stabilizes $-G_{22}$ in the standard (negative) feedback configuration.

There exists a controller that internally stabilizes G if and only if (A, B_2, C_2) is stabilizable and detectable.

If K and F are such that $A - B_2K$ and $A - FC_2$ are Hurwitz then

$$H(s) = \left[\begin{array}{c|c} A - B_2K - FC_2 + FE_{22}K & F \\ \hline -K & 0 \end{array}\right] \tag{7.38}$$

internally stabilizes G.

□ *Proof:* Let (A_H, B_H, C_H, E_H) be a state-space realization of a controller H. The closed-loop "A" matrix of G with H (7.37) is identical to that of $-G_{22}$ with H in the negative feedback configuration.

Suppose (A, B_2, C_2) is stabilizable and detectable and choose K and F so $A - B_2K$ and $A - FC_2$ are Hurwitz. The controller H defined in (7.38) is the negative of the output feedback controller derived in Chapter 5. It was shown there that the closed loop is stable. We will show it directly here. The closed-loop has "A" matrix

$$\hat{A} = \begin{bmatrix} A & -B_2K \\ FC_2 & A - B_2K - FC_2 \end{bmatrix}. \tag{7.39}$$

Using the state transformation

$$P = \begin{bmatrix} I & -I \\ 0 & I \end{bmatrix}, \tag{7.40}$$

we obtain

$$P\hat{A}P^{-1} = \begin{bmatrix} A - FC_2 & 0 \\ FC_2 & A - B_2K \end{bmatrix}. \tag{7.41}$$

Thus, the closed-loop eigenvalues are the union of those of $A - B_2K$ and $A - FC_2$. The closed loop is internally stable.

Now suppose that the closed loop is internally stable. That is, there exists

$$H = \left[\begin{array}{c|c} A_H & B_H \\ \hline C_H & E_H \end{array}\right] \tag{7.42}$$

so that the closed-loop "A" matrix is Hurwitz. The closed-loop "A" matrix (see (7.37)) can be written

$$\begin{aligned} \hat{A} &= \begin{bmatrix} A & B_2L_2E_H \\ 0 & A_H + B_HL_1E_{22}C_H \end{bmatrix} + \begin{bmatrix} B_2E_HL_1C_2 & 0 \\ B_HL_1C_2 & 0 \end{bmatrix} \\ &= \begin{bmatrix} A & B_2L_2E_H \\ 0 & A_H + B_HL_1E_{22}C_H \end{bmatrix} + \begin{bmatrix} B_2E_HL_1 \\ B_HL_1 \end{bmatrix} [C_2 \;\; 0]. \end{aligned}$$

It follows that

$$\left(\begin{bmatrix} A & B_2L_2E_H \\ 0 & A_H + B_HL_1E_{22}C_H \end{bmatrix}, [C_2 \;\; 0] \right) \tag{7.43}$$

is detectable. This means that for all Re $\lambda \geq 0$, the system of equations

$$\begin{bmatrix} A - \lambda I & B_2L_2E_H \\ 0 & A_H + B_HL_1E_{22}C_H - \lambda I \\ C_2 & 0 \end{bmatrix} \begin{bmatrix} x \\ y \end{bmatrix} = \begin{bmatrix} 0 \\ 0 \\ 0 \end{bmatrix} \tag{7.44}$$

has only the trivial solution $x = 0$, $y = 0$. This implies that for all Re $\lambda \geq 0$

$$\begin{bmatrix} A - \lambda I \\ 0 \\ C_2 \end{bmatrix} x = \begin{bmatrix} 0 \\ 0 \\ 0 \end{bmatrix} \tag{7.45}$$

has only the trivial solution. It follows that (A, C_2) is detectable.

We will now show that (A, B_2) is stabilizable. Write the closed-loop "A" matrix

$$\hat{A} = \begin{bmatrix} A & 0 \\ B_H L_1 C_2 & A_H + B_H L_1 E_{22} C_H \end{bmatrix} + \begin{bmatrix} B_2 \\ 0 \end{bmatrix} [E_H L_1 C_2 \quad L_2 C_H]. \tag{7.46}$$

This implies that

$$\left(\begin{bmatrix} A & 0 \\ B_H L_1 C_2 & A_H + B_H L_1 E_{22} C_H \end{bmatrix}, \begin{bmatrix} B_2 \\ 0 \end{bmatrix} \right) \tag{7.47}$$

is stabilizable. This means that for all Re $\lambda \geq 0$,

$$[x \quad y] \begin{bmatrix} A - \lambda I & 0 & B_2 \\ B_H L_1 C_2 & A_H + B_H L_1 E_{22} C_H - \lambda I & 0 \end{bmatrix} = [0 \ \ 0 \ \ 0] \tag{7.48}$$

has only the trivial solution $x = 0$, $y = 0$. Thus, also

$$[x][A - \lambda I \quad B_2] = [0 \quad 0] \tag{7.49}$$

has only the trivial solution $x = 0$. It follows that (A, B_2) is stabilizable. ■

Stabilizability and detectability of a realization of G does not imply stabilizability and detectability of the corresponding realization for G_{22}. For instance (A, B_1) may be stabilizable, in which case $(A, [B_1 \ B_2])$ is stabilizable. This does not imply that (A, B_2) is stabilizable.

One important consequence of this theorem is that a stabilizing controller for the generalized plant G can be obtained by considering stabilizing controllers for G_{22}. Since G_{22} is of lower dimension, this may simplify the problem. Also, there are many results establishing which controllers will stabilize a plant G_{22}.

THEOREM 7.12: Assume that (A, B_2, C_2) is stabilizable and detectable. Choose K and F so $A - B_2 K$, $A - F C_2$ are Hurwitz. The set of all controllers that internally stabilize G is $\mathcal{F}_L(H_0, Q)$ where

$$H_0 = \left[\begin{array}{c|cc} A - B_2 K - F C_2 + F E_{22} K & F & B_2 - F E_{22} \\ \hline -K & 0 & I \\ -C_2 + E_{22} K & I & -E_{22} \end{array} \right] \tag{7.50}$$

and Q is any element of $M(R\mathcal{H}_\infty)$ such that $I + E_{22} E_Q$ is invertible where $E_Q = \lim_{s \to \infty} Q(s)$. (See Fig. 7.10.)

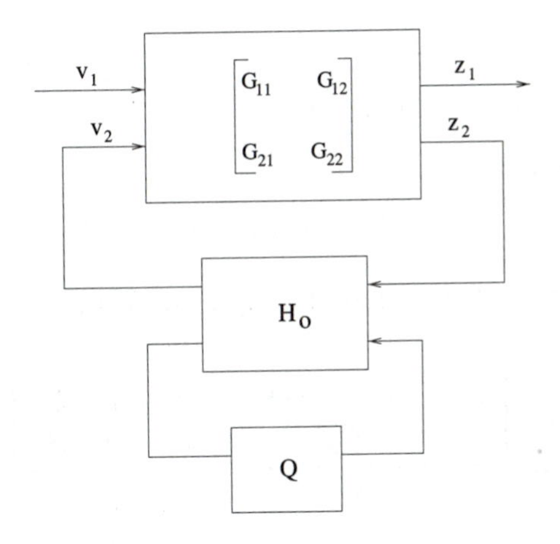

FIGURE 7.10 Family of all internally stable closed loops.

The set of all closed-loop systems from v_1 to z_1 that is internally stable is

$$\mathcal{F}_L(T, Q) = T_{11} + T_{12} Q T_{21} \tag{7.51}$$

where

$$T = \left[\begin{array}{cc|cc} A - B_2 K & -B_2 K & B_1 & B_2 \\ 0 & A - F C_2 & -B_1 + F E_{21} & 0 \\ \hline C_1 - E_{12} K & -E_{12} K & E_{11} & E_{12} \\ 0 & -C_2 & E_{21} & 0 \end{array}\right] \tag{7.52}$$

and $Q \in M(R\mathcal{H}_\infty)$, $I + E_{22} E_Q$ is invertible.

□ *Proof:* Two proofs are given here. First, the family of all stabilizing controllers for G is given by the Youla parametrization (Chapter 6) of the family of all stabilizing controllers for $-G_{22}$. This was written as a linear fractional transformation $\mathcal{F}_L(H_o, Q)$ in Example 7.8. Obtaining a state-space representation for H_o leads to (7.50). The details are outlined in the exercises.

Another proof that does not involve the Youla parametrization may be used. Let H be any controller that internally stabilizes G. Define

$$S = \left[\begin{array}{c|cc} A & F & B_2 \\ \hline K & 0 & I \\ C_2 & I & 0 \end{array}\right]. \tag{7.53}$$

Since

$$S_{22} = G_{22} - E_{22},$$

H internally stabilizes S. Define

$$Q = \mathcal{F}_L(S, H). \tag{7.54}$$

Because H stabilizes S, it follows that $Q \in M(R\mathcal{H}_\infty)$.

Define

$$R = \left[\begin{array}{cc|cc} A - FC_2 - B_2K & B_2K & F & B_2 \\ -FC_2 & A & F & B_2 \\ \hline -K & K & 0 & I \\ -C_2 & C_2 & I & 0 \end{array}\right]. \tag{7.55}$$

Routine calculations show that

$$\mathcal{F}_L(H_0, Q) = \mathcal{F}_L(R, H) \tag{7.56}$$

where using a state-space transformation, we obtain

$$R = \left[\begin{array}{cc|cc} A - FC_2 & B_2K & F & B_2 \\ 0 & A - B_2K & 0 & 0 \\ \hline 0 & K & 0 & I \\ 0 & C_2 & I & 0 \end{array}\right].$$

Observing the structure of this realization, we can simplify to obtain

$$R = \left[\begin{array}{c|cc} 0 & 0 & 0 \\ \hline 0 & 0 & I \\ 0 & I & 0 \end{array}\right].$$

It follows that

$$\mathcal{F}_L(R, H) = H \tag{7.57}$$

and

$$H = \mathcal{F}_L(H_0, Q), \tag{7.58}$$

as was to be shown. Thus, any stabilizing controller can be written $\mathcal{F}_L(H_o, Q)$ where $Q \in M(R\mathcal{H}_\infty)$.

We now show that any stable closed loop can be written $\mathcal{F}_L(T, Q)$ where T is defined in (7.52). To simplify the calculations, we describe only the case where $E_{22} = 0$ in the generalized plant. The generalization is straightforward. The generalized plant G has a state-space realization

$$G = \left[\begin{array}{c|cc} A & B_1 & B_2 \\ \hline C_1 & E_{11} & E_{12} \\ C_2 & E_{21} & 0 \end{array}\right],$$

while for convenience we write the state-space realization (7.50) for H_o as

$$H_o = \left[\begin{array}{c|cc} A_H & B_{H1} & B_{H2} \\ \hline C_{H1} & 0 & I \\ C_{H2} & I & 0 \end{array}\right].$$

Writing out the equations (see Fig. 7.10):

$$\begin{aligned} G: \quad \dot{x}(t) &= Ax(t) + B_1 v_1(t) + B_2 v_2(t) \\ z_1(t) &= C_1 x(t) + E_{11} v_1(t) + E_{12} v_2(t) \\ z_2(t) &= C_2 x(t) + E_{21} v_1(t) \end{aligned}$$

$$\begin{aligned} H_o: \quad \dot{x}_H(t) &= A_H x_H(t) + B_{H1} z_2(t) + B_{H2} v_3(t) \\ v_2(t) &= C_{H1} x_H(t) + v_3(t) \\ z_3(t) &= C_{H2} x_H(t) + z_2(t). \end{aligned}$$

The connection of G with H_o, Q in the configuration shown yields a system with inputs v_1, v_3 and outputs z_1, z_3. Denote this system as T. Any stable closed loop can be written $\mathcal{F}_L(T, Q)$ for some $Q \in M(R\mathcal{H}_\infty)$. We now develop a state-space realization for T. Using the preceding equations to eliminate v_2 and z_2, we obtain

$$\begin{bmatrix} \dot{x}(t) \\ \dot{x}_H(t) \end{bmatrix} = \begin{bmatrix} A & B_2 C_{H1} \\ B_{H1} C_2 & A_H \end{bmatrix} \begin{bmatrix} x(t) \\ x_H(t) \end{bmatrix} + \begin{bmatrix} B_1 & B_2 \\ B_{H1} E_{21} & B_{H2} \end{bmatrix} \begin{bmatrix} v_1(t) \\ v_3(t) \end{bmatrix}$$

$$\begin{bmatrix} z_1(t) \\ z_3(t) \end{bmatrix} = \begin{bmatrix} C_1 & E_{12} C_{H1} \\ C_2 & C_{H2} \end{bmatrix} \begin{bmatrix} x(t) \\ x_H(t) \end{bmatrix} + \begin{bmatrix} E_{11} & E_{12} \\ E_{21} & 0 \end{bmatrix} \begin{bmatrix} v_1(t) \\ v_3(t) \end{bmatrix}.$$

Use the state transformation

$$\begin{bmatrix} x \\ x_H - x \end{bmatrix} = \begin{bmatrix} I & 0 \\ -I & I \end{bmatrix} \begin{bmatrix} x \\ x_H \end{bmatrix}.$$

We obtain the state-space realization (7.52) for T. Since $T_{22} = 0$, (7.51) follows. ■

Once one internally stabilizing controller is found, the family of all stabilizing controllers is found by varying Q over stable transfer matrices.

The family of all possible internally stable transfer matrices is affine in this parameter Q.

As noted earlier, stabilizability and detectability of the generalized plant G do not imply that G_{22} is stabilizable and detectable. The foregoing results state that internally stable closed loops can be obtained by considering controllers that internally stabilize G_{22} (with the positive connection or $-G_{22}$ with the negative connection).

However, even if both G_{22} and H have stabilizable and detectable realizations, external stability of the closed loop does not imply internal stability.

Example 7.13:

$$G = \left[\begin{array}{cc|cc} -1 & 0 & b_1 & 0 \\ 0 & -2 & b_2 & 1 \\ \hline 0 & -6 & 0 & 1 \\ 1 & 0 & 1 & 0 \end{array}\right], \qquad H = \left[\begin{array}{c|c} 4 & 1 \\ \hline 1 & 0 \end{array}\right]. \tag{7.59}$$

The generalized plant G is internally stable and so (A, B_2, C_2) is trivially stabilizable and detectable. The controller H is first-order. It is unstable but stabilizable and detectable. The entries b_1 and b_2 are arbitrary.

The closed loop is externally stable. It has realization

$$\mathcal{F}_L(G, H) = \left[\begin{array}{ccc|c} -1 & 0 & 0 & b_1 \\ 0 & -2 & 1 & b_2 \\ 1 & 0 & 4 & 1 \\ \hline 0 & -6 & 1 & 0 \end{array}\right]. \tag{7.60}$$

This system is detectable if and only if

$$\begin{bmatrix} -1-\lambda & 0 & 0 \\ 0 & -2-\lambda & 1 \\ 1 & 0 & 4-\lambda \\ 0 & -6 & 1 \end{bmatrix} \begin{bmatrix} x \\ y \\ z \end{bmatrix} = \begin{bmatrix} 0 \\ 0 \\ 0 \\ 0 \end{bmatrix} \tag{7.61}$$

has only the trivial solution for all $\operatorname{Re} \lambda \geq 0$. However, setting $\lambda = 4$ and

$$\begin{bmatrix} x \\ y \\ z \end{bmatrix} = \begin{bmatrix} 0 \\ y \\ 6y \end{bmatrix} \tag{7.62}$$

where $y \neq 0$ is arbitrary, we obtain a nontrivial solution to

$$\begin{bmatrix} -5 & 0 & 0 \\ 0 & -6 & 1 \\ 1 & 0 & 0 \\ 0 & -6 & 1 \end{bmatrix}. \tag{7.63}$$

This shows that the closed loop is not detectable.

We could add fictitious inputs to the generalized plant. With stabilizability and detectability defined for this plant, we can show that internal and external stability are equivalent in the closed loop. This result would be entirely analogous to the corresponding result in Chapter 3. (This approach is outlined in the exercises.) However, we usually use the following easily checked sufficient conditions that guarantee stabilizability and detectability of the closed loop. If the closed loop is stabilizable and detectable, then of course external stability implies internal stability.

Theorem 7.14: Assume that G and H both have stabilizable and detectable realizations. Then $\mathcal{F}_L(G, H)$ is

1. Detectable if $\left[\begin{smallmatrix} A-\lambda I & B_2 \\ C_1 & E_{12} \end{smallmatrix}\right]$ has full column rank for all Re $\lambda \geq 0$.

2. Stabilizable if $\left[\begin{smallmatrix} A-\lambda I & B_1 \\ C_2 & E_{21} \end{smallmatrix}\right]$ has full row rank for all Re $\lambda \geq 0$.

□ *Proof:* The realization for the closed loop system is given in (7.37). The closed loop is detectable if and only if

$$\begin{bmatrix} A + B_2 E_H L_1 C_2 - \lambda I & B_2 L_2 C_H \\ B_H L_1 C_2 & A_H + B_H L_1 E_{22} C_H - \lambda I \\ C_1 + E_{12} L_2 E_H C_2 & E_{12} L_2 C_H \end{bmatrix} \begin{bmatrix} x \\ y \end{bmatrix} = \begin{bmatrix} 0 \\ 0 \\ 0 \end{bmatrix} \tag{7.64}$$

where $L_1 = (I - E_{22}E_H)^{-1}$, $L_2 = (I - E_H E_{22})^{-1}$, has only the trivial solution for all Re $\lambda \geq 0$. Now,

$$\begin{aligned} L_2 E_H &= (I - E_H E_{22})^{-1} E_H \\ &= (I - E_H E_{22})^{-1} E_H (I - E_{22} E_H)(I - E_{22} E_H)^{-1} \\ &= (I - E_H E_{22})^{-1} (E_H - E_H E_{22} E_H)(I - E_{22} E_H)^{-1} \\ &= (I - E_H E_{22})^{-1} (I - E_H E_{22}) E_H (I - E_{22} E_H)^{-1} \\ &= E_H L_1. \end{aligned} \tag{7.65}$$

Using this, we rewrite (7.64) as

$$\begin{bmatrix} A-\lambda I & B_2 \\ C_1 & E_{12} \end{bmatrix} \begin{bmatrix} x \\ L_2 E_H C_2 x + L_2 C_H y \end{bmatrix} = \begin{bmatrix} 0 \\ 0 \end{bmatrix} \tag{7.66}$$

$$B_H L_1 (C_2 x + E_{22} C_H y) + (A_H - \lambda I) y = 0. \tag{7.67}$$

If

$$\begin{bmatrix} A-\lambda I & B_2 \\ C_1 & E_{12} \end{bmatrix} \tag{7.68}$$

has full column rank for all Re $\lambda \geq 0$, then the only solution to (7.64) (or equivalently (7.66), (7.67) is

$$\begin{aligned} x &= 0 \\ C_H y &= 0 \\ (A_H - \lambda I) y &= 0. \end{aligned} \tag{7.69}$$

Since (A_H, C_H) is detectable, the only solution to the last two equations is $y = 0$.

Thus we have shown that if (7.68) has full column rank, the system of equations (7.64) has only the trivial solution. It follows that the closed loop system is detectable. This proves point 1.

Point 2 is proven using an identical argument, or by duality. ■

The rank condition just given can be replaced by an equivalent condition on unobservable modes.

THEOREM 7.15: Assume that E has full column rank and define $R = E^* E$. The following conditions are equivalent:

1. $\begin{bmatrix} A-\lambda I & B \\ C & E \end{bmatrix}$ has full column rank for all Re $\lambda \geq 0$.
2. $((I - ER^{-1}E^*)C, A - BR^{-1}E^*C)$ is detectable.

□ *Proof:* Suppose λ, Re $\lambda \geq 0$ is an unobservable mode of $((I - ER^{-1}E^*)C, A - BR^{-1}E^*C)$, that is, there is $x \neq 0$ such that

$$(A - BR^{-1}E^*C - \lambda I)x = 0, (I - ER^{-1}E^*)Cx = 0. \tag{7.70}$$

We can rewrite this as

$$\begin{bmatrix} A-\lambda I & B \\ C & E \end{bmatrix} \begin{bmatrix} I & 0 \\ -R^{-1}E^*C & I \end{bmatrix} \begin{bmatrix} x \\ 0 \end{bmatrix} = \begin{bmatrix} 0 \\ 0 \end{bmatrix}. \tag{7.71}$$

This implies that

$$\begin{bmatrix} A - \lambda I & B \\ C & E \end{bmatrix} \tag{7.72}$$

does not have full column rank.

Conversely, suppose (7.72) does not have full column rank for some λ, Re $\lambda \geq 0$. That is, for x and y not both zero,

$$\begin{bmatrix} A - \lambda I & B \\ C & E \end{bmatrix} \begin{bmatrix} x \\ y \end{bmatrix} = \begin{bmatrix} 0 \\ 0 \end{bmatrix}. \tag{7.73}$$

Define

$$\begin{bmatrix} \bar{x} \\ \bar{y} \end{bmatrix} = \begin{bmatrix} I & 0 \\ R^{-1}E^*C & I \end{bmatrix} \begin{bmatrix} x \\ y \end{bmatrix}.$$

Note that if

$$\begin{bmatrix} x \\ y \end{bmatrix} \neq \begin{bmatrix} 0 \\ 0 \end{bmatrix}, \quad \text{then} \quad \begin{bmatrix} \bar{x} \\ \bar{y} \end{bmatrix} \neq \begin{bmatrix} 0 \\ 0 \end{bmatrix}.$$

Also,

$$\begin{bmatrix} x \\ y \end{bmatrix} = \begin{bmatrix} I & 0 \\ -R^{-1}E^*C & I \end{bmatrix} \begin{bmatrix} \bar{x} \\ \bar{y} \end{bmatrix} = \begin{bmatrix} \bar{x} \\ -R^{-1}E^*C\bar{x} + \bar{y} \end{bmatrix}.$$

This, with (7.73), leads to

$$(A - \lambda I)\bar{x} + B(-R^{-1}E^*C - \bar{x} + \bar{y}) = 0 \tag{7.74}$$

$$C\bar{x} + E(-R^{-1}E^*C\bar{x} + \bar{y}) = 0. \tag{7.75}$$

Multiply the second equation on the left by E^* and recall that $R = E^*E$ to obtain

$$E^*E\bar{y} = 0. \tag{7.76}$$

This implies that $\bar{y} = 0$ and so $\bar{x} \neq 0$. Equations (7.74) and (7.75) become

$$\begin{aligned} (A - BR^{-1}E^*C)\bar{x} &= \lambda\bar{x} \\ (I - ER^{-1}E^*)C\bar{x} &= 0. \end{aligned} \tag{7.77}$$

This completes the proof. ■

7.3 REDHEFFER'S LEMMA

In this section we prove Redheffer's Lemma. This result is needed to prove the main result for $\mathcal{H}_\infty$ controller synthesis in the next chapter. It is also important in other contexts.

Definition 7.16: A transfer function matrix G is *inner* if $G \in M(R\mathcal{H}_\infty)$ and $\tilde{G}G = I$ where $\tilde{G}(s) = [G(-s)]^*$.

Let $\|M\|$ indicate the usual matrix norm (Appendix B). An important property of inner functions is that on the imaginary axis we have $G(j\omega)^*G(j\omega) = I$, and so for functions $v(jw) \in \bar{L}_2$, $\|(Gv)(j\omega)\| = \|v(j\omega)\|$ at each ω and also $\|Gv\|_2 = \|v\|_2$.

The following theorem provides a state-space characterization of inner functions.

Theorem 7.17: Suppose $G = (A, B, C, E)$ where A is Hurwitz and let X be the self-adjoint positive semidefinite solution to the algebraic Riccati equation

$$A^*X + XA + C^*C = 0.$$

Then G is inner if and only if

1. $E^*E = I$,
2. $B^*X + E^*C = 0$ or $B = 0$.

□ *Proof:* $\tilde{G}$ has realization $(-A^*, -C^*, B^*, E^*)$ and a simple calculation (Theorem C.1) yields that $\tilde{G}G$ has realization

$$\left[\begin{array}{cc|c} A & 0 & B \\ -C^*C & -A^* & -C^*E \\ \hline E^*C & B^* & E^*E \end{array}\right].$$

Use the state transformation

$$\tilde{z} = \begin{bmatrix} I & 0 \\ X & I \end{bmatrix} z$$

to obtain

$$\tilde{G}G = \left[\begin{array}{cc|c} A & 0 & B \\ -(A^*X + XA + C^*C) & -A^* & -(XB + C^*E) \\ \hline B^*X + E^*C & B^* & E^*E \end{array}\right].$$

Using $A^*X + XA + C^*C = 0$ and (1), (2), we obtain

$$\tilde{G}G = I$$

as required. ■

Theorem 7.18 (*Redheffer's Lemma*): Let G be a generalized plant. Suppose that $\tilde{G}G = I$, $G \in M(R\mathcal{H}_\infty)$ and $G_{21}^{-1} \in M(R\mathcal{H}_\infty)$. Let

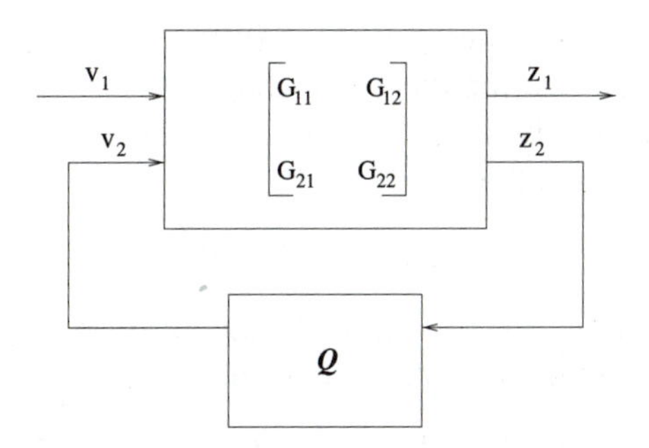

FIGURE 7.11 Diagram for Redheffer's Lemma.

Q be a proper transfer matrix. The following conditions are equivalent:

1. $Q \in M(R\mathcal{H}_\infty)$ and $\|Q\|_\infty < 1$.
2. Q internally stabilizes G and $\|\mathcal{F}_L(G, Q)\|_\infty < 1$.

□ *Proof:* $(1 \to 2)$ Since $G, Q \in M(R\mathcal{H}_\infty)$, the closed-loop transfer function is well-posed and internally stable if $(I - G_{22}Q)^{-1} \in M(R\mathcal{H}_\infty)$. This follows immediately from $\|G_{22}\|_\infty \le 1$ and $\|Q\|_\infty < 1$. We now show that $\|\mathcal{F}_L(G, Q)\|_\infty < 1$. First note that since G is inner, it preserves norms and

$$\|z_1\|_2^2 + \|z_2\|_2^2 = \|v_1\|_2^2 + \|v_2\|_2^2.$$

Define the scalar $M_Q = \|Q\|_\infty < 1$. The transfer function from z_2 to v_1 (Fig. 7.11) is

$$G_{21}^{-1}(I - G_{22}Q)$$

and by assumption, it is stable. Write

$$M_T = \left\|G_{21}^{-1}(I - G_{22}Q)\right\|_\infty.$$

We have

$$\begin{aligned} \|z_1\|_2^2 &= \|v_1\|_2^2 + \|v_2\|_2^2 - \|z_2\|_2^2 \\ &\le \|v_1\|_2^2 - (1 - M_Q^2)\|z_2\|_2^2 \\ &\le \left[1 - \frac{(1 - M_Q^2)}{M_T^2}\right]\|v_1\|_2^2. \end{aligned}$$

Thus, $\|z_1\|_2^2 < \|v_1\|_2^2$ and

$$\|\mathcal{F}_L(G, Q)\|_\infty < 1.$$

(2 → 1). We first show that $Q \in M(R\mathcal{H}_\infty)$. This proof relies on the stable coprime factorization theory in Chapter 6.

Let $Q = N_Q D_Q^{-1}$ be a right coprime factorization (r.c.f.) of Q, and $X, Y \in M(R\mathcal{H}_\infty)$ such that

$$XN_Q + YD_Q = I.$$

We will show that $D_Q^{-1} \in M(\mathcal{H}_\infty)$ and hence $Q \in M(\mathcal{H}_\infty)$.

Since the closed loop is internally stable,

$$Q(I - G_{22}Q)^{-1} = N_Q(D_Q - G_{22}N_Q)^{-1} \in M(R\mathcal{H}_\infty)$$

and

$$(I - G_{22}Q)^{-1} = D_Q(D_Q - G_{22}N_Q)^{-1} \in M(R\mathcal{H}_\infty).$$

Thus,

$$\begin{aligned} &XN_Q(D_Q - G_{22}N_Q)^{-1} + YD_Q(D_Q - G_{22}N_Q)^{-1} \\ &= (D_Q - G_{22}N_Q)^{-1} \in M(R\mathcal{H}_\infty). \end{aligned}$$

This implies that $\det(D_Q - G_{22}N_Q)$ has no zeros in the closed right half-plane and by Cauchy's Principle of the Argument (Theorem 3.42),

$$\text{wind}_\mathcal{C} \det(D_Q - G_{22}N_Q) = 0$$

where $\mathcal{C}$ is the Nyquist contour. Also, since $(I - G_{22}Q)^{-1} \in M(R\mathcal{H}_\infty)$,

$$\text{wind}_\mathcal{C} \det(I - G_{22}Q) = 0.$$

Now,

$$\det(D_Q - G_{22}N_Q) = \det(I - G_{22}Q)\det(D_q),$$

and so

$$\text{wind}_\mathcal{C} \det(D_Q - G_{22}N_Q) = \text{wind}_\mathcal{C} \det(I - G_{22}Q) + \text{wind}_\mathcal{C} \det(D_q).$$

It follows that

$$\text{wind}_\mathcal{C} \det(D_Q) = 0.$$

This implies that $D_Q^{-1} \in M(R\mathcal{H}_\infty)$. Hence, $Q \in M(R\mathcal{H}_\infty)$.

We complete the proof by showing that if $\|Q\|_\infty \geq 1$, then $\|\mathcal{F}(G, Q)\| \geq 1$. In this case, for some $\tilde{z}_2$, $\|Q\tilde{z}_2\| \geq \|\tilde{z}_2\|$. Define

$$\begin{aligned} v_1 &= G_{21}^{-1}(I - G_{22}Q)\tilde{z}_2 \\ \tilde{v}_2 &= Q\tilde{z}_2. \end{aligned}$$

Since G is inner,

$$\|\tilde{z}_1\|^2 + \|\tilde{z}_2\|^2 = \|\tilde{v}_1\|^2 + \|\tilde{v}_2\|^2.$$

Since $\|\tilde{v}_2\| \geq \|\tilde{z}_2\|$,

$$\|\tilde{z}_1\| \geq \|\tilde{v}_1\|,$$

and we obtain $\|\mathcal{F}_L(G, Q)\|_\infty \geq 1$. ∎

NOTES AND REFERENCES

The theory in this chapter can be found in more detail in, for instance, [7, 9, 53]. Zames and Owen [52] have shown that solving the robust performance problem is an infinite-dimensional convex optimization problem. The noise control example (Example 7.3) is from [32], the two-compensator problem (Example 7.4) is in [7], and the uncertainty example (Example 7.9) is in [53].

The MATLAB routine "augment" constructs the generalized plant for the special (but important) case of reducing mixed sensitivity where the generalized plant G is

$$\begin{array}{lll} W_1 & \text{weights} & S \\ W_2 & \text{weights} & H(I+PH)^{-1} \\ W_3 & \text{weights} & T. \end{array}$$

and

$$z_1 = \begin{bmatrix} W_1 S \\ W_2 H S \\ W_3 T \end{bmatrix}$$

(or any subset of these). The controller H is assumed placed in the standard negative feedback connection with P.

EXERCISES

1. Consider Example 7.4 of two-compensator design. Let (A_P, B_P, C_P, E_P) be a realization for P and $(A_{W1}, B_{W1}, C_{W1}, E_{W1})$ a realization for W_1. Construct a state-space realization for the generalized plant that describes the two-compensator design problem.

2. Rewrite the problem

$$\left\| \begin{bmatrix} W_1 S \\ \delta C S \\ W_3 T \end{bmatrix} \right\|_\infty < 1$$

(where $\delta > 0$ is a scalar) in the generalized plant framework. Obtain a state-space realization.

3. Consider the "shifted" $\mathcal{H}_\infty$ problem for some $\alpha > 0$: Find C so the closed loop is α-stable and

$$\sup_{\mathrm{Re}(s) > -\alpha} \sigma_{\max} \left\{ \begin{bmatrix} W_1 S \\ W_2 C S \\ W_3 T \end{bmatrix} (s) \right\} < 1.$$

Rewrite this as a generalized plant so that solving it is a standard $\mathcal{H}_\infty$ problem. Assume that W_1 and P are strictly proper, to simplify the calculations.

4. In Example 7.8 the set of all stabilizing controllers for a plant was obtained as a linear fractional transformation $\mathcal{F}_L(H_o, Q)$. Use the results in Section 6.4 and Exercise 6.10 to obtain a state-space realization for H_o. Verify that this is the same as that given in Theorem 7.12.

5. **(a)** Write the Youla parametrization for the right coprime factors of the stabilizing controllers of a plant:

$$H = (\tilde{X} + DQ)(\tilde{Y} - NQ)^{-1}$$

as a linear fractional transformation $\mathcal{F}_L(\tilde{H}_o, Q)$. That is, calculate $\tilde{H}_o$ in transfer function form. Assume that the factorizations are bicoprime.

(b) Obtain a state-space realization for $\tilde{H}_o$.

6. Let G have the usual realization. Assume that the controller H is stabilizable. Prove that the closed loop $\mathcal{F}_L(G, H)$ is stabilizable if

$$\left[\begin{array}{c|c} A - \lambda I & B_1 \\ \hline C_2 & E_{21} \end{array} \right]$$

has full row rank for all Re $\lambda \geq 0$.

7. **(a)** Assume that $E_{12}^* C_1 = 0$. Show that condition (1) in Theorem 7.14 is satisfied if (A, C_1) is detectable.

(b) Assume that $B_1 E_{21}^* = 0$. Show that (2) in Theorem 7.14 is satisfied if (A, B_1) is stabilizable.

(c) Assume that (A, B_1) and (A, B_2) are stabilizable; that (A, C_1) and (A, C_2) are detectable; and that $E_{12}^* C_1 = 0$ and $B_1 E_{21}^* = 0$. Show that (1) an internally stabilizing controller exists and (2) if the controller realization is minimal, then the closed loop is externally stable if and only if it is internally stable.

8. Consider the standard generalized plant configuration in Fig. 7.2. Denote the controller output by y_H and the controller input by u_H. Add fictitious inputs d_1, d_2 so that $v_2 = d_1 + y_H$ and $u_H = z_2 + d_2$. Draw a diagram. We now have nine transfer functions from inputs v_1, d_1 and d_2 to outputs z_1, u_H, and v_2. The closed loop is externally stable if all nine transfer functions are in $M(R\mathcal{H}_\infty)$. Now show that the closed loop is stabilizable and detectable if the realizations of G and H are stabilizable and detectable. Hence, show that external stability implies internal stability for the closed loop.

9. Assume that P has a stabilizable/detectable realization $(A_P, C_P, C_P, 0)$. Suppose that A_P has an eigenvalue λ with $\text{Re}\,\lambda \geq 0$. Consider the mixed sensitivity problem

$$\left\| \begin{bmatrix} W_1 S \\ W_2 H S \end{bmatrix} \right\|_\infty < 1.$$

Let W_1 and W_2 be arbitrary stable functions with minimal realizations. Let H be any controller with a minimal realization.

(a) Write the realization for the corresponding generalized plant. (See Example 7.2) Are the conditions in Theorem 7.14 satisfied?

(b) Is the closed loop detectable? Is the closed loop stabilizable? Justify your answers.

(c) Does a controller exist so that the closed loop is internally stable?

(d) Repeat parts (a)–(c) for

$$\left\| \begin{bmatrix} W_1 S \\ W_2 H S \\ W_3 S P \end{bmatrix} \right\|_\infty < 1.$$

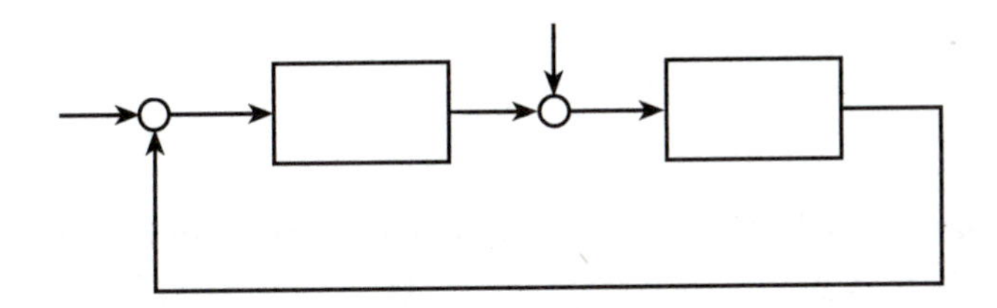

VIII
ESTIMATOR-BASED $\mathcal{H}_\infty$ CONTROLLER DESIGN

In this chapter we will construct controllers to reduce the $\mathcal{H}_\infty$ norm of a closed loop. We will use the generalized plant-controller configuration from Chapter 7. Suppose we have the generalized plant

$$G = \begin{bmatrix} G_{11} & G_{12} \\ G_{21} & G_{22} \end{bmatrix}.$$

For admissible controllers H,

$$\begin{aligned} z_2 &= (I - G_{22}H)^{-1}G_{21}v_1 \\ z_1 &= G_{11}v_1 + G_{12}(Hz_2) \\ &= [G_{11} + G_{12}H(I - G_{22}H)^{-1}G_{21}]v_1 \\ &= \mathcal{F}_L(G, H)v_1. \end{aligned}$$

The controller design problem is to find a stabilizing H so that

$$\|z_1\|_2 < \gamma \quad \text{for all} \quad v_1 \text{ with } \quad \|v_1\|_2 \leq 1.$$

In other words, find H so that

$$\|G_{11} + G_{12}H(I - G_{22}H)^{-1}G_{21}\|_\infty < \gamma \tag{8.1}$$

or

$$\|\mathcal{F}_L(G, H)\|_\infty < \gamma.$$

As discussed in Chapter 7, many important control problems fit into this framework.

Definition 8.1: If a controller H stabilizes the generalized plant G so that (8.1) is satisfied, then we say that H provides *attenuation* γ.

In this chapter we develop a state-space-based method for solving this problem and obtaining a controller. Several frequency-domain methods, minimum-distance and interpolation, are discussed briefly in Chapter 9.

8.1 RICCATI EQUATIONS AND L_2 GAIN

Consider a linear system,

$$\begin{aligned} \dot{x}(t) &= Ax(t) + Bu(t), \quad x(0) = x_0 \\ y(t) &= Cx(t), \end{aligned} \tag{8.2}$$

and assume that A is Hurwitz. Indicate the transfer function by $G(s)$. Since A is Hurwitz, $G(s) \in M(R\mathcal{H}_\infty)$. The norm $\|G\|_\infty \leq \gamma$ for some γ if and only if

$$\int_0^\infty - \|Cx(\tau)\|^2 + \gamma^2\|u(\tau)\|^2 d\tau \geq 0, \tag{8.3}$$

with initial condition $x(0) = 0$.

Theorem 8.2: Assume that A is Hurwitz. Indicate the system order by n. We have L_2 gain less than γ if and only if there is a continuous nonnegative function $S : R^n \to R^+$ such that $S(0) = 0$ and

$$S(x(0)) + \int_0^t - \|Cx(\tau)\|^2 + \gamma^2\|u(\tau)\|^2 d\tau \geq S(x(t)) \tag{8.4}$$

for all t, u, $x(0)$, and $x(t)$ satisfying (8.2).

□ *Proof:* Suppose that (8.4) holds. Since A is Hurwitz, if $u \in L_2(0, \infty; R^m)$ then $y \in L_2(0, \infty; R^p)$. Taking $x(0) = 0$ and letting $t \to \infty$, we obtain (8.3) as required.

Now, suppose that (8.3) holds and define the "available storage"

$$S_a(x_0) = \sup_{\substack{u \in L_2 \\ T>0 \\ x(0)=x_0}} \int_0^T \|Cx(\tau)\|^2 - \gamma^2\|u(\tau)\|^2 d\tau.$$

Clearly, $S_a(x_0) \geq 0$ since $u \equiv 0$ yields $S_a(x_0) \geq 0$. Also, since A is Hurwitz, $S_a(x_0) < \infty$ for all x_0.

Suppose we take the system from $x(0)$ at time 0 to $x(t)$ at time t, and then extract the available storage. This is clearly a suboptimal procedure and so

$$S_a(x_0) \geq \int_0^t \|Cx(\tau)\|^2 - \gamma^2\|u(\tau)\|^2 d\tau + S_a(x(t)).$$

Rearranging, we obtain (8.4). ∎

Inequality (8.4) is known as a dissipation inequality (DIE). The function S is a generalized energy function, often called a *storage function*. The integrand

$$-\|Cx(t)\|^2 + \gamma^2\|u(t)\|^2$$

is a special case of a *supply rate*.

If S is differentiable, we can differentiate the DIE (8.4) with respect to t and rearrange to obtain

$$\frac{dS}{dt} + \|y(t)\|^2 - \gamma^2\|u(t)\|^2 \leq 0. \tag{8.5}$$

If we indicate the gradient of S with respect to x as the column vector ∇S_x,

$$(\nabla S_x)^*(\dot{x}(t)) + \|y(t)\|^2 - \gamma^2\|u(t)\|^2 \leq 0.$$

Using (8.2),

$$(p)^*(Ax(t) + Bu(t)) + \|y(t)\|^2 - \gamma^2\|u(t)\|^2 \leq 0 \tag{8.6}$$

where $p \in R^n$ denotes ∇S_x. If whenever the DIE holds the storage function S can be chosen differentiable, then the preceding differentiable form of the DIE (DDIE) is equivalent to the DIE.

Theorem 8.3: Assume that for all x there exists p such that

$$p^*Ax + \frac{1}{4\gamma^2}p^*BB^*p + x^*C^*Cx \leq 0. \tag{8.7}$$

Then the L_2 gain of (8.2) is less than γ.

□ *Proof:* Define

$$H(x, p, u) = p^*(Ax + Bu) + x^*C^*Cx - \gamma^2 u^* u.$$

This function H is the left-hand side of the DDIE. The function H is quadratic in u and has a unique maximum at

$$u_*(x, p) = \frac{1}{2\gamma^2} B^* p.$$

This maximum is

$$\begin{aligned} H_*(x, p) &= H(x, p, u^*) \\ &= p^* Ax + \frac{1}{4\gamma^2} p^* BB^* p + x^* C^* Cx \end{aligned} \tag{8.8}$$

and

$$H(x, p, u) = H_*(x, p) - \gamma^2 \|u - u_*(x, p)\|^2. \tag{8.9}$$

Inequality (8.7) can be written $H_*(x, p) \leq 0$ for all x.

If (8.7) holds, then from (8.9)

$$H(x, p, u) \leq 0$$

and the DDIE (8.6) is satisfied. The DIE (8.4) follows, and hence the L_2 gain is less than γ. ■

The following theorem will be used in proof of the next theorem on the relationship between algebraic Riccati equations and L_2 gain. It displays that there is a close connection between the solution to an ARE and the eigenspaces of an associated matrix called the *Hamiltonian*.

THEOREM 8.4: Let Q and R be real symmetric matrices and let A be a real matrix. Suppose that the Hamiltonian matrix

$$Z = \begin{bmatrix} A & R \\ -Q & -A^* \end{bmatrix}$$

has no eigenvalues on the imaginary axis. For matrices X, Y define the matrix

$$\begin{bmatrix} X \\ Y \end{bmatrix}$$

whose columns are the (generalized) eigenvectors associated with the negative real eigenvalues of Z. Also define

$$\mathcal{V} = \text{Range}\begin{bmatrix} X \\ Y \end{bmatrix}.$$

If

$$\mathcal{V} \cap \text{Range}\begin{bmatrix} 0 \\ I \end{bmatrix} = \{0\}, \tag{8.10}$$

then the ARE

$$A^*P + PA + PRP + Q = 0 \tag{8.11}$$

has a solution $P = YX^{-1}$ with the property that $A + RP$ is Hurwitz.

Conversely, if the ARE has a stabilizing solution, then (8.10) holds and the Hamiltonian matrix has no imaginary axis eigenvalues.

If a stabilizing solution P does exist, P is real and symmetric, and it is the only solution with the property that $A + RP$ is Hurwitz.

□ *Proof:* The key step in this result is to rewrite the ARE in the equivalent form

$$\begin{bmatrix} A & R \\ -Q & -A^* \end{bmatrix}\begin{bmatrix} I \\ P \end{bmatrix} = \begin{bmatrix} I \\ P \end{bmatrix}(A + RP). \tag{8.12}$$

Thus, P is a solution of the ARE if and only if

$$\text{Range}\begin{bmatrix} I \\ P \end{bmatrix}$$

is an n-dimensional invariant subspace of Z. Also, $A + RP$ characterizes the restriction of Z to this subspace.

First assume that Z has no imaginary axis eigenvalues and that (8.10) holds. Since the eigenvalues of Z are symmetric about the imaginary axis (Lemma B.4), this means that Z has an n-dimensional invariant subspace $\mathcal{V}$ on which the restriction of Z has all eigenvalues with negative real parts. The statement (8.10) implies that X is nonsingular and so

$$\mathcal{V} = \text{Range}\begin{bmatrix} I \\ YX^{-1} \end{bmatrix}.$$

Define $P = YX^{-1}$. Then (8.12) implies that P solves the ARE. Since $\mathcal{V}$ is the invariant subspace of Z associated with its negative real eigenvalues, (8.12) implies that $A + RP$ is Hurwitz.

Now, assume that a solution P to the ARE exists such that $A + RP$ is Hurwitz. Then, from (8.12),

$$\text{Range}\begin{bmatrix} I \\ P \end{bmatrix}$$

is an n-dimensional invariant subspace of Z. Also, $A + RP$ characterizes the restriction of Z to this subspace. Thus, Z has an n-dimensional eigenspace associated with eigenvalues with negative real values. Since the spectrum of Z is symmetric about the imaginary axis, this implies that Z has n eigenvalues with positive real parts, n with negative real parts, and thus no imaginary eigenvalues. Clearly,

$$\text{Range}\begin{bmatrix} I \\ P \end{bmatrix} \cap \text{Range}\begin{bmatrix} 0 \\ I \end{bmatrix} = \{0\}.$$

It only remains to show that a stabilizing solution P is real, symmetric, and unique. We first show that P is real. The columns of

$$\begin{bmatrix} X \\ Y \end{bmatrix}$$

can be chosen complex conjugate in pair so that if X_c, Y_c are the complex conjugates of X and Y, respectively, then

$$\begin{bmatrix} X_c \\ Y_c \end{bmatrix} M = \begin{bmatrix} X \\ Y \end{bmatrix}$$

where M is a permutation matrix. Then

$$P = YX^{-1} = Y_c M M^{-1} X_c^{-1} = Y_c X_c^{-1} = P_c$$

and so $P = P_c$. The matrix P is real.

To show that P is symmetric, multiply (8.12) on the left by $[I \quad P^*]J$ where

$$J = \begin{bmatrix} 0 & -I \\ I & 0 \end{bmatrix}.$$

We obtain

$$[I \quad P^*]JZ\begin{bmatrix} I \\ P \end{bmatrix} = \begin{bmatrix} P^* & -P \end{bmatrix}(A + RP).$$

We take transposes and note that $JZ = Z^*J^*$,

$$(P^* - P)(A + RP) + (A + RP)^*(P^* - P) = 0.$$

By a result in linear algebra (Theorem B.5), the fact that $A + RP$ is Hurwitz implies that the only solution is $P - P^* = 0$. Thus, P is symmetric.

Uniqueness follows by a similar argument. Suppose that an ARE has two stabilizing solutions, P_1 and P_2. Then

$$A^* P_1 + P_1 A + P_1 R P_1 + Q = 0$$
$$A^* P_2 + P_2 A + P_2 R P_2 + Q = 0.$$

Subtracting the second equation from the first and rearranging,

$$(P_1 - P_2)(A + RP_1) + (A + RP_2)^*(P_1 - P_2) = 0.$$

Again, since $A + RP_1$ and $A + RP_2$ are Hurwitz, $P_1 - P_2 = 0$, proving that the stabilizing solution is unique. ■

This theorem is useful from a computational viewpoint, since it shows that the solution to an ARE can be found by calculating the eigenvectors of the associated matrix Z.

THEOREM 8.5: Consider the system (8.2). The following are equivalent.

1. A is Hurwitz and the L_2 gain of the system is strictly less than γ.
2. A is Hurwitz and the Hamiltonian matrix

$$Z = \begin{bmatrix} A & \frac{1}{\gamma^2} BB^* \\ -C^*C & -A^* \end{bmatrix}$$

   has no eigenvalues on the imaginary axis.
3. The algebraic Riccati equation (ARE)

$$A^*X + XA + C^*C + \frac{1}{\gamma^2} XBB^*X = 0 \tag{8.13}$$

   has a unique symmetric positive semidefinite solution X such that $A + \frac{1}{\gamma^2} BB^*X$ is Hurwitz.
4. There exists a symmetric positive definite matrix $P > X$ such that

$$A^*P + PA + \frac{1}{\gamma^2} PB^*BP + C^*C < 0. \tag{8.14}$$

□ *Proof:* ($1 \leftrightarrow 2$). We showed the equivalence of (8.2) and (8.3) in Chapter 3 (Theorem 3.7).

($2 \rightarrow 3$) Since A is Hurwitz, the pair $(A, \frac{1}{\gamma^2} BB^*)$ is trivially stabilizable and (A, C^*C) is detectable. By Theorem 5.16 there exists a unique symmetric stabilizing solution $X \geq 0$ of the ARE (8.13). We now show an

alternative proof that $X \geq 0$. Since

$$A^*X + XA = \frac{1}{-\gamma^2} X^*B^*BX - C^*C,$$

the quadratic form $V^-(x) = x^*Xx$ satisfies, along trajectories $\dot{x}(t) = Ax(t)$,

$$\frac{dV^-}{dt} = x^*(A^*X + XA)x = -\left\| \frac{1}{\gamma} BXx(t) \right\|^2 - \|Cx(t)\|^2 \leq 0.$$

Therefore, $V^-(x(t)) \leq V^-(x(0))$ for each $x(0), t \geq 0$. Since $\lim_{t\to\infty} V^-(x(t)) = 0$ (A is Hurwitz), it follows that $V(x) \geq 0$ for all x and $X \geq 0$. (If $V(x_0) = 0$ for some x_0, then the trajectory $\dot{x}(t) = Ax(t), x(0) = x_0$ yields $V(x(t)) = 0$ for all $t \geq 0$ and so $Cx(t) = 0, t \geq 0$. Hence, any such x_0 is an unobservable state. Thus, if (A, C) is observable, then X is positive definite.)

(3 → 1) Suppose there is $X \geq 0$ such that

$$A^*X + XA + \frac{1}{\gamma^2} XB^*BX + C^*C = 0.$$

Define W to be a system with state-space realization

$$W(s) = \left[\begin{array}{c|c} A & -B \\ \hline B^*X & \gamma^2 I \end{array}\right].$$

Then (Theorem C.1)

$$W^{-1}(s) = \left[\begin{array}{c|c} A + \frac{1}{\gamma^2} BB^*X & \frac{1}{\gamma^2} B \\ \hline \frac{1}{\gamma^2} B^*X & \frac{1}{\gamma^2} I \end{array}\right].$$

The transfer function W^{-1} has no poles on the imaginary axis and so $W(s)$ has no zeros on the imaginary axis. Adding and subtracting $j\omega X$ to the Riccati equation,

$$-X(j\omega I - A) - (j\omega I - A)^*X + \frac{1}{\gamma^2} XBB^*X + C^*C = 0.$$

Multiply by $B^*[(j\omega I - A)^{-1}]^*$ on the left and $(j\omega I - A)^{-1}B$ on the right to obtain

$$\begin{aligned} &-B^*[(j\omega I - A)^{-1}]^*XB - B^*X(j\omega I - A)^{-1}B \\ &\quad + \frac{1}{\gamma^2} B^*[(j\omega I - A)^{-1}]^*XBB^*X(j\omega I - A)^{-1}B + G(j\omega)^*G(j\omega) = 0. \end{aligned}$$

where $G(j\omega) = C(j\omega I - A)^{-1}B$. Rearranging, we obtain

$$G^*(jw)G(jw) - \gamma^2 I + W^*(jw)\frac{1}{\gamma^2}W(jw) = 0$$

$$G^*(jw)G(jw) = \gamma^2 I - \frac{1}{\gamma^2}W^*(jw)W(jw).$$

Since W has no zeros on the imaginary axis,

$$\sup_w \|G(jw)\|^2 < \gamma^2.$$

($4 \to 1$) If (8.14) holds, define the storage function

$$S(x) = x^* P x.$$

Since $P > 0$, the DDIE holds with strict inequality and it can be argued that the L_2 gain is strictly less than γ.

Alternatively, note that the strict inequality in (8.14) continues to hold if γ is replaced by $\gamma_0 < \gamma$, provided that $\gamma - \gamma_0$ is sufficiently small. The DIE (8.4) then holds with $\gamma_0 < \gamma$, and hence the L_2 gain is strictly less than γ.

($1 \to 4$) Suppose the L_2-gain is strictly less than γ:

$$\|C(sI - A)^{-1}B\| < \gamma.$$

Then for sufficiently small $\varepsilon > 0$ we have

$$\left\| \begin{pmatrix} C \\ \varepsilon I \end{pmatrix} (sI - A)^{-1}B \right\|_\infty < \gamma.$$

This implies that the ARE has a solution $P \geq 0$

$$A^*P + PA + \frac{1}{\gamma^2}PBB^*P + (C^* + \varepsilon I)\begin{bmatrix} C \\ \varepsilon I \end{bmatrix} = 0.$$

Since $(A, [\begin{smallmatrix} C \\ \varepsilon I \end{smallmatrix}])$ is observable, $P > 0$ (see Chapter 5, Exercise 3, or the argument for $2 \to 3$). Also,

$$A^*P + PA + \frac{1}{\gamma^2}PBB^*P + C^*C = -\varepsilon^*\varepsilon < 0$$

as required. ∎

8.2 STANDARD ASSUMPTIONS

We now return to the problem of obtaining a stabilizing controller H so that $\|\mathcal{F}_L(G, H)\|_\infty < \gamma$. Let G have the state-space realization

$$\left[\begin{array}{c|cc} A & B_1 & B_2 \\ \hline C_1 & E_{11} & E_{12} \\ C_2 & E_{21} & E_{22} \end{array}\right].$$

The following standard assumptions will be used in the remainder of this chapter:

(A1a) The matrix E_{12} has full column rank. The assumption that E_{12} has full column rank ensures a nonsingular penalty z_1 on the control v_2.

(A1b) The matrix E_{21} has full row rank. This assumption is dual to (A1a) and it relates to effect of the disturbance on the controller input. It ensures that the effect of the exogenous input v_1 on z_2 is nonsingular.

(A2a) (A, B_2) is stabilizable.

(A2b) (A, C_2) is detectable.

As shown in Chapter 7, assumption (A2) is necessary and sufficient for existence of an internally stabilizing controller.

(A3a) $\begin{bmatrix} A - \lambda I & B_2 \\ C_1 & E_{12} \end{bmatrix}$ has full column rank, $\mathrm{Re}(\lambda) \geq 0$.

(A3b) $\begin{bmatrix} A - \lambda I & B_1 \\ C_2 & E_{21} \end{bmatrix}$ has full row rank for all $\mathrm{Re}(\lambda) \geq 0$.

Assumption (A3) ensures that the closed loop is stabilizable and detectable (Theorem 7.14). Thus, if the closed loop is externally stable, then it is internally stable.

We will now make some additional assumptions solely to obtain simpler formulas. Relaxation of these assumptions involves routine manipulations.

(a) $E_{12}^*[E_{12} \;\; C_1] = [I \;\; 0]$ and

$$\begin{bmatrix} E_{21} \\ B_1 \end{bmatrix} E_{21}^* = \begin{bmatrix} I \\ 0 \end{bmatrix}.$$

As long as E_{12} has full column rank and E_{21} has full row rank (standard assumption A1), a simple transformation will put any system in this form.

(b) We will also assume that $E_{11} = 0$, $E_{22} = 0$. Any system can be transformed into one in this form.

Once the simplifying assumptions have been made, the standard assumptions become the following *standard simplified assumptions*: In addition to $E_{11} = 0$, E_{22}=0, we assume

(S1) $E_{12}^*[E_{12} \;\; C_1] = [I \;\; 0]$ and

$$\begin{bmatrix} E_{21} \\ B_1 \end{bmatrix} E_{21}^* = \begin{bmatrix} I \\ 0 \end{bmatrix}.$$

(S2a) (A, B_2) is stabilizable.
(S2b) (A, C_2) is detectable.
(S3a) (A, C_1) is detectable and
(S3b) (A, B_1) is stabilizable.

8.3 SPECIAL PROBLEMS

In this section we first consider the simplest problem where the full state x and disturbance v_1 are available. This is called the *full information* (FI) problem. The solution to the full information problem is then used to obtain solutions to some other special problems. The solutions to these special problems will be used in the next section to obtain a controller for the general case of output feedback.

8.3.1 Full Information

In a full information problem, the plant has the form

$$\dot{x}(t) = Ax + B_1 v_1 + B_2 v_2 \tag{8.15}$$

$$z_1 = C_1 x + E_{12} v_2 \tag{8.16}$$

$$z_2 = \begin{bmatrix} x \\ v_1 \end{bmatrix}.$$

Indicating plants of this form by G_{FI}, the state-space realization can be written compactly:

$$G_{FI} = \left[\begin{array}{c|cc} A & B_1 & B_2 \\ \hline C_1 & 0 & E_{12} \\ \begin{bmatrix} I \\ 0 \end{bmatrix} & \begin{bmatrix} 0 \\ I \end{bmatrix} & \begin{bmatrix} 0 \\ 0 \end{bmatrix} \end{array}\right].$$

The relevant simplified assumptions for the full information problem are that $E_{12}^*[E_{12}\ \ C_1] = [I\ \ 0], (A, B_2)$ is stabilizable, and (A, C_1) is detectable. The control problem is to find H so $\|\mathcal{F}_L(G_{FI}, H)\|_\infty < \gamma$.

THEOREM 8.6: Assume that $E_{12}^*[E_{12}\ \ C_1] = [I\ \ 0]$, (A, B_2) is stabilizable and (A, C_1) is detectable. If there exists a controller such that $\|\mathcal{F}_L(G, H)\|_\infty < \gamma$, then there exists a static state feedback K so that $\|\mathcal{F}_L(G, K)\|_\infty < \gamma$.

□ *Proof:* Suppose that there exists a suitable controller and let (A_H, B_H, C_H, E_H) be a minimal realization. Writing

$$\bar{E}_H = E_H \begin{bmatrix} I \\ 0 \end{bmatrix}, \quad \bar{B}_H = B_H \begin{bmatrix} I \\ 0 \end{bmatrix}, \quad \tilde{E}_H = E_H \begin{bmatrix} 0 \\ I \end{bmatrix}, \quad \tilde{B}_H = B_H \begin{bmatrix} 0 \\ I \end{bmatrix}$$

we obtain that the closed loop has realization

$$\left[\begin{array}{cc|c} A + B_2\bar{E}_H & B_2C_H & B_1 + B_2\tilde{E}_H \\ \bar{B}_H & A_H & \tilde{B}_H \\ \hline C_1 + E_{12}\bar{E}_H & E_{12}C_H & E_{12}\tilde{E}_H \end{array}\right].$$

To simplify the argument, assume that $E_H = 0$. Let $(A_0, B_0, C_0, 0)$ indicate the state-space realization for the closed loop. Since the closed loop is internally stable, A_0 is Hurwitz. From the theorem on L_2-gain (Theorem 8.5), there exists a self-adjoint positive definite matrix P so

$$A_0^*P + PA_0 + \frac{1}{\gamma^2}PB_0B_0^*P + C_0^*C_0 < 0.$$

Set $S = P^{-1}$. Multiply on the left and on the right by S. Let n be the order of the full-information system and n_H the order of the controller. For all $x \in R^n, x_H \in R^{n_H}$,

$$[x^*\ \ x_H^*]\left[SA_0^* + A_0S + \frac{1}{\gamma^2}B_0B_0^* + SC_0^*C_0S\right]\begin{bmatrix} x \\ x_H \end{bmatrix} < 0. \tag{8.17}$$

Partition

$$S = \begin{bmatrix} S_{11} & S_{21}^* \\ S_{21} & S_{22} \end{bmatrix}.$$

Since (8.17) is true for all

$$\begin{bmatrix} x \\ x_H \end{bmatrix},$$

it is true for any

$$\begin{bmatrix} x \\ 0 \end{bmatrix}, \qquad x \in R^n.$$

Using the simplifying assumption $E_{12}^* C_1 = 0$ and $E_{12}^* E_{12} = I$, we obtain that for all $x \in R^n$,

$$x^* \Bigg[S_{11}A^* + S_{21}^* C_H^* B_2^* + AS_{11} + B_2 C_H S_{21} + \frac{1}{\gamma^2} B_1 B_1^*$$
$$+ S_{11} C_1^* C_1 S_{11} + S_{21}^* C_H^* C_H S_{21} \Bigg] x < 0.$$

Now $S = P^{-1}$ and so S is symmetric positive definite. Thus, S_{11}^{-1} exists. Use this to rearrange the preceding inequality to obtain, defining $Q = (S_{11})^{-1}$,

$$x^* S_{11} \Bigg[A^* Q + Q S_{21}^* C_H^* B_2^* Q + QA + QB_2 C_H S_{21} Q$$
$$+ \frac{1}{\gamma^2} Q B_1 B_1^* Q + C_1^* C_1 + Q S_{21}^* C_H^* C_H S_{21} Q \Bigg] S_{11} x < 0.$$

Define $K = C_H S_{21} Q$ and obtain

$$(A + B_2 K)^* Q + Q(A + B_2 K) + \frac{1}{\gamma^2} Q B_1 B_1^* Q$$
$$+ (C_1 + E_{12} K)^* (C_1 + E_{12} K) < 0.$$

Thus (Theorem 8.5), the system with realization

$$\left[\begin{array}{c|c} A + B_2 K & B_1 \\ \hline C_1 + E_{12} K & 0 \end{array}\right]$$

has L_2 gain less than γ. But this is precisely the full-information system with state feedback $v_2 = Kx$. ■

The foregoing theorem shows that not only can the control be chosen static without loss of generality, it can be chosen to be a state feedback. It is not necessary to include any feedthrough from the disturbance v_1. The next theorem is constructive. A technical lemma, similar to part of Theorem 8.5, is required. The proof is omitted here.

Lemma 8.7: Assume that

$$\begin{bmatrix} A - j\omega I \\ C^*C \end{bmatrix}$$

has full column rank for all real ω. If there is a symmetric positive definite matrix X such that

$$A^*X + XA + XRX + C^*C < 0,$$

then there exists a unique symmetric, positive semidefinite matrix P such that

$$A^*P + PA + PRP + C^*C = 0$$

and $A + RP$ is Hurwitz.

Theorem 8.8: Assume that $E_{12}^*[E_{12} \;\; C_1] = [I \;\; 0]$, (A, B_2) is stabilizable and (A, C_1) is detectable. There exists a stabilizing controller for the full information problem so that $\|\mathcal{F}_L(G_{FI}, H)\|_\infty < \gamma$ if and only if there exists a symmetric solution $P \geq 0$ of the ARE

$$A^*P + PA + P\left(\frac{1}{\gamma^2}B_1B_1^* - B_2B_2^*\right)P + C_1^*C_1 = 0 \qquad (8.18)$$

such that $A + (\frac{1}{\gamma^2}B_1B_1^* - B_2B_2^*)P$ is Hurwitz. If so, one control is

$$v_2 = [-B_2^*P \quad 0]z_2$$

and $A - B_2B_2^*P$ is Hurwitz.

□ *Proof:* First, assume that there exists a symmetric stabilizing solution $P \geq 0$ to the ARE. Then, for $x(t)$ satisfying $\dot{x}(t) = Ax(t) + B_1 v_1(t) + B_2 v_2(t)$,

$$\begin{aligned}
\frac{d}{dt}(x^* P x) &= (x^* A^* + v_1^* B_1^* + v_2^* B_2^*) P x + x^* P (Ax + B_1 v_1 + B_2 v_2) \\
&= x^*(AP + PA)x + v_1^* B_1^* P x + v_2^* B_2^* P X + x^* P B_1 v_1 \\
&\quad + x^* P B_2 v_2 \\
&= x^* \left(-C_1^* C_1 - \frac{1}{\gamma^2} P B_1 B_1^* P + P B_2 B_2^* P \right) x + v_1^* B_1^* P x \cdots \\
&\quad + v_2^* B_2^* P x + x^* P B_1 v_1 + x^* P B_2 v_2 \\
&= -\|C_1 x\|^2 + \|B_2^* P x\|^2 - \frac{1}{\gamma^2} \|B_1^* P x\|^2 + \langle v_2, B_2^* P x \rangle \cdots \\
&\quad + \langle v_1, B_1^* P x \rangle + \langle B_1^* P x, v_1 \rangle + \langle B_2^* P x, v_2 \rangle \\
&= -\|z_1\|^2 + \langle v_2 + B_2^* P x, v_2 + B_2^* P x \rangle + \gamma^2 \|v_1\|^2 \\
&\quad - \gamma^2 \left\langle v_1 - \frac{1}{\gamma^2} B_1^* P x, v_1 - \frac{1}{\gamma^2} B_1^* P x \right\rangle \\
&= -\|z_1(t)\|^2 + \gamma^2 \|v_1(t)\|^2 + \|v_2(t) + B_2^* P x(t)\|^2 \\
&\quad - \gamma^2 \left\| v_1(t) - \frac{1}{\gamma^2} B_1^* P x(t) \right\|^2 .
\end{aligned}$$

Choosing $v_2(t) = -B_2^* P x(t)$, we obtain that

$$\frac{d}{dt}(x^* P x) + \|z_1(t)\|^2 \leq \gamma^2 \|v_1(t)\|^2.$$

Set $x(0) = 0$ and integrate from $t = 0$ to T to obtain

$$x^*(T) P x(T) + \int_0^T \|z_1(t)\|^2 dt \leq \gamma^2 \int_0^T \|v_1(t)\|^2 dt.$$

Since P is positive semidefinite,

$$\int_0^T \|z_1(t)\|^2 dt \leq \gamma^2 \int_0^\infty \|v_1(t)\|^2 dt.$$

Since the right-hand side is independent of T we obtain that for any $v_1 \in L_2$,

$$\int_0^\infty \|z_1(t)\|^2 dt \leq \gamma^2 \int_0^\infty \|v_1(t)\|^2 dt.$$

Thus, the control $v_2 = -B_2^* P x(t)$ yields a closed loop with L_2 gain less than γ.

We now show that $A - B_2 B_2^* P$ is Hurwitz. With the control $v_2 = -B_2^* P$, the system has state-space equations

$$\dot{x}(t) = (A - B_2 B_2^* P)x(t) + B_1 v(t)$$
$$z_1(t) = C_1 x(t) + E_{12} v_2(t).$$

The simplifying assumptions on E_{12} imply that the L_2 norm of the output z_1 is equal to the L_2 norm of

$$z_1 = \begin{bmatrix} C_1 x \\ v_2 \end{bmatrix},$$

and so we consider the equivalent system

$$\dot{x}(t) = (A - B_2 B_2^* P)x(t) + B_1 v_1(t)$$
$$z_1(t) = \begin{bmatrix} C_1 x \\ E_{12} v_2 \end{bmatrix}.$$

The output z_1 is in L_2 for all disturbances v_1 in L_2, and so $C_1 x(t)$ is in L_2 for all v_1 in L_2. Lemmas 5.14 and 5.15 then imply that $A - B_2 B_2^* P$ is Hurwitz.

Rearranging the ARE (8.18), and defining $K = B_2^* P$, we obtain

$$(A - B_2 K)^* P + P(A - B_2 K) + \frac{1}{\gamma^2} P B_1 B_1^* P$$
$$+ (C_1 + E_{12} K)^* (C_1 + E_{12} K) = 0.$$

Since $A - B_2 K$ is Hurwitz, Theorem 8.5 implies that the L_2 gain of the closed loop is strictly less than γ.

Now assume that there is a controller so that the closed loop has L_2 gain strictly less than γ. By the previous theorem, we can assume without loss of generality the form

$$v_2 = Kx$$

for some matrix K. By Theorem 8.5 this implies the existence of a symmetric matrix $Q > 0$ where

$$(A + B_2 K)^* Q + Q(A + B_2 K) + \frac{1}{\gamma^2} Q B_1 B_1^* Q$$
$$+ (C_1 + E_{12} K)^* (C_1 + E_{12} K) < 0.$$

Rewriting the left-hand side of this inequality and using the simplifying assumptions, we obtain

$$A^*Q + QA + \frac{1}{\gamma^2}QB_1B_1^*Q + (QB_2 + K^*)(QB_2 + K^*)^*$$
$$- QB_2B_2^*Q + C_1^*C_1 < 0,$$

Or,

$$A^*Q + QA + Q\left(\frac{1}{\gamma^2}B_1B_1^* - B_2B_2^*\right)Q + C_1^*C_1 < 0.$$

The existence of a stabilizing solution to the ARE (8.18) now follows from Lemma 8.7. The fact that $A - B_2B_2^*P$ is Hurwitz follows by the same argument used before. ■

Notice that both signals are given by state feedback. Also, although controller input z_2 includes the signal v_1, this information is not needed in the feedback control. The controller derived in this theorem provides the required attenuation for all disturbances v_1.

8.3.2 Full Control

The full control problem is G_{FC} with state-space realization

$$\begin{cases} \dot{x} = Ax + B_1v_1 + [I \quad 0]v_2 \\ z_1 = C_1x + [0 \quad I]v_2 \\ z_2 = C_2x + E_{21}v_1 \end{cases}.$$

This can be written compactly as

$$G_{FC} = \left[\begin{array}{c|cc} A & B_1 & [I \;\; 0] \\ \hline C_1 & 0 & [0 \;\; I] \\ C_2 & E_{21} & [0 \;\; 0] \end{array}\right].$$

THEOREM 8.9: Assume that

$$\begin{bmatrix} E_{21} \\ B_1 \end{bmatrix} E_{21}^* = \begin{bmatrix} I \\ 0 \end{bmatrix},$$

(A, C_2) is detectable and (A, B_1) is stabilizable. There exists a stabilizing solution to the full control problem $\|\mathcal{F}_L(G_{FC}, H)\|_\infty < \gamma$ if and only if there exists a symmetric solution $Q \geq 0$ to the ARE

$$QA^* + AQ + Q\left(\frac{1}{\gamma^2}C_1^*C_1 - C_2^*C_2\right)Q + B_1B_1^* = 0 \tag{8.19}$$

such that $A + Q(\frac{1}{\gamma^2}C_1^*C_1 - C_2C_2^*)$ is Hurwitz. If so, one control is

$$v_2 = \begin{bmatrix} -QC_2^* \\ 0 \end{bmatrix} z_2$$

and $A - QC_2^*C_2$ is Hurwitz.

□ *Proof:* This problem is dual to the full information problem:

$$G_{FC}^* = \left[\begin{array}{c|cc} A^* & C_1^* & C_2^* \\ \hline B_1^* & 0 & E_{21}^* \\ \begin{bmatrix} I \\ 0 \end{bmatrix} & \begin{bmatrix} 0 \\ I \end{bmatrix} & \begin{bmatrix} 0 \\ 0 \end{bmatrix} \end{array}\right].$$

Using the full information result to find H so $\|\mathcal{F}_L(G_{FC}^*, H)\|_\infty < \gamma$, we obtain

$$AQ + QA^* + B_1B_1^* + Q\left(\frac{1}{\gamma^2}C_1^*C_1 - C_2^*C_2\right)Q = 0,$$

and the controller is $[-C_2Q \ \ 0]$. The transpose of this is as before. ■

8.3.3 Output Estimation

The output estimation problem is

$$G_{OE} = \begin{cases} \dot{x} = Ax + B_1v_1 + B_2v_2 \\ z_1 = C_1x + v_2 \\ z_2 = C_2x + E_{21}v_1 \end{cases}.$$

In this problem, finding a controller H so that $\|\mathcal{F}_L(G_{OE}, H)\|_\infty < \gamma$ is equivalent to finding a good estimate v_2 of C_1x. The state space realization of G_{OE} can be written compactly as

$$\left[\begin{array}{c|cc} A & B_1 & B_2 \\ \hline C_1 & 0 & I \\ C_2 & E_{21} & 0 \end{array}\right].$$

The realizations for the full control problem and the output estimation problem are very similar and differ only in B_2 and E_{12}. The two problems are equivalent in that a controller for one problem can be obtained from a controller for the other problem. This is stated precisely and proven in the following theorem.

THEOREM 8.10: Let the full control problem G_{FC} and the output estimation problem G_{OE} be as defined previously. Assume that

$$\begin{bmatrix} E_{21} \\ B_1 \end{bmatrix} E_{21}^* = \begin{bmatrix} I \\ 0 \end{bmatrix},$$

(A, C_2) is detectable, (A, B_1) is stabilizable, and (A, C_1) is detectable.

1. Suppose that H_{OE} internally stabilizes G_{OE}. Then

$$H_{FC} = \begin{bmatrix} B_2 \\ I \end{bmatrix} H_{OE}$$

internally stabilizes G_{FC}. Also,

$$\|\mathcal{F}_L(G_{FC}, H_{FC})\|_\infty = \|\mathcal{F}_L(G_{OE}, H_{OE})\|_\infty.$$

2. Assume that $A - B_2C_1$ is stable. Define

$$P_{OE} = \left[\begin{array}{c|cc} A - B_2C_1 & 0 & [I \ \ -B_2] \\ \hline C_1 & 0 & [0 \ \ I] \\ C_2 & I & [0 \ \ 0] \end{array}\right].$$

If H_{FC} internally stabilizes G_{FC}, then

$$H_{OE} = \mathcal{F}_L(P_{OE}, H_{FC})$$

internally stabilizes G_{OE}. Also,

$$\|\mathcal{F}_L(G_{FC}, H_{FC})\|_\infty = \|\mathcal{F}_L(G_{OE}, H_{OE})\|_\infty.$$

(See Fig. 8.1.)

□ *Proof:* (1) The solution for the full control problem can be obtained from the output estimation problem by defining

$$\tilde{v}_2 = \begin{bmatrix} B_2 \\ I \end{bmatrix} v_2.$$

The proof of (1) is thus a straightforward substitution.

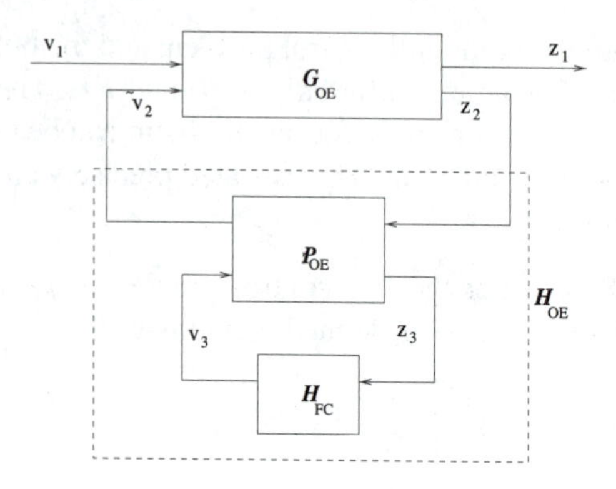

FIGURE 8.1 Equivalence of controller design problems for output estimation and full control.

(2) Let (A_H, B_H, C_H, E_H) indicate a minimal realization for the controller H_{FC}. The state-space equations for G_{OE} with controller $\mathcal{F}_L(P_{OE}, H_{FC})$ are

$$\begin{aligned}
\dot{x} &= Ax + B_1 v_1 + B_2 v_2 \\
z_1 &= C_1 x + v_2 \\
z_2 &= C_2 x + E_{21} v_1 \\
\dot{x}_p &= (A - B_2 C_1) x_p + [I \quad -B_2] v_3 \\
v_2 &= C_1 x_p + [0 \quad I] v_3 \\
z_3 &= C_2 x_p + z_2
\end{aligned}$$

with the controller equations

$$\begin{aligned}
\dot{x}_H &= A x_H + B_H z_3 \\
v_3 &= C_H x_H + E_H z_3.
\end{aligned}$$

Eliminating the intermediate signals z_2, v_2 we obtain

$$\begin{bmatrix} \dot{x} \\ \dot{x}_p \end{bmatrix} = \begin{bmatrix} A & B_2 C_1 \\ 0 & A - B_2 C_1 \end{bmatrix} \begin{bmatrix} x \\ x_p \end{bmatrix} + \begin{bmatrix} B_1 \\ 0 \end{bmatrix} v_1 + \begin{bmatrix} 0 & B_2 \\ I & -B_2 \end{bmatrix} v_3$$

$$z_1 = [C_1 \quad C_1] \begin{bmatrix} x \\ x_p \end{bmatrix} + [0 \quad I] v_3$$

$$z_3 = [C_2 \quad C_2] \begin{bmatrix} x \\ x_p \end{bmatrix} + E_{21} v_1.$$

Use the state transformation

$$\tilde{x} = \begin{bmatrix} I & I \\ 0 & I \end{bmatrix} \begin{bmatrix} x \\ x_p \end{bmatrix}$$

to obtain

$$\frac{d\tilde{x}}{dt} = \begin{bmatrix} A & 0 \\ 0 & A - B_2 C_1 \end{bmatrix} \tilde{x} + \begin{bmatrix} B_1 \\ 0 \end{bmatrix} v_1 + \begin{bmatrix} I & 0 \\ I & -B_2 \end{bmatrix} v_3$$

$$z_1 = [C_1 \quad 0]\tilde{x} + [0 \quad I]v_3$$

$$z_3 = [C_2 \quad 0]\tilde{x} + E_{21} v_1.$$

Denote this system by $\tilde{G}$. We have shown so far that the connection of the controller $H_{OE} = \mathcal{F}_L(P_{OE}, H_{FC})$ with G_{OE} is equivalent to the connection of H_{FC} with $\tilde{G}$. That is,

$$\mathcal{F}_L(G_{OE}, H_{OE}) = \mathcal{F}_L(\tilde{G}, H_{FC}).$$

The second set of states in $\tilde{G}$ is unobservable and so $\tilde{G}$ has the equivalent realization

$$\left[\begin{array}{c|cc} A & B_1 & [I \quad 0] \\ \hline C_1 & 0 & [0 \quad I] \\ C_2 & E_{21} & [0 \quad 0] \end{array}\right].$$

This is G_{FC}. Thus,

$$\mathcal{F}_L(G_{OE}, H_{OE}) = \mathcal{F}_L(G_{FC}, H_{FC}).$$

Assume now that $A - B_2 C_1$ is Hurwitz. Then the detectability of (A, C_1) and the stabilizability of (A, B_1) imply that $\tilde{G}$ satisfies assumption (A3). Thus, the closed loop

$$\mathcal{F}_L(\tilde{G}, H_{FC}) = \mathcal{F}_L(G_{OE}, H_{OE})$$

is internally stable if and only if it is externally stable.

The conclusion follows. ■

The following corollary follows immediately from Theorems 8.9 and 8.10.

Corollary 8.11: Assume that

$$\begin{bmatrix} E_{21} \\ B_1 \end{bmatrix} E_{21}^* = \begin{bmatrix} I \\ 0 \end{bmatrix},$$

(A, C_2) is detectable and (A, B_1) is stabilizable. Also assume that $A - B_2 C_1$ is Hurwitz.

There exists a stabilizing solution to the output estimation problem if and only if there exists a symmetric solution $Q \geq 0$ to the ARE (8.19) such that $A - Q(C_2C_2^* - \frac{1}{\gamma^2}C_1^*C_1)$ is Hurwitz. If so, defining $F = QC_2^*$, one controller is

$$H_{OE} = \left[\begin{array}{c|c} A - B_2C_1 - FC_2 & -F \\ \hline C_1 & 0 \end{array}\right].$$

8.4 OUTPUT FEEDBACK

We now use the solutions to the special problems established in the previous section to calculate a controller H that internally stabilizes G and solves the problem $\|\mathcal{F}_L(G, H)\|_\infty < \gamma$. In order to simplify the formulas and the proofs, we assume that the problem has been scaled so $\gamma = 1$. We are trying to find a stabilizing controller such that

$$\|\mathcal{F}_L(G, H)\|_\infty < 1.$$

Consider a generalized plant G satisfying the standard simplified assumptions (S1)–(S3) with state-space equations

$$\begin{aligned} \dot{x} &= Ax + B_1v_1 + B_2v_2 \\ z_1 &= C_1x + E_{12}v_2 \\ z_2 &= C_2x + E_{21}v_1. \end{aligned}$$

We will separate this into a full control problem and an output estimation problem. The problem will first be redefined so that $G = \mathcal{F}_L(G_i, \tilde{G})$ for some G_i where $\tilde{G}$ is an output estimation problem. We will then show that solving the output estimation problem $\|\mathcal{F}_L(\tilde{G}, H)\|_\infty < 1$ is equivalent to solving $\|\mathcal{F}_L(G, H)\|_\infty < 1$. Calculation of the controller formula will then follow.

We first redefine the problem so $G = \mathcal{F}_L(G_i, \tilde{G})$. If we have full information, then the control $-B_2^*Px$ where P solves the ARE (8.18) provides the required attenuation. This suggests defining the cost for $\tilde{G}$ to be

$$\tilde{v} = B_2^*Px + v_2.$$

Then, if $\|\tilde{v}\|_\infty$ is small, v_2 will be a good estimate of $-B_2^*Px$.

Choose the uncontrolled input $\tilde{z}$ to $\tilde{G}$ to be the difference between v_1 and the "worst" disturbance in the full information case:

$$\tilde{z} = v_1 - B_1^*Px.$$

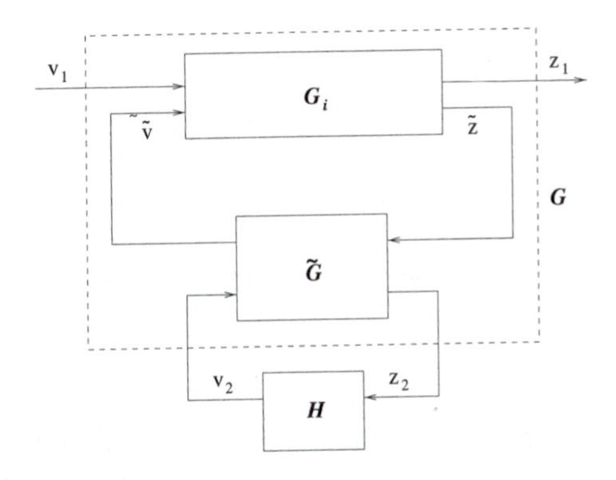

FIGURE 8.2 Output feedback problem.

This construction is shown in Fig. 8.2. These definitions yield

$$G_i \begin{cases} \dot{x} = (A - B_2 B_2^* P)x + B_1 v_1 + B_2 \tilde{v} \\ z_1 = (C_1 - E_{12} B_2^* P)x + E_{12} \tilde{v} \\ \tilde{z} = -B_1^* P x + v_1, \end{cases}$$

and, defining $\tilde{A} = A + B_1 B_1^* P$,

$$\tilde{G} \begin{cases} \dot{x} = \tilde{A}x + B_1 \tilde{z} + B_2 v_2 \\ \tilde{v} = B_2^* P x + v_2 \\ z_2 = C_2 x + E_{21} \tilde{z} \end{cases} .$$

Note that by construction, $G = \mathcal{F}_L(G_i, \tilde{G})$. Also, finding a controller H so that

$$\|\mathcal{F}_L(\tilde{G}, H)\|_\infty < 1$$

is an output estimation problem. Since v_2 will then be a good estimate of $-B_2^* P x$, the controller H should also provide

$$\|\mathcal{F}_L(G, H)\|_\infty < 1.$$

The following theorem is a formal proof of this heuristic argument. It relies on Redheffer's Lemma (Theorem 7.18). We first show that the problem is now in a form where Redheffer's Lemma can be applied.

Lemma 8.12: Define as before

$$G_i = \left[\begin{array}{c|cc} A - B_2B_2^*P & B_1 & B_2 \\ \hline C_1 - E_{12}B_2^*P & 0 & E_{12} \\ -B_1^*P & I & 0 \end{array}\right].$$

If $-B_2^*P$ solves the full information problem with attenuation 1, then the transfer function matrix $G_i \in M(R\mathcal{H}_\infty)$, G_i is inner, and $(G_i)_{21}^{-1} \in M(R\mathcal{H}_\infty)$.

□ *Proof:* It was shown previously (Theorem 8.8) that $A - B_2B_2^*P$ is Hurwitz. It can easily be verified that conditions (1)–(2) in Theorem 7.17 are satisfied. Hence, G_i is inner. Also,

$$(G_i)_{21} = \left[\begin{array}{c|c} A - B_2B_2^*P & B_1 \\ \hline -B_1^*P & I \end{array}\right]$$

and $(G_i)_{21}^{-1}$ has state matrix $A - B_2B_2^*P + B_1B_1^*P$ (Theorem C.1). By assumption, this matrix is Hurwitz (Theorem 8.8). ■

Lemma 8.13: Assume that the full information problem is solvable with attenuation 1. The standard assumption (A3) for G implies that the closed loop of a controller H with $\tilde{G}$ is stabilizable and detectable.

□ *Proof:* We assume as usual that the realization of H is stabilizable and detectable. Now,

$$\begin{bmatrix} \tilde{A} - \lambda I & B_1 \\ C_2 & E_{21} \end{bmatrix} = \begin{bmatrix} A - \lambda I & B_1 \\ C_2 & E_{21} \end{bmatrix} \begin{bmatrix} I & 0 \\ B_1^*P & I \end{bmatrix}.$$

Assumption (A3b) implies that the matrix on the left-hand side has full row rank for all $\text{Re}(s) \geq 0$.

We also want to show that

$$\begin{bmatrix} \tilde{A} - \lambda I & B_2 \\ B_2^*P & I \end{bmatrix}$$

has full column rank for all $\text{Re}(s) \geq 0$. By Theorem 7.15, this would be implied by the fact that $\tilde{A} - B_2B_2^*P$ is Hurwitz. But this follows from the assumption and Theorem 8.8. ■

The following key theorem proves that the output feedback problem can be approached by solving the full information problem, followed by an output estimation problem.

THEOREM 8.14: Assume that full information problem is solvable (with attenuation 1). Then H stabilizes G with attenuation 1 if and only if H stabilizes $\tilde{G}$ with attenuation 1.

□ *Proof:* The previous lemma implies that internal stability is equivalent to external stability for each closed loop. Thus we only need to consider external stability. Theorem 7.17 and Lemma 8.12 show that the assumptions of Redheffer's Lemma (Theorem 7.18) are satisfied. By Redheffer's Lemma, $\|\mathcal{F}_L(G, H)\|_\infty < 1$ if and only if $\|\mathcal{F}_L(\tilde{G}, H)\|_\infty < 1$. (Set $Q = \mathcal{F}_L(\tilde{G}, H)$ in Redheffer's Lemma.) ■

This is a key result. It states that an output feedback controller can be designed by separately designing a full-information (or state-feedback) controller and an estimator. The following theorem is constructive, and it shows that an output feedback controller can be designed by solving two AREs.

THEOREM 8.15: Assume that the standard simplified assumptions (S1)–(S3) hold. There exists a controller H such that the closed loop $\mathcal{F}_L(G, H) \in M(R\mathcal{H}_\infty)$ and

$$\|\mathcal{F}_L(G, H)\|_\infty < 1$$

if and only if the following two conditions are satisfied:

1. The algebraic Riccati equation

$$A^*P + PA + P(B_1B_1^* - B_2B_2^*)P + C_1^*C_1 = 0 \qquad (8.20)$$

has a symmetric positive semidefinite solution P with $A + (B_1B_1^* - B_2B_2^*)P$ Hurwitz.

2. Define $\tilde{A} = A + B_1B_1^*P$ and $K = B_2^*P$. The algebraic Riccati equation

$$\tilde{A}\tilde{Q} + \tilde{Q}\tilde{A}^* + \tilde{Q}(K^*K - C_2^*C_2)\tilde{Q} + B_1B_1^* = 0 \qquad (8.21)$$

has a symmetric positive semidefinite solution $\tilde{Q}$ such that $\tilde{A} + \tilde{Q}(K^*K - C_2^*C_2)$ is Hurwitz.

Moreover, if the above conditions are satisfied, then defining $\tilde{F} = \tilde{Q}C_2^*$, one controller that provides attenuation 1 is

$$H = \left[\begin{array}{c|c} \tilde{A} - B_2K - \tilde{F}C_2 & -\tilde{F} \\ \hline K & 0 \end{array}\right]. \qquad (8.22)$$

□ *Proof:* First, suppose that a stabilizing controller H exists so that

$$\|\mathcal{F}_L(G, H)\|_\infty < 1.$$

Then the controller $H[C_2 \;\; E_{21}]$ solves the full information problem. It follows that condition (1) is necessary. We will complete the proof by showing that if condition (1) holds, condition (2) is both necessary and sufficient.

Assume then that (1) holds. With this assumption, Theorem 8.14 shows that solution of the output feedback problem is equivalent to solving the output estimation problem $\|\mathcal{F}_L(\tilde{G}, H)\|_\infty < 1$. Now, by assumption $(A + B_1 B_1^* P) - B_2 B_2^* P$ is Hurwitz and so (Theorem 8.10) this output estimation problem is equivalent to a full control problem. The conclusion now follows immediately from Corollary 8.11, as does the controller formula. ■

As for the linear quadratic regulator/estimator design in Chapter 5, the controller H has the same order as the plant G, and calculation of a state-space formula for the controller requires solving two Riccati equations.

A major difference is that the solution $\tilde{Q}$ to the estimation ARE (8.21) involves the solution P to the ARE (8.20) for the full information problem. (This is reasonable since the estimation problem was to find an estimate of the full information control.) This coupling is undesirable, since errors in the solution of (8.20) will affect the solution to (8.21). This next theorem shows that we can instead solve two independent AREs plus an additional third condition involving the size of the solutions to the AREs.

THEOREM 8.16: Assume that the standard simplified assumptions (S1)–(S3) hold. There exists a stabilizing controller H such that $\|\mathcal{F}_L(G, H)\|_\infty < 1$ if and only if the following three conditions all hold:

(i) The algebraic Riccati equation

$$A^* P + PA + P(B_1 B_1^* - B_2 B_2^*)P + C_1^* C_1 = 0 \qquad (8.23)$$

has a symmetric positive semidefinite solution P with $A + (B_1 B_1^* - B_2 B_2^*)P$ Hurwitz;

(ii) The algebraic Riccati equation

$$AQ + QA^* + Q(C_1^* C_1 - C_2^* C_2)Q + B_1 B_1^* = 0 \qquad (8.24)$$

has a symmetric positive semidefinite solution Q with $A + Q(C_1^* C_1 - C_2^* C_2)$ Hurwitz; and

(iii) $r(PQ) < 1$ where $r(PQ) = \max_i |\lambda_i(PQ)|$.

Moreover, if these conditions are satisfied, then defining $\tilde{F} = (I - QP)^{-1}QC_2^*$, and K as in Theorem 8.15 one stabilizing controller is (8.22) as before.

□ *Proof:* Comparing this theorem to Theorem 8.15 we see that it is only required to show that conditions (ii) and (iii) are equivalent to condition (2) in Theorem 8.15. The key step is to relate the Hamiltonian matrix for each ARE. Recalling that $K = B_2^* P$, define the Hamiltonian matrix associated with (8.21),

$$\tilde{Z} = \begin{bmatrix} \tilde{A}^* & K^*K - C_2^*C_2 \\ -B_1^*B_1 & -\tilde{A} \end{bmatrix},$$

and the Hamiltonian matrix associated with (8.24);

$$Z = \begin{bmatrix} A^* & C_1^*C_1 - C_2^*C_2 \\ -B_1^*B_1 & -A \end{bmatrix}.$$

Defining

$$T = \begin{bmatrix} I & -P \\ 0 & I \end{bmatrix},$$

$$Z = T^{-1}\tilde{Z}T.$$

Since the two matrices are similar, they have the same eigenspaces.

First, assume that (i)–(iii) hold. We need to show that (2) is satisfied. Let $\mathcal{V}_-$ be the invariant subspace on which the restriction of Z has all eigenvalues with negative real parts. By Theorem 8.4, we can write

$$\mathcal{V}_- = \text{Range}\begin{bmatrix} I \\ Q \end{bmatrix}.$$

In other words,

$$Z\left(\text{Range}\begin{bmatrix} I \\ Q \end{bmatrix}\right) \subset \text{Range}\begin{bmatrix} I \\ Q \end{bmatrix}.$$

Thus,

$$\tilde{Z}T\left(\text{Range}\begin{bmatrix} I \\ Q \end{bmatrix}\right) \subset T\ \text{Range}\begin{bmatrix} I \\ Q \end{bmatrix},$$

and the subspace

$$\text{Range}\begin{bmatrix} I & -P \\ 0 & I \end{bmatrix}\begin{bmatrix} I \\ Q \end{bmatrix} = \text{Range}\begin{bmatrix} I - PQ \\ Q \end{bmatrix}$$

is $\tilde{Z}$-invariant. It is also complementary to

$$\text{Range}\begin{bmatrix} 0 \\ I \end{bmatrix},$$

since (iii) implies that $I - PQ$ is invertible. It follows from Theorem 8.4 that $\tilde{Q} = Q(I - PQ)^{-1}$ solves the ARE (8.21). But $(I - QP)^{-1}Q = Q(I - PQ)^{-1}$ (Lemma B.8). Also, since Z and $\tilde{Z}$ are similar, $A + R\tilde{Q}$ is Hurwitz. Thus, (2) holds and from Theorem 8.15, (i)–(iii) are sufficient for the existence of a stabilizing controller providing attenuation 1.

We now show that (i)–(iii) are necessary. It was shown in the first part of the proof of Theorem 8.15 that (i) is necessary. The necessity of (ii) follows by a duality argument. It only remains to show that (iii) is necessary. Assume that the control problem is solvable. Then (i) and (ii) hold, as does (2). Thus, there exists a positive semidefinite stabilizing solution $\tilde{Q}$ to the ARE (8.21) and a positive semidefinite stabilizing solution Q to the ARE (8.24). By the preceding argument, $I - PQ$ is non-singular and

$$\tilde{Q} = Q(I - PQ)^{-1} = (I - QP)^{-1}Q.$$

By Lemma B.8, the fact that $Q \geq 0$ and $\tilde{Q} \geq 0$ implies that $r(PQ) = r(QP) < 1$.

The controller formula follows. ∎

Conditions (i)–(iii) in the preceding theorem (or (1) and (2) in Theorem 8.15) are necessary and sufficient. If any of the three conditions (i)–(iii) fail to hold, there is no stabilizing feedback controller H such that

$$\|\mathcal{F}_L(G, H)\|_\infty < 1.$$

Consider the optimal control problem

$$\inf_{H \in \mathcal{S}(G)} \|\mathcal{F}_L(G, H)\|_\infty.$$

One way to solve this is to attempt to solve

$$\left\| \mathcal{F}_L\left(\frac{1}{\gamma}G, H\right) \right\|_\infty < 1$$

for decreasing values of γ. The infinum γ_* is the smallest value of γ for which the problem $\|\mathcal{F}_L(\frac{1}{\gamma}G, H)\|_\infty < 1$ is solvable. Unfortunately, this approach is numerically sensitive, since the Riccati equations are difficult to solve accurately for γ close to optimal. The other approaches discussed in Chapter 9 may be preferable.

8.5 DISCUSSION OF ASSUMPTIONS

As mentioned previously, assumptions (a) and (b) are merely simplifying assumptions. Any problem that satisfies all the assumptions except (a) and (b) can be transformed into a problem that satisfies the simplifying assumptions. The only difference is that the formulas for the controller are more complicated. We give here a theorem corresponding to Theorem 8.16 for which the simplifying assumptions are weakened.

THEOREM 8.17: Assume that assumptions (A1)–(A3) hold, that $E_{11} = 0$, $E_{22} = 0$, $E_{12}^* E_{12} = I$, and $E_{21} E_{21}^* = I$. Define $A_c = A - B_2 E_{12}^* C_1$ and $A_f = A - B_1 E_{21}^* C_2$. There exists a stabilizing controller H such that $\|\mathcal{F}_L(G, H)\|_\infty < 1$ if and only if the following three conditions all hold:

(i) The algebraic Riccati equation

$$A_c^* P + P A_c + P(B_1 B_1^* - B_2 B_2^*)P + C_1^*(I - E_{12} E_{12}^*)C_1 = 0$$

has a self-adjoint positive semidefinite solution P with $A_c + (B_1 B_1^* - B_2 B_2^*)P$ Hurwitz;

(ii) The algebraic Riccati equation

$$A_f Q + Q A_f^* + Q(C_1^* C_1 - C_2^* C_2)Q + B_1(I - E_{21}^* E_{21})B_1^* = 0$$

has a self-adjoint positive semidefinite solution Q with $A_f + Q(C_1^* C_1 - C_2^* C_2)$ Hurwitz; and

(iii) $r(PQ) < 1$.

Moreover, if the listed conditions are satisfied, then if we define $K = B_2^* P + E_{12}^* C_1$, $\tilde{F} = (I - QP)^{-1}(QC_2^* + B_1 E_{21}^*)$, one stabilizing controller H is

$$H = \left[\begin{array}{c|c} A_c + B_1 B_1^* P - B_2 B_2^* P - \tilde{F}(C_2 + E_{21} B_1^* P) & -\tilde{F} \\ \hline K & 0 \end{array} \right].$$

The assumption that $E_{22} = 0$ can easily be removed (see exercises). The assumption $E_{11} = 0$ can also be removed, but the calculations and resulting controller formula are lengthy.

8.5.1 Assumption (A1)

The first assumption is essentially that the penalty of the control effort is nonsingular.

For the common mixed sensitivity problem where the design objective is to find a controller so that

$$\left\| \begin{matrix} W_1 S \\ W_3 T \end{matrix} \right\|_\infty < 1,$$

we have the generalized plant

$$z_1 = \begin{bmatrix} W_1 \\ 0 \end{bmatrix} v_1 + \begin{bmatrix} -W_1 P \\ W_3 P \end{bmatrix} v_2.$$

If P is strictly proper and W_1, W_3 are proper, then

$$E_{12} = 0.$$

There are a number of ways to redefine the problem so that the control cost is nonsingular.

1. Change the plant slightly to

$$P + \epsilon I$$

where ϵ is some small nonzero number. If W_3 is nonstrictly proper, then E_{12} will have full column rank. This is not a common solution.

2. If we explicitly weight the control cost by some weight W_2, the problem becomes

$$\left\| \begin{matrix} W_1 S \\ W_2 H S \\ W_3 T \end{matrix} \right\|_\infty < 1.$$

If W_2 is nonstrictly proper, then E_{12} has full column rank. A constant function, for instance,

$$W_2 = 0.01,$$

is a common choice. Earlier, we showed that

$$\|W_2 H S\|_\infty < 1$$

was a criterion for robust stability with respect to additive uncertainty. It can also be regarded as a penalty on the control effort. If $W_2(j\omega)$ is large where input $r(j\omega)$ is large, then $\|W_2 H S\|_\infty$ small will ensure that the controller outputs are kept small. This is important when there are limits on the actuator inputs, as is common.

3. We can choose an improper weight W_3, as long as $W_3 P$ is proper. If $W_3 P$ is nonstrictly proper, then E_{12} will have full column rank. Recall that for large ω, strictly proper P and proper H,

$$T \approx PH.$$

Therefore (omitting the dependence of the transfer functions on s), $|W_3 T| < 1$ implies that $|PC| < |W_3^{-1}|$ or

$$|H| < |(PW_3)^{-1}|.$$

Ensuring that $W_3 P$ is nonstrictly proper means that there is a penalty on the controller response $H(j\omega)$ for large ω.

Assumption (A1) also includes an assumption that E_{21} has full row rank. This is dual to the assumption on E_{12}. It ensures that the effect of the disturbance on the controller input is non-singular.

In the standard feedback system (Fig. 3.2), we usually include a reference input or a plant output disturbance r so

$$z_2 = r - Pv_2 = v_1 - Pv_2,$$

and it follows that

$$E_{21} = I.$$

Thus, the fourth assumption is satisfied for all plants, weights, and choices of z_1. However, if the uncontrolled input consists only of a plant input disturbance d, then

$$z_2 = -P(d + v_2) = -Pv_1 - Pv_2.$$

The matrix $E_{21} = 0$ if P is strictly proper. If a plant output disturbance r is included, then

$$v_1 = \begin{bmatrix} r \\ d \end{bmatrix}$$

and

$$z_2 = [I \ -P] \begin{bmatrix} r \\ d \end{bmatrix} + [-P]v_2.$$

Now, E_{21} has full row rank and the disturbance weighting is nonsingular.

8.5.2 Standard Assumption (A2)

Standard assumption (A2) on the stabilizability and detectability (A, B_2, C_2) is necessary and sufficient for the existence of an internally stabilizing controller.

8.5.3 Standard Assumption (A3)

Standard assumption (A3) is sufficient for stabilizability and detectability of the closed loop. With this assumption we are assured that a controller H that provides

$$\|\mathcal{F}_L(G, H)\|_\infty < \gamma$$

also internally stabilizes the plant. Assumption (A3) can be weakened to an assumption that the relevant matrices have full rank on the imaginary axis. This weaker condition is necessary for the existence of stabilizing solutions to the Riccati equations.

Suppose our cost is

$$\left\| \begin{matrix} W_1 S \\ W_2 H S \\ W_3 T \end{matrix} \right\|_\infty$$

and we have the standard control configuration. What happens if A has poles on the imaginary axis? For instance, consider the case where our original plant P has a pole at the origin. If $W_2 = \delta \cdot I$ and P, W_1 are strictly proper, we have the generalized plant

$$A = \begin{bmatrix} A_P & 0 & 0 \\ B_{w1} C_P & A_{w1} & 0 \\ B_{w3} C_P & 0 & A_{w3} \end{bmatrix}$$

$$B_1 = \begin{bmatrix} 0 \\ -B_{w1} \\ 0 \end{bmatrix}, \; B_2 = \begin{bmatrix} B_P \\ 0 \\ 0 \end{bmatrix}$$

$$C_1 = \begin{bmatrix} 0 & -C_{w1} & 0 \\ E_{w3} C_P & 0 & C_{w3} \\ 0 & 0 & 0 \end{bmatrix}, \quad C_2 = [-C_P \quad 0 \quad 0]$$

$$E_{11} = 0, \; E_{12} = \begin{bmatrix} 0 \\ 0 \\ \delta \cdot I \end{bmatrix}, \; E_{21} = I, \;\; E_{22} = 0.$$

For this problem,

$$E_{12}^*[C_1 \quad E_{12}] = [0 \quad 0 \quad 0 \quad \delta^2].$$

Assumptions (A1) and (A2) are satisfied. Assumption (A3a) is implied by detectability of (A_P, C_P) and the usual construction that A_{w1} and A_{w2} are Hurwitz.

Now consider the remaining assumption (A3b). For this problem (A3b) is that the matrix

$$\begin{bmatrix} A_P - \lambda I & 0 & 0 & 0 \\ -B_1 C_P & A_{w1} - \lambda I & 0 & B_{w1} \\ B_3 C_P & 0 & A_{w3} - \lambda I & 0 \\ -C_P & 0 & 0 & I \end{bmatrix}$$

has full row rank for all $\mathrm{Re}(\lambda) \geq 0$. It is necessary that this matrix have full row rank for all imaginary λ in order for the related Riccati equation to have a stabilizing solution. However, for $\lambda = 0$, A_P loses rank and even the weakened assumption is not satisfied.

There are a number of ways to cope with this. One is to shift the poles of the plant to the right by a small amount $\epsilon > 0$, by replacing A_P by $A_P + \epsilon I$. After a controller is designed with the modified plant, the matrix A_H is replaced by $A_H - \epsilon I$ and implemented with the original system. Another method is to replace only the imaginary axis pole(s) $j\omega_i$ by $-\epsilon + j\omega_i$. It must be verified that the controller satisfies the design criteria with the original plant. (This is a wise practice in all design problems.) It is also possible to redefine the controller design problem, so that all the standard assumptions are satisfied for the generalized plant.

8.6 EXAMPLE

In this example we consider control of a double spring–mass system. (See Fig. 8.3.) This problem has application to the design of active suspension systems. The fixed spring with stiffness k_1 is a model for a tire, while the second spring and mass can be regarded as the passive suspension and the vehicle mass. The system has the same mathematical model as Example 2.2, with the exception that we now include a force on the first mass. Letting x_1 indicate the deviation from equilibrium position of the first mass and x_2 the deviation from equilibrium of the second, we obtain

$$\begin{aligned} m_1 \ddot{x}_1(t) &= -k_1 x_1(t) + k_2(x_2(t) - x_1(t)) - d_1 \dot{x}_1(t) + d_2(\dot{x}_2(t) \\ &\quad - \dot{x}_1(t)) + f_1(t) \\ m_2 \ddot{x}_2(t) &= -k_2(x_2(t) - x_1(t)) - d_2(\dot{x}_2(t) - \dot{x}_1(t)) + f_2(t) \end{aligned}$$

where $f_1(t)$ is external forces applied to the mass m_1 and $f_2(t)$ is external forces applied to the mass m_2. Values of the spring–mass system parameters for a car suspension model are given in Table 8.1. To write in standard first-order state-space form, define the state vector $\mathbf{x}$ of length

TABLE 8.1 Plant parameters

Parameter	Value
m_1	36 kg
m_2	240 kg
k_1	160,000 N/m
k_2	16,000 N/m
d_1	100 Ns/m
d_2	980 Ns/m

4 as $\mathbf{x}^* = [x_1, \dot{x}_1, x_2, \dot{x}_2]^*$. Rewriting the two second-order differential equations just given, we get

$$\dot{\mathbf{x}}(t) = A_P \mathbf{x}(t) + B_{P1} f_1(t) + B_{P2} f_2(t),$$

where

$$A_P = \begin{bmatrix} 0 & 1 & 0 & 0 \\ -\frac{k_1+k_2}{m_1} & -\frac{d_1+d_2}{m_1} & \frac{k_2}{m_1} & \frac{d_2}{m_1} \\ 0 & 0 & 0 & 1 \\ \frac{k_2}{m_2} & \frac{d_2}{m_2} & -\frac{k_2}{m_2} & -\frac{d_2}{m_2} \end{bmatrix}, \; B_{P1} = \begin{bmatrix} 0 \\ \frac{1}{m_1} \\ 0 \\ 0 \end{bmatrix}, \; B_{P2} = \begin{bmatrix} 0 \\ 0 \\ 0 \\ \frac{1}{m_2} \end{bmatrix}.$$

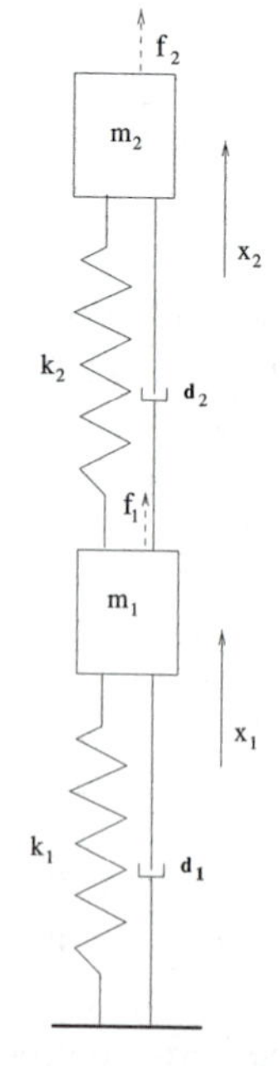

FIGURE 8.3 Double spring–mass system.

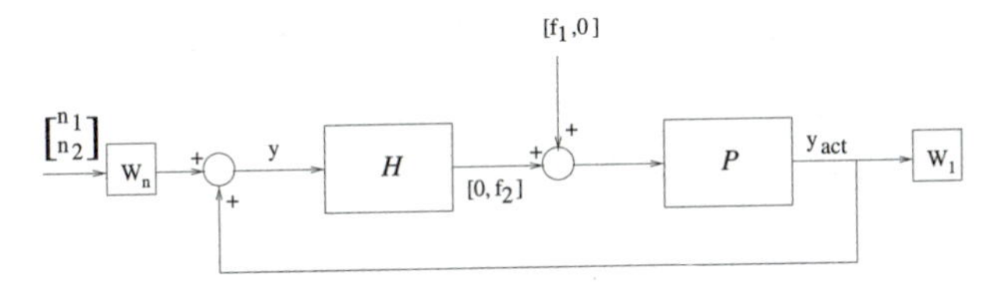

FIGURE 8.4 Block diagram for control of double spring–mass system.

We are measuring the position of each mass, but these measurements are corrupted by sensor noise,

$$y(t) = C_P \mathbf{x}(t) + \begin{bmatrix} \tilde{n}_1(t) \\ \tilde{n}_2(t) \end{bmatrix},$$

where

$$C_P = \begin{bmatrix} 1 & 0 & 0 & 0 \\ 0 & 0 & 1 & 0 \end{bmatrix}.$$

The first sensor noise is described by $W_{n1}n_1$ where n_1 is white noise and W_{n1} has realization $(A_{n1}, B_{n1}, C_{n1}, E_{n1})$, and the second sensor noise is described similarly by W_{n2} where W_{n2} has realization $(A_{n2}, B_{n2}, C_{n2}, E_{n2})$. The state matrices A_{n1} and A_{n2} are Hurwitz. The controller will apply a force f_2 to the second mass. The aim of the controller is to reject the effect of a disturbance f_1 on the position of the first mass. We will attempt to reject disturbances of frequencies up to the natural frequency of the first mass. We need to limit the controller effort for practical reasons, and to include some robustness in the design to account for modeling and parameter error. Before listing the system parameters and precisely formulating the controller objectives, we will write this problem in the generalized plant framework. Figure 8.4 is a block diagram describing the control problem. Note that we have used the positive feedback convention here.

Let P_1 indicate the transfer function from f_1 to the 2×1 position vector y and P_2 the transfer function from f_2 to y. Let W_1 indicate the weight on the position x_1. The part of the cost due to restricting the position of x_1 is

$$[0 \;\; W_1]P_1 f_1 + [0 \;\; W_1]P_2 f_2.$$

Let W_2 indicate the weight on the controller output (due to restricting controller gain and/or additive uncertainty). The part of the cost due to restricting the controller output is

$$W_2 f_2.$$

The uncontrolled input is

$$v_1 = \begin{bmatrix} f_1 \\ n_1 \\ n_2 \end{bmatrix},$$

whereas the controlled input is

$$v_2 = f_2.$$

Thus, we can write the cost z_1 in terms of controlled input v_2 and uncontrolled input v_1 as

$$z_1 = \begin{bmatrix} [0 \ \ W_1]P_1 & 0 & 0 \\ 0 & 0 & 0 \end{bmatrix} v_1 + \begin{bmatrix} [0 \ \ W_1]P_2 \\ W_2 \end{bmatrix} v_2.$$

The controller input z_2 is the measured plant output y,

$$\begin{aligned} z_2 &= y_{act} + \begin{bmatrix} \tilde{n}_1 \\ \tilde{n}_2 \end{bmatrix} \\ &= \begin{bmatrix} P_1 & \begin{matrix} W_{n1} & 0 \\ 0 & W_{n2} \end{matrix} \end{bmatrix} v_1 + [P_2]\, v_2. \end{aligned}$$

Thus, we have the generalized plant

$$G_{11} = \begin{bmatrix} [0 \ \ W_1]P_1 & 0 & 0 \\ 0 & 0 & 0 \end{bmatrix}, \quad G_{12} = \begin{bmatrix} [0 \ \ W_1]P_2 \\ W_2 \end{bmatrix}$$

$$G_{21} = \begin{bmatrix} P_1 & \begin{matrix} W_{n1} & 0 \\ 0 & W_{n2} \end{matrix} \end{bmatrix}, \quad G_{22} = P_2.$$

We now derive a state-space realization for the generalized plant. Let $C_{P1} = [1 \ \ 0 \ \ 0 \ \ 0]$, $C_{P2} = [0 \ \ 0 \ \ 1 \ \ 0]$, so that

$$C_P = \begin{bmatrix} C_{P1} \\ C_{P2} \end{bmatrix}.$$

The transfer function $[0 \ \ I]P_1$ has realization $(A_P, B_{P1}, C_{P2}, 0)$ and $[0 \ \ I]P_2$ has realization $(A_P, B_{P2}, C_{P2}, 0)$. The weight on performance W_1 is in $\mathcal{RH}_\infty$, and since we are only attempt to reject low-frequency disturbances, W_1 is chosen strictly proper. Let $(A_{W1}, B_{W1}, C_{W1}, 0)$ be a minimal realization of W_1. Since we do not have a precise description of the weighting on the control cost, and we wish to penalize the controller gain at high frequencies, let W_2 be a constant: $W_2(s) = \delta > 0$.

A state-space realization of $[0 \;\; W_1]P_1$ is (Theorem. C.1)

$$\left[\begin{array}{cc|c} A_{W1} & B_{W1}C_{P2} & 0 \\ 0 & A_P & B_{P1} \\ \hline C_{W1} & 0 & 0 \end{array}\right].$$

The realization of $[0 \;\; W_1]P_2$ will be the same, except that B_{P1} is replaced by B_{P2}. Thus,

$$G_{11} = \left[\begin{array}{cc|ccc} A_{W1} & B_{W1}C_{P2} & 0 & 0 & 0 \\ 0 & A_P & B_{P1} & 0 & 0 \\ \hline C_{W1} & 0 & 0 & 0 & 0 \\ 0 & 0 & 0 & 0 & 0 \end{array}\right],$$

$$G_{12} = \left[\begin{array}{cc|c} A_{W1} & B_{W1}C_{P2} & 0 \\ 0 & A_P & B_{P2} \\ \hline C_{W1} & 0 & 0 \\ 0 & 0 & \delta \end{array}\right].$$

Also,

$$G_{21} = \left[\begin{array}{ccc|ccc} A_P & 0 & 0 & B_{P1} & 0 & 0 \\ 0 & A_{N1} & 0 & 0 & B_{N1} & 0 \\ 0 & 0 & A_{N2} & 0 & 0 & B_{N2} \\ \hline C_{P1} & C_{N1} & 0 & 0 & E_{N1} & 0 \\ C_{P2} & 0 & C_{N2} & 0 & 0 & E_{N2} \end{array}\right],$$

$$G_{22} = \left[\begin{array}{c|c} A_P & B_{P2} \\ \hline C_{P1} & 0 \\ C_{P2} & 0 \end{array}\right].$$

We need a composite state that includes all the subsystem states. Define

$$A = \begin{bmatrix} A_{W1} & B_{W1}C_{P2} & 0 & 0 \\ 0 & A_P & 0 & 0 \\ 0 & 0 & A_{N1} & 0 \\ 0 & 0 & 0 & A_{N2} \end{bmatrix}.$$

We obtain

$$G = \left[\begin{array}{cccc|ccc:c} A_{W1} & B_{W1}C_{P2} & 0 & 0 & 0 & 0 & 0 & 0 \\ 0 & A_P & 0 & 0 & B_{P1} & 0 & 0 & B_{P2} \\ 0 & 0 & A_{N1} & 0 & 0 & B_{N1} & 0 & 0 \\ 0 & 0 & 0 & A_{N2} & 0 & 0 & B_{N2} & 0 \\ \hline C_{W1} & 0 & 0 & 0 & 0 & 0 & 0 & 0 \\ 0 & 0 & 0 & 0 & 0 & 0 & 0 & \delta \\ \cdots & \cdots & \cdots & \cdots & \cdots & \cdots & \cdots & \cdots \\ 0 & C_{P1} & C_{N1} & 0 & 0 & E_{N1} & 0 & 0 \\ 0 & C_{P2} & 0 & C_{N2} & 0 & 0 & E_{N2} & 0 \end{array}\right].$$

We now check that the standard assumptions are satisfied. First, check the simplified assumptions:

$$E_{12}^*[E_{12} \quad C_1] = \begin{bmatrix} 0 \\ \delta \end{bmatrix}^* \begin{bmatrix} 0 & C_{W1} & 0 & 0 & 0 \\ \delta & 0 & 0 & 0 & 0 \end{bmatrix}$$
$$= \begin{bmatrix} \delta^2 & 0 & 0 & 0 & 0 \end{bmatrix}.$$

So, E_{12} has full column rank and E_{12} and C_1 are orthogonal. Also,

$$\begin{bmatrix} E_{21} \\ B_1 \end{bmatrix} E_{21}^* = \begin{bmatrix} 0 & E_{N1} & 0 \\ 0 & 0 & E_{N2} \\ B_{P1} & 0 & 0 \\ 0 & B_{N1} & 0 \\ 0 & 0 & B_{N2} \end{bmatrix} \begin{bmatrix} 0 & E_{N1} & 0 \\ 0 & 0 & E_{N2} \end{bmatrix}^*$$

$$= \begin{bmatrix} E_{N1}E_{N1}^* & 0 \\ 0 & E_{N2}E_{N2}^* \\ 0 & 0 \\ B_{N1}E_{N1}^* & 0 \\ 0 & B_{N2}E_{N2}^* \end{bmatrix}.$$

As long as E_{N1} and E_{N2} have full row rank, standard assumption (A1) is satisfied, but the simplifying orthogonality assumption on B_1 and E_{21} does not hold in general.

Since all the subsystems are stable, the state matrix is Hurwitz. Assumption (A2) on the stabilizability and detectability of (A, B_2, C_2) follows trivially; verifying (A3) is straightforward. Since $E_{11} = 0$, $E_{22} = 0$,

Theorem 8.17 can be used to calculate the controller after normalization of E_{12} and E_{21}.

The parameter values in Table 8.1 yield a natural frequency for the first mass of approximately

$$\sqrt{\frac{k_1}{m_1}} = 67.$$

In order to reject disturbances up to this frequency, choose W_1 to be a first-order system with a pole beyond this,

$$W_1(s) = \frac{\gamma}{1 + s/100},$$

with a minimal realization

$$W_1 = \left[\begin{array}{c|c} -100 & 100 \\ \hline \gamma & 0 \end{array}\right].$$

The value of γ will be increased as much as possible, in order to improve disturbance rejection. We set

$$W_2(s) = 0.1.$$

The sensor noise weights have minimal realizations

$$W_{n1} = \left[\begin{array}{c|c} -200 & 1 \\ \hline -10 & .2 \end{array}\right], \quad W_{n2} = \left[\begin{array}{c|c} -200 & 3 \\ \hline -3 & .1 \end{array}\right].$$

Using the formulas, we can now, for each γ in the performance weight W_1, attempt to calculate a controller H so that

$$\|\mathcal{F}_L(G, H)\|_\infty < 1.$$

This will yield a closed loop that rejects low frequency disturbances and also has some robustness. For large γ, this problem will not be solvable. Table 8.2 shows the result of iteration on γ as produced by the MATLAB routine *hinfopt*. The final controller, with $\gamma = 30.25$, has a realization

$$\left[\begin{array}{ccccccc|cc}
-289.54 & 0.73125 & -0.10144 & 1.957 & 3.2416 & 1.3453 & -13.544 & 2.1011 & 0.0325 \\
0.7506 & -372.66 & 45.361 & -50.327 & -1295.5 & 21.162 & 5583.1 & 0.1167 & 1.1551 \\
2.9095 & 6.9311 & -142.57 & 7.3124 & 135.08 & -3.5359 & -584.2 & 0.4774 & 7.3853 \\
-138.9 & -2.9485 & 6.8692 & -356.23 & -34.559 & 126.52 & 185.58 & -27.92 & -0.1340 \\
-44.244 & 57.033 & -7.2778 & -100.19 & -282.52 & -11.026 & 901.32 & -8.9198 & -0.1044 \\
-511.51 & 43.796 & 29.709 & -1274.9 & -18.871 & -353.51 & -30.057 & -103.19 & 0.0127 \\
5.5302 & -0.234 & 29.449 & 13.564 & -90.521 & 9.2202 & -355.59 & 1.0958 & 1.3503 \\
\hline
-0.0086 & 28.238 & -4.8068 & 4.1198 & 99.963 & -1.7063 & -430.42 & 0 & 0
\end{array}\right].$$

TABLE 8.2 Controller design for suspension system

		≪ H-Infinity Optical Control Synthesis ≫						
No	Gamma	D11 <= 1	P-Exist	P >= 0	Q-Exist	Q >= 0	lam(PQ) < 1	C.L.
1	1.0000e+00	OK	OK	OK	OK	OK	OK	STAB
2	2.0000e+00	OK	OK	OK	OK	OK	OK	STAB
3	4.0000e+00	OK	OK	OK	OK	OK	OK	STAB
4	8.0000e+00	OK	OK	OK	OK	OK	OK	STAB
5	1.6000e+01	OK	OK	OK	OK	OK	OK	STAB
6	3.2000e+01	OK	OK	OK	FAIL	OK	OK	STAB
7	2.4000e+01	OK	OK	OK	OK	OK	OK	STAB
8	2.8000e+01	OK	OK	OK	OK	OK	OK	STAB
9	3.0000e+01	OK	OK	OK	OK	OK	OK	STAB
10	3.1000e+01	OK	OK	OK	FAIL	OK	OK	STAB
11	3.0500e+01	OK	OK	OK	FAIL	OK	OK	STAB
12	3.0250e+01	OK	OK	OK	OK	OK	OK	STAB

Iteration no. 12 is your best answer under the tolerance: 0.0100.

The generalized closed loop is shown in Fig. 8.5. Notice that the maximum gain is close to 1 as it should be. Comparisions of various aspects of open- and closed-loop behavior are shown in Figs. 8.6–8.8. There is a significant reduction in the gain of the map from the disturbance force

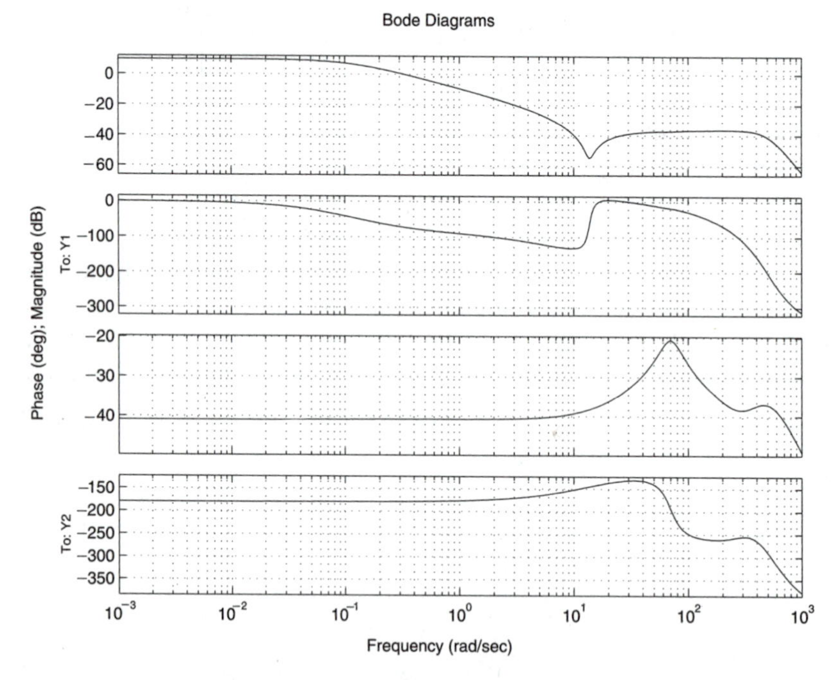

FIGURE 8.5 Bode diagram for generalized closed loop.

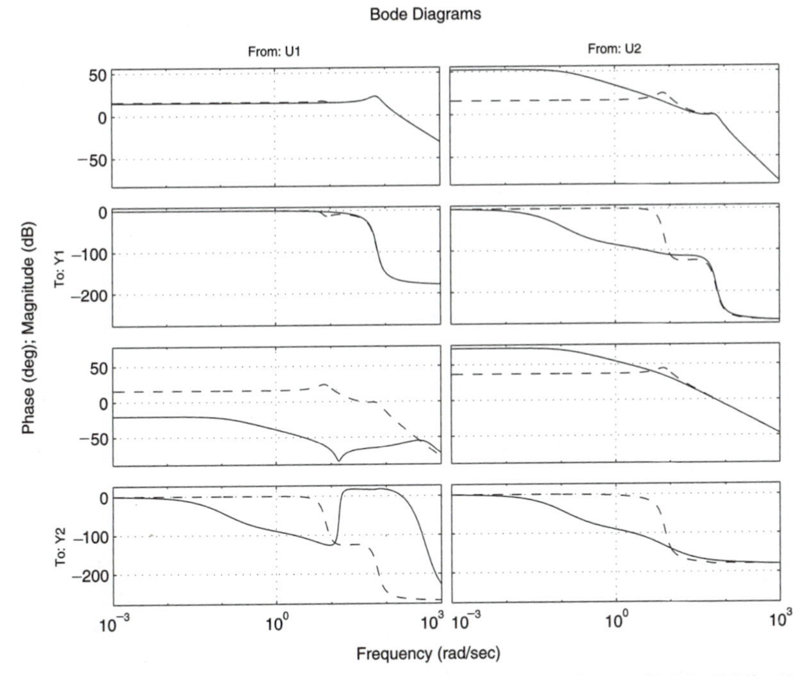

FIGURE 8.6 Bode diagrams for uncontrolled (dashes) and controlled (solid lines) position transfer functions.

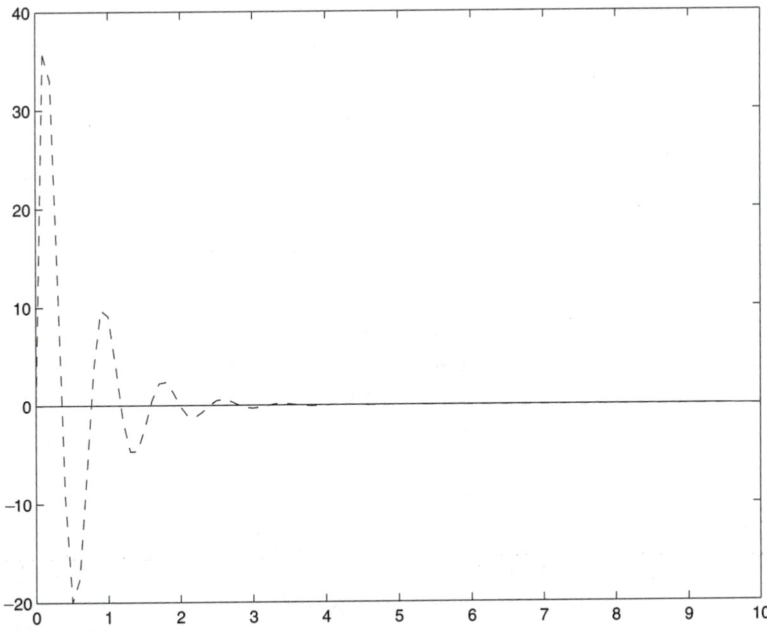

FIGURE 8.7 Open (dashes) and closed loop (solid line) responses of $x_2(t)$ to an impulse f_1.

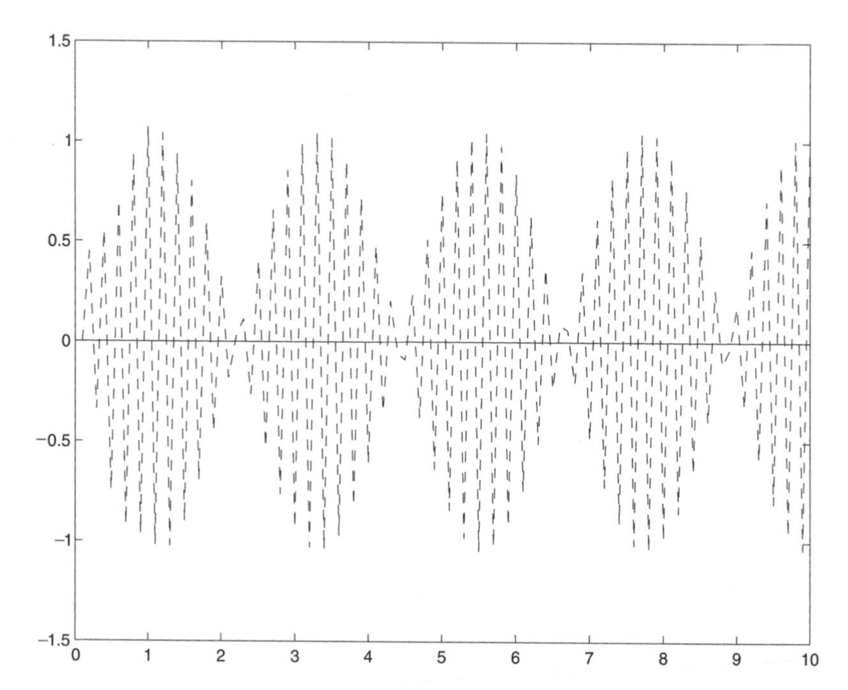

FIGURE 8.8 Open (dashes) and closed loop responses (solid line) of $x_2(t)$ to $f_1(t) = \sin(30t)$.

f_1 to the position of the second mass x_2. There is little improvement in the other maps, and in some cases the closed-loop behavior is worse than the open-loop behavior. This is not surprising, since the controller design objective was only to reduce the map from f_1 to x_2.

NOTES AND REFERENCES

The result on state feedback (Theorem 8.6) relies on [24]. The proof for the full information problem (Theorem 8.8) is based on the energy arguments in [27]. Algebraic proofs may be found in [9, 53]. A proof of Lemma 8.7 can be found in, for example, [27]. Doyle, Glover, Khargonekar, and Francis [14] first obtained a solution to the $\mathcal{H}_\infty$ problem in terms of the solution to two Riccati equations. This is given here as Theorem 8.16. Details on how to rewrite a problem for which the simplifying assumptions do not hold to one for which the simplifying assumptions do hold, and the resulting controller formula, can be found in [9, 53]. MATLAB routines in the "robust control" toolbox will solve the controller design problem for a plant that satisfies the standard assumptions (A1)–(A3), returning a suitable controller for the original problem if one exists. The singular problem for which the first standard assumption does not hold is covered in [41]. The parameters for the suspension system controller design are from [6]. A detailed design example is in [29] and several are given in [21]. The idle speed control design (exercise 4) is based on [45]. Exercise 2 is concerned with the connection between $\mathcal{H}_2$ (or linear quadratic) control and $\mathcal{H}_\infty$ control. This is covered in a number of the texts just cited.

One drawback of the controller design method in this chapter is that the controller is of order equal to the generalized plant. Satisfactory performance can usually be obtained with a controller of lower order. Common order reduction method are balanced truncations, Hankel norm approximation, and approximation of the controller coprime factors. Care needs to be taken so that closed-loop stability and performance is maintained with the reduced-order controller. Details can be found in, for example [21, 53].

EXERCISES

1. The full information problem can be restated as find a $v_2 \in L_2(0, \infty; R^m)$ where m is the number of columns in B_2 so that

$$\begin{aligned} J_1 &= \int_0^\infty z_1^*(t)z_1(t) - \gamma^2 v_1(t)^* v_1(t)\,dt \\ &= \int_0^\infty \begin{bmatrix} x^* & v_2 \end{bmatrix} \begin{bmatrix} C_1^*C_1 & 0 \\ 0 & I \end{bmatrix} \begin{bmatrix} x \\ v_2 \end{bmatrix} - \gamma^2 v_1(t)^* v_1(t) dt \\ &< 0. \end{aligned}$$

The optimal solution of this problem requires finding the controller v_2 that stabilizes the system and minimizes J_1 for the worst disturbance v_1. Thus, we wish to solve

$$\inf_{v_2 \in L_2} \sup_{v_1 \in L_2} J_1.$$

Show that the disturbance with the "worst" effect on the L_2 norm of the cost z_1 is

$$v_1^{worst}(t) = +\frac{1}{\gamma^2} B_1^* P x(t)$$

and the "best" control v_2 is

$$v_2^{best}(t) = -B_2^* P x(t).$$

(Hint: Read the proof of Theorem 8.8 carefully.)

2. For some problems, only a specific disturbance input with Laplace transform W_1 needs to be considered. Using the generalized plant framework, the cost

$$\begin{aligned} z_1 &= [G_{11} + G_{12}H(I - G_{22}H)^{-1}G_{21}]W_1 \\ &= [\tilde{G}_{11} + G_{12}H(I - \tilde{G}_{22}H)^{-1}\tilde{G}_{21}] \end{aligned}$$

where $\tilde{G}_{ij} = G_{ij}W_1$. Let $\|z_1\|_2$ indicate the $\mathcal{H}_2$-norm of z_1. (This is equal to the L_2-norm of the inverse Laplace transform of z_1 in the time domain.) We want a stable closed and also a $\mathcal{H}_2$ norm that is as small as possible. One example where this occurs is in reducing the reponse in a duct to a particular acoustic disturbance (Example 7.3). The controller design problem is to find a stabilizing controller H so that

$$\|\mathcal{F}_L(\tilde{G}, H)\|_2$$

is as small as possible. Consider a generalized plant that satisfies all the simplified assumptions of this chapter. Assume that it is a full information problem.

(a) Write the controller design objective in the time domain. Show that this is a linear quadratic optimal control problem.

(b) Hence show that the solution to the full-information $\mathcal{H}_2$-design problem can be obtained by solving (8.18) with $\gamma \to \infty$.

3. Redo the controller design for the suspension system done in this chapter, except include the weight W_1 on x_1 and x_2. Compare the closed-loop response of your controller to that in the example.

4. A schematic for a model for a controlled engine is shown in Fig. 8.9. The controller design objective is to reduce the sensitivity of the engine speed to changes in the load (for instance, air conditioning). The controller outputs are a valve setting, $u_1(t)$ and the spark advance timing $u_2(t)$. The external disturbance is indicated by $d(t)$. A simplified linear model of order 4 about the usual operating speed of 800 rpm is

$$\begin{aligned} \dot{x}(t) &= A_p x(t) + B_p u(t) \\ y(t) &= C_p x(t) + d(t) \end{aligned}$$

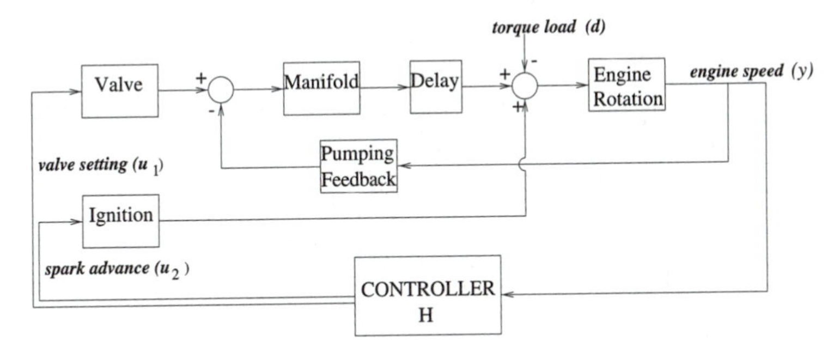

FIGURE 8.9 Block diagram for engine speed control.

where $C_p = [0\ \ 0\ \ 0\ \ 10.6]$,

$$A_p = \begin{bmatrix} -6.25 & 0 & 0 & 0 \\ 0 & -25 & 0 & 0 \\ 6.25 & -733.33 & -2.4 & -5.05 \\ 0 & 0 & 1 & 0 \end{bmatrix}, \quad B_p = \begin{bmatrix} 1 & 0 \\ 0 & 1 \\ 0 & 33.3 \\ 0 & 0 \end{bmatrix}.$$

(a) The controller objectives are as follows: (1) reduce the response of the engine speed to low frequency disturbances $d(t)$ and (2) constrain the controller outputs. The constraint on the controller outputs also improves robustness. (Why?) The engine speed is weighted by W_1 and the control variables u_1 and u_2 are weighted by W_v and W_s, respectively. Calculate the generalized plant G so that the controller design problem is to find the stabilizing controller H where

$$\|\mathcal{F}_L(G, H)\|_\infty < \gamma,$$

where γ is as small as possible. (Hint: First write the problem as a standard mixed sensitivity problem.)

(b) Verify that the problem satisfies the standard assumptions, provided that the weighting functions satisfy some simple conditions.

(c) Using weights

$$W_1(s) = \frac{8.5}{s + 0.00085}, \quad W_v(s) = \frac{1000s + 10}{s + 4000},$$

$$W_s(s) = \frac{0.57s + 2}{s + 0.002},$$

design a controller for the system so that γ is as small as possible. Compare the open- and closed-loop responses to a step disturbance and to a periodic disturbance with period 1 rad/s.

(d) Repeat part (c) for the weights

$$W_1(s) = \frac{17}{s + 0.0017}, \quad W_v(s) = \frac{1000s + 10}{s + 10,000},$$

$$W_s(s) = \frac{0.85s + 30}{s + 0.002}.$$

(e) Compare the responses of the two controlled systems. Also compare the controller outputs.

5. In Exercise 5.7, linear quadratic control was used to design a controller for an inverted pendulum attached to a moving cart. Let H_o indicate the transfer function of the output feedback controller designed in Exercise 5.7.

(a) Calculate the closed loop sensitivity S_o with this controller and use $1/|S_o|$ as a guideline to calculate the performance weight on sensitivity W_1. Calculate the robustness to additive uncertainty of this closed loop, and use this as a guideline to choose the controller output weight W_2.

(b) Use these weights to formulate the controller design problem as an $\mathcal{H}_\infty$-controller design problem, using the linearized model as the nominal plant. Assume that only position and angle can be measured.

(c) Design an $\mathcal{H}_\infty$ controller.

(d) How well does the controller work with the nonlinear model?

(e) Compare the performance of the $\mathcal{H}_\infty$ controller with the controller H_o.

6. In Section 8.5 it was shown that standard assumption (A3) does not hold for the standard mixed sensitivity problem if the plant has imaginary axis poles. Change the problem to include a plant input disturbance so that the uncontrolled input

$$v_1 = \begin{bmatrix} r \\ d \end{bmatrix}.$$

Show that the standard assumptions are satisfied for a new generalized plant if the plant is strictly proper with stabilizable and detectable realization $(A_p, B_p, C_p, 0)$. Also assume that W_1 and

W_3 are stable with minimal realizations. Assume that W_1 is strictly proper and W_3 is non-strictly proper.

7. Consider the flexible beam in Example 5.17. The natural frequencies w_i each may be in error by $\pm 10\%$ and the damping ξ is 0.05 ± 0.01. The objective is to design a controller so that the tip position y should respond to a step reference input with no steady-state error. The closed loop should reject disturbances with frequencies up to the first natural frequency w_1.

(a) First, find a weight W_3 so that the range of plants included in the variation in natural frequency and damping is included in $\mathcal{M}(P, W_3)$. To do this, plot

$$\left| \frac{\tilde{P}(j\omega)}{P(j\omega)} - 1 \right|$$

for a variety of perturbed plants $\tilde{P}$ and find W_3 so that

$$|W_3(j\omega)| > \left| \frac{\tilde{P}(j\omega)}{P(j\omega)} - 1 \right|.$$

(This approach was used in Example 3.55.) To keep the controller order low, W_3 should have order less than 3.

(b) Choose W_1 to be a function that is large up to frequency w_1, and then drops off. In order to keep the controller order low, W_1 should have order less than 3.

(c) Write down a generalized plant G that satisfies all the standard assumptions, and so that a controller that provides attenuation 1 for G should provide satisfactory closed loop performance for the beam. (This not straightforward, since the beam transfer function is strictly proper and has a pole at 0.)

(d) Design the controller, reducing the weighted sensitivity $W_1 S$ as much as possible.

(e) Check whether the controller satisfies all the design objectives.

(f) It is desirable that the closed loop have settling time less than $2s$ and no overshoot during the transient. Is this satisfied?

8. **(a)** Repeat Exercise 7, except do the design using a plant model with only the first frequency (i.e., a model of order 4). Regard the given model with the additional frequencies as a perturbation. This will require a different choice of uncertainty weight W_3. Use the same weight W_1 on sensitivity.

(b) Check whether the controller satisfies all the design objectives.

(c) Implement the controller with the higher order model from Exercise 7. Compare the sensitivity obtained with the two controllers. Compare the other performance criteria.

9. Suppose that a given generalized plant G does not have $E_{22} = 0$. Let H provide attenuation γ for G with E_{22} set equal to zero. Show that $H(I + E_{22}H)^{-1}$ provides attenuation γ for G.

10. Derive a parametrization of all stabilizing controllers for the full information problem. (Hint: Use the controller given in this chapter and Redheffer's Lemma.)

11. Derive a parametrization of all stabilizing controllers for the output feedback problem.

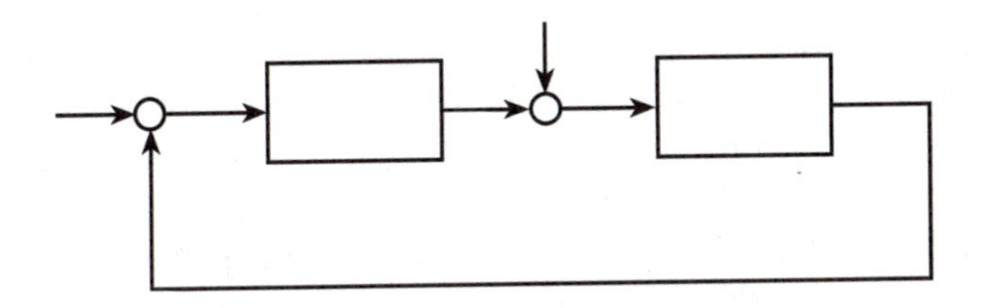

IX
MODEL MATCHING

Consider the model-matching problem

$$\gamma = \inf_{Q \in \mathcal{H}_\infty} \|T_1 - T_2 Q\|_\infty$$

where $T_1, T_2 \in M(R\mathcal{H}_\infty)$. Because of the parametrization of all stabilizing controllers, many control problems can be put in this form. (This was shown in Chapter 6, and also Chapter 7.) The most general form is to minimize

$$\|T_1 - T_2 Q T_3\|_\infty,$$

but this will not be considered in this chapter.

If there is a matrix $T_2^{-R} \in M(R\mathcal{H}_\infty)$ so that $T_2 T_2^{-R} = I$, then obviously with

$$Q = T_2^{-R} T_1 \in M(R\mathcal{H}_\infty),$$

$\gamma = 0$. The problem is find a solution in the more general, and more interesting, case where T_2 is not right-invertible in $R\mathcal{H}_\infty$.

In the previous chapter controllers to minimize the $\mathcal{H}_\infty$ norm of a system were constructed using a state-feedback and -estimation approach. We discuss two different approaches to this problem in this chapter. One is based on viewing the problem as a minimum distance problem. The other approach formulates the model-matching problem as an interpolation problem. For most problems the state-space method covered in the previous chapter is computationally preferable, but there are some situations where the methods from this chapter may be advantageous. Also, the different viewpoints provide additional insight into $\mathcal{H}_\infty$ design.

We will only consider scalar problems where $T_1, T_2 \in R\mathcal{H}_\infty$. References that deal with the generalization to matrix-valued problems are given at the end of the chapter.

9.1 MINIMUM-DISTANCE PROBLEM

Do an inner–outer factorization of T_2 (Section 4.6):

$$T_2 = T_{2i} T_{2o}.$$

For any $Q \in \mathcal{H}_\infty$, we have

$$\begin{aligned}
\|T_1 - T_2 Q\|_\infty &= \|T_1 - T_{2i} T_{2o} Q\|_\infty \\
&= \left\| T_{2i}\left(T_{2i}^{-1} T_1 - T_{2o} Q\right)\right\|_\infty \\
&= \sup_\omega \left| T_{2i}\left(T_{2i}^{-1} T_1 - T_{2o} Q\right)\right| \\
&\quad \text{(the argument } \omega \text{ is not indicated)} \\
&= \sup_\omega \left| T_{2i}^{-1} T_1 - T_{2o} Q\right|.
\end{aligned}$$

Define $R = T_{2i}^{-1} T_1$, and $X = T_{2o} Q$. Assume for now that $T_{2o}^{-1} \in M(R\mathcal{H}_\infty)$. (In other words, T_2 is nonstrictly proper with no imaginary axis zeros.) We will remove this assumption later. Since $Q = T_{2o}^{-1} X$, our problem is

$$\begin{aligned}
\gamma &= \inf_{X \in \mathcal{H}_\infty} \sup_\omega |R(jw) - X(jw)| \\
&= \inf_{X \in \mathcal{H}_\infty} \|R - X\|_{\bar{L}_\infty}
\end{aligned}$$

where

$$\|f\|_{\bar{L}_\infty} = \sup_\omega |f(j\omega)|.$$

Here

$$\bar{L}_\infty = \{g(jw), \sup |g(jw)| < \infty\}.$$

If $R \in R\mathcal{H}_\infty$ (T_2 has no r.h.p. zeros), then obviously $X = R$ will yield $\gamma = 0$. In general, decompose R into its stable and unstable parts:

$$R = \underbrace{R_u}_{\notin \mathcal{H}_\infty} + \underbrace{R_s}_{\in \mathcal{H}_\infty}.$$

Now,

$$\gamma = \inf_{X \in \mathcal{H}_\infty} \|R_u + (R_s - X)\|_{L_\infty}.$$

Only the unstable part of R will affect the infinum of our problem. We want to find the closest stable transfer function to an unstable transfer function. Distance is measured in the L_∞ norm. We would like an optimal solution in $R\mathcal{H}_\infty$, since R is rational. Clearly

$$\inf_{X \in \mathcal{H}_\infty} \|R - X\|_\infty \leq \inf_{X \in R\mathcal{H}_\infty} \|R - X\|_\infty.$$

It will turn out that, because R is rational, the preceding quantities are equal. The solution to this problem as a minimum distance problem requires a result from functional analysis known as the *Nehari Theorem.*

9.1.1 Nehari Theorem

Define

$$\mathcal{H}_2 = \{f(s) \text{ analytic in } \operatorname{Re}(s) > 0; \underbrace{\sup_{x>0} \frac{1}{\sqrt{2\pi}} \int_{-\infty}^{\infty} |f(x + jw)|^2 \, dw^{1/2}}_{\|f\|_2} < \infty\}$$

$$\bar{L}_2 = \{g(jw), \underbrace{\frac{1}{\sqrt{2\pi}} \int_{-\infty}^{\infty} |g(jw)|^2 \, dw^{1/2}}_{\|g\|_2} < \infty\}.$$

The norms on both spaces will be indicated by $\| \ \|_2$. Also, if $f \in \mathcal{H}_2$, for almost all x, the limit

$$\hat{f}(jw) = \lim_{x \to 0} f(x + jw)$$

exists, and $\hat{f} \in \bar{L}_2$. Moreover, $\|\hat{f}\|_2 = \|f\|_2$. Thus, we may identify $\mathcal{H}_2$

with a subspace of $\bar{L}_2$. The space $\mathcal{H}_2$ is the space of Laplace transforms of functions $u \in L_2(0, \infty)$. (See Section 3.1.)

The orthogonal complement $\mathcal{H}_2^\perp$ of $\mathcal{H}_2$ in L_2 is the space

$$\mathcal{H}_2^\perp = \left\{ f \text{ analytic in } \mathrm{Re}(s) < 0 ; \sup_{x<0} \frac{1}{\sqrt{2\pi}} \int_{-\infty}^{\infty} |f(x + jw)|^2 dw^{1/2} < \infty \right\}.$$

We identify $\mathcal{H}_2^\perp$ via its values on the imaginary axis with a subspace of $\bar{L}_2$. The space $\mathcal{H}_2^\perp$ consists of the Laplace transforms of functions $u \in L_2(-\infty, 0)$.

Also, we have a map between functions in $\mathcal{H}_\infty$ and their boundary values in $\bar{L}_\infty$. For $f \in \mathcal{H}_\infty$, the limit

$$\hat{f}(jw) = \lim_{x \to 0} f(x + jw)$$

exists for almost all x and $\hat{f} \in \bar{L}_\infty$. This map is also one-to-one and norm preserving. We can regard $\mathcal{H}_\infty$ as a closed subspace of $\bar{L}_\infty$.

Now, for $f \in \bar{L}_\infty$ define the operator from $\bar{L}_2$ to $\bar{L}_2$ by

$$\Lambda_f g = fg.$$

The operator Λ_f is called a *Laurent operator*, and f is its symbol.

A related operator is the restriction of Λ_f to $\mathcal{H}_2$, denoted $\Lambda_f|_{\mathcal{H}_2}$, which maps $\mathcal{H}_2$ to $\bar{L}_2$.

Let Π_u indicate the projection of $\bar{L}_2$ onto $\mathcal{H}_2^\perp$. The Hankel operator with symbol f is defined as

$$\Gamma_f g = \Pi_u \Lambda_f|_{\mathcal{H}_2}.$$

That is, for $g \in \mathcal{H}_2$,

$$\Gamma_f g = \Pi_u f g.$$

Note that a Hankel operator maps $\mathcal{H}_2$ to $\mathcal{H}_2^\perp$.

Example 9.1:

$$f = \frac{s-1}{(s+2)(s-2)} = \frac{3/4}{(s+2)} + \frac{1/4}{(s-2)}$$

For any $g \in \mathcal{H}_2$,

$$\begin{aligned}
\Gamma_f g &= \Pi_u f_s g + \Pi_u f_u g \\
&= \Pi_u f_u g \\
&= \Pi_u \frac{\left(\frac{1}{4}\right) g}{(s-2)} \\
&= \Pi_u \left[\frac{\left(\frac{1}{4}\right) g(2)}{(s-2)} + \frac{\left(\frac{1}{4}\right)(g(s) - g(2))}{(s-2)} \right] \\
&= \frac{\left(\frac{1}{4}\right) g(2)}{(s-2)}.
\end{aligned}$$

For instance, if $g = 1/(s+1)$,

$$\begin{aligned}
\Gamma_f g &= \Pi_u \frac{1/4}{(s+1)(s-2)} \\
&= \Pi_u \left[\frac{1}{4} \left(\frac{1}{3} \frac{1}{s-2} - \frac{1}{3} \frac{1}{s+1} \right) \right] \\
&= \frac{1}{12} \frac{1}{s-2}.
\end{aligned}$$

Notice that only the unstable part of f affects Γ_f.

THEOREM 9.2: Decompose any $f \in \bar{L}_\infty$ into its stable and unstable parts,

$$f = f_u + f_s,$$

where $f_s(s) \in \mathcal{H}_\infty$ and $f_u(-s) \in \mathcal{H}_\infty$. Then

$$\|\Gamma_f\| = \|f_u\|_{\bar{L}_\infty}.$$

□ *Proof:*

$$\begin{aligned}
\Gamma_f g &= \Pi_u (f g) \\
&= \Pi_u f_u g + \Pi_u f_s g \\
&= \Gamma_{f_u} g \\
&= f_u g.
\end{aligned}$$

■

The Hankel operator can be analyzed in the time domain. Let $(A, B, C, 0)$ be a realization for a "totally unstable" function f,

$$f(s) = C(s - A)^{-1} B,$$

where A has all its eigenvalues in the right half-plane. The inverse bilateral Laplace transform of f is

$$f(t) = \begin{cases} -Ce^{At}B & t < 0 \\ 0 & t \geq 0 \end{cases}.$$

The time domain analog of the Hankel operator maps a function u in $L_2(0, \infty)$ to a function y in $L_2(-\infty, 0)$ defined by

$$\begin{aligned} y(t) &= \int_{-\infty}^{\infty} f(t-\tau)u(\tau)d\tau = \int_0^{\infty} f(t-\tau)u(\tau)d\tau, \quad t < 0 \\ &= -Ce^{At}\int_0^{\infty} e^{-A\tau}Bu(\tau)d\tau, \quad t < 0. \end{aligned}$$

Define the controllability operator $\Psi_c : L_2(0, \infty) \to R$ as

$$\Psi_c u := -\int_0^{\infty} e^{-A\tau}Bu(\tau)d\tau$$

and the observability operator $\Psi_0 : R \to L_2(-\infty, 0)$ as

$$(\Psi_0 x)(t) := Ce^{At}x \quad t < 0.$$

Thus,

$$y = \Psi_0\Psi_c u,$$

and in the time domain,

$$\Gamma_f = \Psi_0\Psi_c.$$

There is another interpretation of the Hankel operator. Let $(A, B, C, 0)$ be a realization for the unstable transfer function f and consider the usual state-space equations

$$\dot{x} = Ax(t) + Bu(t), \quad x(0) = x_0 \tag{9.1}$$

$$y = Cx. \tag{9.2}$$

Apply an input $u \in L_2(0, \infty)$, such that $x(t) \to 0$. Then

$$x(t) = e^{At}x_0 + e^{At}\int_0^t e^{-A\tau}Bu(\tau)d\tau, \quad t \geq 0.$$

Thus, since $x(t) \to 0$,

$$\begin{aligned} x_0 &= -\int_0^{\infty} e^{-A\tau}Bu(\tau)d\tau \\ &= \Psi_c u. \end{aligned}$$

Now solve (9.1) and (9.2) backwards in time starting at $t = 0$. Noting that $u(t) = 0$ for $t < 0$,

$$\begin{aligned} y(t) &= Ce^{At}x_0 \\ &= (\Psi_o x_0)(t), \quad t < 0 \\ &= \Psi_o \Psi_c u. \end{aligned}$$

The Hankel operator can be seen as mapping future inputs to initial state to past output.

Decompose any $R \in \bar{L}_\infty$ into stable and unstable parts:

$$R = \underbrace{R_u}_{\notin R\mathcal{H}_\infty} + \underbrace{R_s}_{\in R\mathcal{H}_\infty}.$$

The function R_u is strictly proper and analytic in Re $s \leq 0$. Thus, R_u has a minimal state-space realization

$$[A, B, C, 0]$$

where Re $\lambda_i(A) \leq 0$.

The controllability grammian is defined as

$$L_c = \int_0^\infty e^{-At} BB^* e^{-A^*t} dt.$$

It is routine to show that L_c is the solution of the Lyapunov equation

$$AL_c + L_c A = BB^*.$$

Define similarly the observability grammian

$$L_o = \int_0^\infty e^{-A^*t} C^* C e^{-At} dt.$$

The operator L_o is the solution of the Lyapunov equation

$$A^* L_o + L_o A = C^* C.$$

Note that L_c is a matrix representation of $\Psi_c \Psi_c^*$ and L_o is a matrix representation of $\Psi_o \Psi_o^*$.

We now state and prove the first of two lemmas needed for Nehari's Theorem.

Lemma 9.3: The operator $\Gamma_f^* \Gamma_f$ and the matrix $L_c L_o$ have the same eigenvalues.

□ *Proof:* Let λ be a nonzero eigenvalue of $\Gamma_f^* \Gamma_f$. It is easy to show that λ is an eigenvalue of the time-domain analogue of $\Gamma_f^* \Gamma_f$.

Thus, for some nonzero u,

$$\Psi_c^* \Psi_o^* \Psi_o \Psi_c u = \lambda u. \tag{9.3}$$

Premultiply by Ψ_c and define $x = \Psi_c u$:

$$L_c L_o x = \lambda x. \tag{9.4}$$

If $x = 0$, then $u = 0$ (see Eq. (9.3).) Thus, λ is an eigenvalue of $L_c L_o$.

The converse is this argument in reverse. Let λ be a nonzero eigenvalue of $L_c L_o$ and x a corresponding eigenvector. Premultiply (9.4) by $\Psi_c^* L_o$ and define $u = \Psi_c^* L_o x$ to obtain (9.3). Since the realization is minimal, Ψ_c and L_o are one-to-one. (See Theorems 2.1, 2.11.) This implies that u is nonzero. ■

LEMMA 9.4: Decompose $R = R_u + R_s$ and let $[A, B, C, 0]$ be a minimal realization of R_u. Let L_c, L_o be the controllability and observability grammians. Let λ^2 be the largest eigenvalue of $L_c L_o$ and w a corresponding eigenvector:

$$L_c L_o w = \lambda^2 w.$$

Defining $v := \lambda^{-1} L_o w$, we have

$$L_c v = \lambda w$$
$$L_O w = \lambda v.$$

Define the real-rational functions

$$f(s) = C(s - A)^{-1} w$$
$$g(s) = B^*(s + A^*)^{-1} v.$$

Then

$$\Gamma_R g = \lambda f \tag{9.5}$$
$$\Gamma_R^* f = \lambda g. \tag{9.6}$$

□ *Proof:*

$$A L_c + L_c A^* = B B^*$$

Add and subtract $s L_c$:

$$-(s - A) L_c + L_c (s + A^*) = B B^*.$$

Premultiply by $C(s - A)^{-1}$ and postmultiply by $(s + A^*)^{-1} v$ to obtain

$$\underbrace{-C L_c (s + A^*)^{-1} v}_{\in R\mathcal{H}_2} + \underbrace{C(s - A)^{-1} L_c v}_{\lambda f} = \underbrace{C(s - A)^{-1} B B^* (s + A^*)^{-1} v}_{R_u g}.$$

Project both sides of the equation onto $\mathcal{H}_2^\perp$:

$$\begin{aligned}\lambda f &= \Pi_u R_u g \\ &= \Gamma_{R_u} g \\ &= \Gamma_R g.\end{aligned}$$

This proves (9.5).

The second statement is proved similarly, starting with

$$A^* L_0 + L_0 A = C^* C.$$ ■

We now return to our minimum distance problem. For any $X \in R\mathcal{H}_\infty$,

$$\|R - X\|_\infty = \|\Lambda_R - \Lambda_X\|$$

where $\| \ \|$ on the right-hand side is the operator norm of an operator from $L_2(-\infty, \infty)$ to $L_2(-\infty, \infty)$. Thus,

$$\begin{aligned}\|R - X\|_\infty &\geq \left\|\Pi_u(\Lambda_R - \Lambda_x)|_{\mathcal{H}_2}\right\| \\ &= \|\Gamma_R - \Gamma_X\| \\ &= \|\Gamma_R\| \\ &= \|\Gamma_{Ru}\|\end{aligned}$$

where R_u is the unstable part of R. Thus, $\|\Gamma_R\|$ is a lower bound on the distance from R to $\mathcal{H}_\infty$.

Theorem 9.5 (*Nehari's Theorem*): Let $R \in \bar{L}_\infty$ be rational with real coefficients. There exists a closest function $X \in \mathcal{H}_\infty$ to R and

$$\|R - X\| = \|\Gamma_R\|.$$

Also, X is rational with real coefficients.

□ *Proof:* We shall prove this for our problem, where R is real rational, by constructing a suitable X.

Now,

$$\begin{aligned}\inf \|R - X\|_\infty &\geq \|\Gamma_{R_u}\| \\ &= \left[\lambda_{\max}\left(\Gamma_{R_u}^* \Gamma_{R_u}\right)\right]^{1/2} \\ &= [\lambda_{\max}(L_c L_o)]^{1/2},\end{aligned}$$

where L_c, L_o are calculated using a minimal realization of R_u. We now compute an X so that

$$\|R - X\|_\infty = \|\Gamma_{R_u}\|.$$

Defining f and g as in Lemma 9.4,

$$\begin{aligned}\Gamma_R^* \Gamma_R g &= \Gamma_R^* \lambda f \\ &= \lambda^2 g.\end{aligned}$$

That is, g is an eigenvector of $\Gamma_R^* \Gamma_R$ corresponding to λ^2, the largest eigenvalue.

Also,

$$\begin{aligned}|f(jw)|^2 &= f(jw)^* f(jw) \\ &= \frac{1}{\lambda^2} (\Gamma_R g)^* (\Gamma_R g) \\ &= \frac{1}{\lambda^2} g^* \Gamma_R^* \Gamma_R g \\ &= g^*(jw) g(jw) \\ &= |g(jw)|^2.\end{aligned}$$

Thus,

$$\left| \frac{f(j\omega)}{g(jw)} \right| = 1.$$

Define

$$X = R - \lambda \frac{f}{g}.$$

$|f/g| = 1$, so X is proper. Also, using (9.5),

$$\begin{aligned}\Pi_u \left(R - \lambda \frac{f}{g} \right) &= \Pi_u R - \Pi_u (\Gamma_R g) \frac{1}{g} \\ &= \Pi_u R - \Pi_u (R g) \frac{1}{g} \\ &= 0.\end{aligned}$$

Therefore, $X \in R\mathcal{H}_\infty$. With this choice of X,

$$\begin{aligned}\|R - X\|_\infty &= \left\| \frac{\lambda f}{g} \right\|_\infty \\ &= \lambda \left\| \frac{f}{g} \right\|_\infty \\ &= \lambda \\ &= \|\Gamma_R\|.\end{aligned}$$

Since

$$\inf_{X \in R\mathcal{H}_\infty} \|R - X\|_\infty \geq \|\Gamma_R\|,$$

we have shown that

$$\inf_{X \in R\mathcal{H}_\infty} \|R - X\|_\infty = \|\Gamma_R\|$$

by constructing a suitable $X \in R\mathcal{H}_\infty$. ■

Note that with the optimal choice of X, $R - X$ is an inner function. The following result is now straightforward.

COROLLARY 9.6: Consider the model-matching problem

$$\inf_{Q \in \mathcal{H}_\infty} \|T_1 - T_2 Q\|$$

where $T_1, T_2 \in R\mathcal{H}_\infty$. Define $R = T_2^{-1} T_1$. The infimal model matching error $\hat{\gamma} = \|\Gamma_R\|$, and if $T_{2o}^{-1} \in R\mathcal{H}_\infty$, the optimal $Q = T_{2o}^{-1} X$ where X is the solution to the Nehari Problem.

If T_{2o} is strictly proper, it does not have a stable inverse, and the optimal model-matching error is not achievable. This situation is discussed later.

Since the preceding proof was constructive, it yields an algorithm that can be used to solve the model-matching problem.

ALGORITHM: Find $Q_{opt} \in \mathcal{H}_\infty$, $\hat{\gamma}$ so

$$\hat{\gamma} = \inf_{Q \in \mathcal{H}_\infty} \|T_1 - T_2 Q\|_\infty = \|T_1 - T_2 Q_{opt}\|_\infty$$

1. Do an inner–outer factorization

$$T_2 = T_{2i} T_{2o}.$$

2. Define $R = T_{2i}^{-1} T_1$.

3. Decompose $R = R_u + R_s$ and find a minimal realization for $R_u = [A, B, C, 0]$.

4. Solve

$$-AL_c - L_c A^* = -BB^*,$$
$$-A^* L_o - L_o A = -C^* C.$$

(This should be done on a computer, using a subroutine, such as "lyap" in MATLAB, to solve Lyapunov equations.)

5. $\hat{\gamma} = [\lambda_{\max}(L_c L_o)]^{1/2}$. If the optimal Q is also wanted, calculate the eigenvector w corresponding to the largest eigenvalue, of $L_c L_o$.

6. Define f to be the transfer function with realization $[A, w, C, 0]$ and $g(s)$ to be the transfer function with realization $[-A^*, \hat{\gamma}^{-1}L_o w, B^*, 0]$. Then define

$$X = R - \hat{\gamma} f/g.$$

7. $Q_{opt} = T_{2o}^{-1}X$.

8. As a check, verify that $T_1 - T_2 Q_{opt} = T_1 - T_{2i}X$ is an inner function, multiplied by $\hat{\gamma}$.

Steps (1)–(5) to calculate the optimal γ can be performed without too much difficulty. Care is needed in step (6) to ensure that the calculated $X \in M(\mathcal{H}_\infty)$. Floating-point calculations often lead to a calculated X that is not stable. Hand or symbolic calculations are preferable. Also, because of the system manipulations required in most steps, purely state-space manipulations typically lead to a Q with high order. We now illustrate this procedure with an example.

Example 9.7:

$$T_1 = \frac{1}{s+1}, \quad T_2 = \frac{(s-1)(s-2)}{(s+1)(s+3)}.$$

Step 1: Calculate an inner–outer factorization of T_2:

$$T_{2i} = \frac{(s-1)(s-2)}{(s+1)(s+2)}, \quad T_{2o} = \frac{(s+2)}{(s+3)}.$$

Step 2: Compute R:

$$R = T_{2i}^{-1}T_1 = \frac{(s+2)}{(s-1)(s-2)}.$$

Step 3: By calculating the residues of R at its r.h.p. poles 1 and 2, calculate R_u and R_s:

$$R = \frac{-3}{(s-1)} + \frac{4}{(s-2)} = R_u$$

so $R_s = 0$. A realization for R_u is

$$A = \begin{bmatrix} 2 & 0 \\ 0 & 1 \end{bmatrix}, \quad B = \begin{bmatrix} 4 \\ -3 \end{bmatrix}, \quad C = [1 \quad 1], \quad E = 0.$$

Step 4: Solve

$$-AL_c - L_c A^* = -BB^*$$
$$-A^*L_0 - L_0 A = -C^*C$$

for L_c and L_o.

Step 5:

$$\hat{\gamma} = (\lambda_{\max}(L_c L_o))^{1/2} = .72871355, \quad w = \begin{bmatrix} 0.61073290 \\ -0.79183668 \end{bmatrix}.$$

The extra digits are required for sufficient accuracy to obtain a stable solution X.

Step 6: Calculate X:

$$f = [A, w, c, 0] = \frac{-0.1811s + 0.9729}{(s-1)(s-2)}$$

$$g = [-A^*, \hat{\gamma}^{-1} L_o w, B^*, 0] = \frac{0.1811s + 0.9729}{(s+1)(s+2)}$$

$$X = R - \hat{\gamma}\frac{f}{g} = \frac{\hat{\gamma}(s+2)}{(s+5.372)}.$$

Step 7: Calculate Q_{opt}:

$$Q_{opt} = T_{2o}^{-1} X = \frac{\hat{\gamma}(s+3)}{(s+5.372)}.$$

Step 8: Check:

$$T_1 - T_2 Q_{opt} = \frac{\hat{\gamma}(s-5.372)}{(s+5.372)}.$$

9.2 NONACHIEVABLE INFIMUM

What if T_{2o} is not invertible? This will happen, for instance, if T_2 is strictly proper. The model-matching problem

$$\inf_{Q \in \mathcal{H}_\infty} \Bigg\| \underbrace{T_{2i}^{-1} T_1}_{R} - \underbrace{T_{2o} Q}_{X} \Bigg\|_\infty$$

has the formal optimal solution

$$Q_{opt} = T_{2o}^{-1} X$$

where X is as calculated earlier. However, if T_{2o}^{-1} is not in $\mathcal{H}_\infty$, the optimal Q does not exist in $\mathcal{H}_\infty$.

If the difficulty is that T_{2o} is strictly proper, as is common, then Q_{opt} can be approximated by a proper function Q_a, by rolling Q_{opt} off at high frequencies. For instance, we can approximate Q_{opt} on the band $[0, \omega_1]$ by

$$Q_a(s) = \frac{Q_{opt}(s)}{\left(.1\frac{s}{\omega_1} + 1\right)^{\ell}}$$

where ℓ is some integer.

The closeness of the performance level

$$\gamma(Q_a) = \left\| T_{2i}^{-1} T_1 - T_{2o} Q_a \right\|_\infty$$

to the optimal value depends on T_1. In most cases T_1 is strictly proper and we can obtain $Q_a \in \mathcal{H}_\infty$ with $\gamma(Q_a)$ arbitrarily close to the optimal value. However, an *optimal* solution does *not* exist in $\mathcal{H}_\infty$.

The case where T_{2o} is not invertible because of imaginary axis zeros is handled similarly.

Example 9.8: Minimize the weighted sensitivity for the plant

$$P(s) = \frac{(s-1)(s-2)}{(s-1)(s^2+s+1)}$$

with weight

$$W(s) = \frac{(s+1)^2}{(10s+1)^2}.$$

Step 1: Write the problem as a model-matching problem:

$$N = P, \quad D = 1, \quad X = 0, \quad Y = 1$$

$$W(s) = \frac{(s+1)^2}{(10s+1)^2}$$

$$T_1 = \left(\frac{s+1}{10s+1}\right)^2, \quad T_2 = \left(\frac{s+1}{10s+1}\right)^2 \frac{(s-1)(s-2)}{(s+1)(s^2+s+1)}.$$

Step 2: Write the model-matching problem as a minimum-distance problem:

$$T_{2i} = \frac{(s-1)(s-2)}{(s+1)(s+2)} \quad T_{2o} = \left(\frac{s+1}{10s+1}\right)^2 \frac{(s+2)}{(s^2+s+1)}$$

$$R = T_{2i}^{-1} T_1, \quad X = T_{2o} Q$$

$$R = \left(\frac{s+1}{10s+1}\right)^2 \frac{(s+1)(s+2)}{(s-1)(s-2)}.$$

Step 3: Calculate R_u and R_s:

$$R = R_u + R_s$$

$$R_u = \frac{a}{(s-1)} + \frac{b}{(s-2)}$$

$$a = \left(\frac{1+1}{10+1}\right)^2 \frac{(1+1)(1+2)}{(1-2)} = \left(\frac{2}{11}\right)^2 \frac{-6}{1};$$

$$b = \left(\frac{3}{21}\right)^2 \left(\frac{3 \cdot 4}{1}\right) = 12 \left(\frac{1}{7}\right)^2.$$

A minimal realization for R_u is

$$A = \begin{bmatrix} 1 & 0 \\ 0 & 2 \end{bmatrix}, \quad B = \begin{bmatrix} 1 \\ 1 \end{bmatrix}, \quad C = \begin{bmatrix} -6\left(\frac{2}{11}\right)^2 & 12\left(\frac{1}{7}\right)^2 \end{bmatrix}, \quad E = 0.$$

Step 4: Solve

$$-AL_c - L_c A^* = -BB^*$$

$$-A^* L_o - L_o A = -C^* C.$$

$$L_c = \begin{bmatrix} .5 & .33 \\ .33 & .5 \end{bmatrix}, \quad L_o = \begin{bmatrix} 0.0197 & -0.0162 \\ -0.0162 & 0.0150 \end{bmatrix}.$$

Step 5:

$$\hat{\gamma} = (\lambda_{\max}(L_c L_0))^{1/2} = 0.051141.$$

Step 6: We now calculate the corresponding optimal X. The corresponding eigenvector of $L_c L_0$ with eigenvalue $\hat{\gamma}^2$ is

$$w = \begin{bmatrix} 0.8619 \\ 0.5071 \end{bmatrix}.$$

$$\begin{aligned}
f(s) &= [A, w, c, 0] \\
&= C \begin{bmatrix} (s-1)^{-1} & 0 \\ 0 & (s-2)^{-1} \end{bmatrix} \omega \\
&= -6\left(\frac{2}{11}\right)^2 (s-1)^{-1}(0.8619) + 12\left(\frac{3}{21}\right)^2 (s-2)^{-1}(0.5071) \\
&= \frac{-0.046756s + 0.217706}{(s-1)(s-2)} \\
g(s) &= \left[-A^*, \hat{\gamma}^{-1} L_o w, B^*, 0\right] \\
&= [1 \quad 1] \begin{bmatrix} (s+1)^{-1} & 0 \\ 0 & (s+2)^{-1} \end{bmatrix} \begin{bmatrix} 0.1715 \\ -0.1246 \end{bmatrix} \\
&= \frac{0.046756s + 0.217706}{(s+1)(s+2)} \\
X &= R - \hat{\gamma}\frac{f}{g} \\
&= R_s - \frac{0.005742}{(s+4.657)}.
\end{aligned}$$

Step 7: Calculate a suitable Youla parameter (and hence a suitable controller). The optimal Youla parameter is

$$Q_{opt} = T_{2o}^{-1} X,$$

which is not proper.

Approximate Q_{opt} on an operating band $[0, \omega_1]$ by rolling off at high frequencies, say, 10 above W_1,

$$Q_a = \frac{Q(s)}{(s/0.1+1)^{\ell}},$$

for some $\ell > 0$.

9.3 INTERPOLATION

We now explain a different approach to solving the optimal model-matching problem (9) where $T_1, T_2 \in \mathcal{RH}_\infty$.

For some "performance level" γ consider the model matching problem:

$$\text{Find } Q \in R\mathcal{H}_\infty; \quad \|T_1 - T_2 Q\|_\infty < \gamma. \tag{9.7}$$

We assume that T_2 is nonstrictly proper with no imaginary axis zeros. If T_2 has no right-hand-plane zeros, then as discussed earlier, the problem is easily solved for any performance level γ.

Suppose first that T_2 has only one zero, z, in the closed right half-plane. Clearly, for any $Q \in R\mathcal{H}_\infty$,

$$\|T_1 - T_2 Q\|_\infty \geq |T_1(z)|.$$

If Q is chosen so that

$$|T_1(j\omega) - T_2(j\omega)Q(j\omega)| = |T_1(z)|$$

for all ω, then this Q will be a minimizing choice. Suppose that

$$T_1(s) - T_2(s)Q(s) = T_1(z)$$

for all s. Rearranging,

$$Q_{opt} = \frac{T_1(s) - T_1(z)}{T_2(s)}.$$

Notice that since the numerator is zero at the only r.h.p. zero of T_2, $Q_{opt} \in R\mathcal{H}_\infty$.

The more general case where T_{2o}^{-1} is nonstrictly proper, but has more than one r.h.p. zero, is similar. Let z_i, $i = 1, \ldots, n$ be the r.h.p. zeros of T_2. We reformulate the model matching problem as an interpolation problem. First, note that (9.7) is equivalent to

$$\text{Find } Q \in R\mathcal{H}_\infty; \quad \left\| \frac{T_1}{\gamma} - \frac{T_2}{\gamma} Q \right\|_\infty \leq 1. \tag{9.8}$$

The foregoing discussion shows that this problem can be solved by solving the following interpolation problem:

$$\text{Find G} \in R\mathcal{H}_\infty \text{ so that } G(z_i) = a_i, \quad i = 1, \ldots, n \text{ and } \|G\|_\infty \leq 1. \tag{9.9}$$

(For our problem, $a_i = T_1(z_i)/\gamma$.) This is known as the Nevanlinna–Pick problem. Once one such function G is found, then $Q_\gamma \in R\mathcal{H}_\infty$ is easily calculated as

$$Q_\gamma = \frac{T_1(s) - G(s)}{T_2(s)}.$$

The optimal performance level $\hat{\gamma}$ is the smallest γ for which the interpolation problem has a solution.

We first solve a simple example for which an *ad hoc* procedure can be used.

Example 9.9: $T_1 = 1/(s+1)$, $T_2 = (s-1)/(s+3)$. For all $Q \in R\mathcal{H}_\infty$,

$$\|T_1 - T_2 Q\|_\infty \geq T_1(1) = \frac{1}{2}.$$

We will attempt to find a Q so that

$$|T_1(j\omega) - T_2(j\omega)Q(j\omega)| = \frac{1}{2}$$

for all ω. The parameter Q needs to interpolate the r.h.p. zero of T_2. Try

$$Q(s) = \frac{T_1(s) - T_1(1)}{T_2(s)} = -\frac{1}{2}\frac{s+3}{s+1}.$$

Substituting this choice of Q,

$$T_1 - T_2 Q = \frac{1}{2},$$

and so

$$Q_{opt} = -\frac{1}{2}\frac{s+3}{s+1} \quad \text{with } \hat{\gamma} = \frac{1}{2}.$$

For more complicated problems a formal technique for interpolation is required. This technique is known as *Nevanlinna–Pick interpolation.* The main result, given next, relies on Sarason's Theorem and some other results in complex analysis. The proof is omitted.

Theorem 9.10 (*Pick's Theorem*): There exists $G \in R\mathcal{H}_\infty$ such that $\|G\|_\infty \leq 1$ and $G(z_i) = a_i$, for $i = 1..n$ where $\text{Re}(z_i) > 0$ if and only if the *Pick Matrix* M is positive semidefinite, where M is defined by its ijth entries to be

$$[M]_{ij} = \frac{1 - a_i\overline{a_j}}{z_i + z_j}.$$

Corollary 9.11: For the problem (9.7) assume that T_2 has no imaginary zeros and no repeated r.h.p. zeros. Let z_i, $i = 1..n$ indicate the r.h.p. zeros of T_2 and define $a_i = T_1(z_i)$, $i = 1..n$. Define the matrix A with entries $[A]_{ij} = 1/(z_i + \overline{z_j})$ and the matrix B with entries

$[B]_{ij} = a_i\overline{a_j}/(z_i + \overline{z_j})$. Then

$$\hat{\gamma} = \sqrt{\lambda_{\max}\left(A^{-\frac{1}{2}} B A^{-\frac{1}{2}}\right)}.$$

□ *Proof:* The Pick matrix for the problem

$$\left\| \frac{T_1}{\gamma} - \frac{T_2}{\gamma} Q \right\|_\infty \leq 1$$

is

$$M = A - \frac{1}{\gamma^2} B.$$

Thus, $\hat{\gamma}$ is the smallest γ so that $A - \frac{1}{\gamma^2}B$ is positive semidefinite. The matrix A is Hermitian and positive definite. Hence, there is a positive definite matrix $A^{1/2}$ such that $A = A^{1/2}A^{1/2}$. Now,

$$A - \frac{1}{\gamma^2} B = \frac{1}{\gamma^2} A^{1/2} \left(\gamma^2 I - A^{-1/2} B A^{-1/2}\right) A^{1/2},$$

and so the Pick matrix is positive semidefinite if and only if $\gamma^2 I - A^{-1/2} B A^{-1/2}$ is positive semidefinite. The smallest γ for which this is true satisfies

$$\hat{\gamma}^2 = \lambda_{\max}\left(A^{-1/2} B A^{-1/2}\right).$$ ■

EXAMPLE 9.12:

$$T_1 = \frac{1}{s+1}, \quad T_2 = \frac{(s-1)(s-2)}{(s+1)(s+3)}.$$

This problem was previously solved as a minimum distance problem (Example 9.7). We will use the previous corollary to determine $\hat{\gamma}$. Calculating the matrices A and B, we find

$$A = \begin{bmatrix} \frac{1}{2} & \frac{1}{3} \\ \frac{1}{3} & \frac{1}{4} \end{bmatrix}, \quad B = \begin{bmatrix} \frac{1}{12} & \frac{1}{18} \\ \frac{1}{18} & \frac{1}{24} \end{bmatrix}.$$

The smallest γ for which the Pick matrix $A - \frac{1}{\gamma^2}B$ is positive semidefinite is

$$\sqrt{\lambda_{\max}\left(A^{-\frac{1}{2}} B A^{-\frac{1}{2}}\right)} = 0.72871355.$$

This is equal to the previously calculated value.

We now discuss calculation of the corresponding Q_{opt}. We need to calculate G that solves the Nevanlinna–Pick problem, with $\gamma = \hat{\gamma}$. Once such a G is found, then

$$Q_{opt} = \frac{T_1 - \hat{\gamma}G}{T_2}.$$

It was shown earlier that if T_2 has only one r.h.p. zero, z_1, then

$$G = \frac{1}{\hat{\gamma}}T_1(z_1)$$

(a constant function) is a solution. The solution for problems with two r.h.p. zeros proceeds by replacing the problem by a problem with one r.h.p. zero, and so on, for problems with greater numbers of r.h.p. zeros.

We first need some definitions.

Definition 9.13: A *Mobius function* has the form $M_b(s) = (s - b)/(1 - s\bar{b})$ for b with $|b| < 1$.

A Mobius function has the important properties that $|M(s)| = 1$ for $|s| = 1$. It maps the unit circle onto the unit circle and also the unit disc onto the unit disc. The inverse

$$\begin{aligned} M_b(s)^{-1} &= \frac{s+b}{1+s\bar{b}} \\ &= M_{-b}(s), \end{aligned}$$

and this is also a Mobius function.

Definition 9.14: A *Blaschke function* $B_a(s) = (s - a)/(s + \bar{a})$ for a with $\mathrm{Re}(a) > 0$.

Clearly all Blaschke functions are inner functions with $|B_a(j\omega)| = 1$.

Definition 9.15: A function $G(s) \in R\mathcal{H}_\infty$ is *all-pass* if $G(s)$ is a constant multiplying an inner function.

In other words, an all-pass function has constant magnitude on the imaginary axis. An inner function is thus a special all-pass function, where the constant is 1.

Lemma 9.16: Consider the Nevanlinna–Pick problem (9.9) with one data point (z, a). There are three cases:

1. $|a| > 1$. There is no solution.
2. $|a| = 1$. There is a unique solution, $G(s) = a$.
3. $|a| < 1$. There are an infinite number of solutions. Any solution can be written

$$G(s) = M_{-a}(p)$$

where $p = G_1(s)B_z(s)$ and $G_1 \in R\mathcal{H}_\infty$, $\|G\|_\infty \leq 1$. If G_1 is an all-pass function, then so is G.

THEOREM 9.17: Consider the Nevanlinna–Pick problem (9.9) with n data points. Assume that all $|a_i| \leq 1$. Define the reduced problem with $n-1$ points: Find $G_{n-1} \in R\mathcal{H}_\infty$ so that

$$G_{n-1}(z_i) = \frac{M_{a_1}(a_i)}{B_{z_1}(a_i)}, \quad i = 2, \ldots, n \text{ and } \|G_{n-1}\|_\infty \leq 1.$$

The set of solutions to the original problem is

$$G(s) = M_{-a_1}(p)$$

where $p = G_{n-1}(s)B_{z_1}(s)$ and G_{n-1} is any solution to the reduced problem. If G_{n-1} is an all-pass function, then so is G.

EXAMPLE 9.18 (*Example 9.12 Continued*): We will now calculate the optimal $Q_{opt.}$ We need to solve the Nevanlinna–Pick problem: Find $G \in R\mathcal{H}_\infty$ with $\|G\|_\infty \leq 1$ so that

$$G(1) = \frac{T_1(1)}{\hat{\gamma}} = \frac{1}{2\hat{\gamma}} = a_1$$

$$G(2) = \frac{T_1(2)}{\hat{\gamma}} = \frac{1}{3\hat{\gamma}} = a_2$$

where $\hat{\gamma} = 0.72871355$. We need the Mobius function corresponding to the first value a_1, and the Blaschke function for the first zero, $z_1 = 1$, to construct the reduced problem:

$$M_{a_1}(p) = \frac{p - a_1}{1 - p\overline{a_1}}, \quad B_1(s) = \frac{s-1}{s+1}.$$

The reduced problem is: Find $G_1 \in R\mathcal{H}_\infty$ with $\|G_1\|_\infty \leq 1$ so that

$$G_1(2) = \frac{M_{a_1}(a_2)}{B_1(a_2)} = -1.$$

The correct choice is the constant function $G_1(s) = -1$.

We now construct G. We have

$$M_{-a_1}(p) = \frac{p + a_1}{1 + p\overline{a_1}},$$

and the solution is

$$G(s) = \frac{p + a_1}{1 + p\overline{a_1}}$$

with

$$\begin{aligned} p &= G_1(s)B_1(s) \\ &= -1\frac{s-1}{s+1}. \end{aligned}$$

Substituting,

$$G(s) = \frac{0.5372 - s}{0.5372 + s}.$$

(Notice that since G_1 was all-pass, so is G.) Now,

$$Q_{opt}(s) = \frac{T_1(s) - \hat{\gamma} G(s)}{T_2(s)}.$$

After some manipulations, we obtain

$$Q_{opt}(s) = \frac{\hat{\gamma}(s+3)}{(s+5.372)}$$

as previously obtained.

Note that calculation of the optimal $\hat{\gamma}$ is straightforward. Calculation of Q and hence the controller requires care so that $Q \in R\mathcal{H}_\infty$. Symbolic or hand calculations are preferable to floating-point calculation.

So far we have assumed that T_2 is strictly proper, with no imaginary axis zeros. This implies that $T_{2o}^{-1} \in R\mathcal{H}_\infty$. In many situations, however, this is not the case. Suppose we attempt to solve the model-matching problem as before. Write the model-matching problem

$$\|T_1 - T_2 Q\|_\infty = \|T_1 - (T_{2i} T_{2o} Q)\|_\infty.$$

Assume that the Nevanlinna–Pick interpolation problem (9.9) has a solution G. Then

$$Q = (T_1 - G)T_2^{-1}$$

appears to be the solution. However, because T_{2o}^{-1} is not in $R\mathcal{H}_\infty$, $Q \notin R\mathcal{H}_\infty$. As discussed in Section 9.2, Q needs to be approximated by a function in $R\mathcal{H}_\infty$.

Example 9.19:

$$P = \frac{s-1}{(s+1)(s-p)}, \quad 0 < p \neq 1.$$

The precise location of the pole p is uncertain. We suspect that this plant may be difficult to control if p is close to 1. Suppose that the uncertainty model is multiplicative uncertainty with weight

$$W_3(s) = \frac{s+0.1}{s+1}.$$

How does the robust stability margin vary with p? The problem is to calculate

$$\inf_{H \in \mathcal{S}(P)} \|W_3 T\|_\infty$$

for different values of p. To convert this to a model-matching problem, we need a coprime factorization for P. Defining

$$N = \frac{(s-1)}{(s+1)^2}, \quad D = \frac{s-p}{s+1}$$

$$X = \frac{(p+1)^2}{(p-1)}, \quad Y = \frac{s-a}{s+1}$$

where

$$a = \frac{(p+3)}{(p-1)}.$$

$$P = \frac{N}{D},$$

$$XN + YD = 1.$$

Any stabilizing controller for P is of the form

$$H = (X + DQ)(Y - NQ)^{-1}, \quad Q \in \mathcal{H}_\infty.$$

Since

$$\begin{aligned} T &= PH(1+PH)^{-1} \\ &= N(X+DQ), \end{aligned}$$

the optimal robust stability problem is

$$\begin{aligned} \gamma &= \inf_{Q \in R\mathcal{H}_\infty} \|W_3 N(X+DQ)\|_\infty \\ &= \inf_{Q \in R\mathcal{H}_\infty} \|T_1 - T_2 Q\|_\infty \end{aligned}$$

where $T_1 = W_3 N X$ and $T_2 = W_3 N D$. Since W_3 is outer,

$$T_{2i} = N_i D_i.$$

Because N appears in both T_1 and T_2, we can simplify this problem before proceeding further, by factoring out N_i:

$$\begin{aligned}\gamma &= \inf_{H\in\mathcal{S}(P)} \|W_3 T\|_\infty \\ &= \inf_{Q\in\mathcal{H}_\infty} \|N_i W_3 N_o(X + DQ)\|_\infty \\ &= \inf_{Q\in\mathcal{H}_\infty} \|W_3 N_o X + (W_3 N_o D)Q\|_\infty.\end{aligned}$$

Since N_i has been factored out, it can be seen that the right-half-plane zeros of P do not affect robust stability:

$$\begin{aligned}T_1(s) &= W_3 N_o X \\ &= \frac{(p+1)^2(s+0.1)}{(p-1)(s+1)^2}, \\ T_2(s) &= W_3 N_o D \\ &= \frac{-(s+0.1)(s-p)}{(s+1)^3}.\end{aligned}$$

Since $T_2(p) = 0$,

$$\begin{aligned}\gamma &> |T_1(p)| \\ &= \left|\frac{p+0.1}{p-1}\right|.\end{aligned}$$

For the optimal Q_{opt},

$$|T_2(j\omega) - T_2(j\omega)Q(j\omega)| = \frac{p+0.1}{p-1}.$$

Since T_2 only has one r.h.p. zero, try

$$T_1(s) - T_2(s)Q(s) = \frac{p+0.1}{p-1}.$$

Rearranging,

$$\begin{aligned}Q_{opt}(s) &= \frac{T_1(s) - \frac{p+0.1}{p-1}}{T_2(s)} \\ &= \frac{-1.2(s+1)(s-1.25)}{s+0.1}.\end{aligned}$$

This function has no r.h.p poles, but is improper. We expected this because T_2 is strictly proper. We can approximate Q_{opt} by a proper $Q_a \in R\mathcal{H}_\infty$ to obtain $\|T_1 - T_2 Q_a\|_\infty$ arbitrarily close to

$$\left| \frac{p + 0.1}{p - 1} \right|.$$

Notice that as p gets close to 1, γ becomes very large, and as expected, the allowable uncertainty is reduced.

The case where T_2 has imaginary axis zeros is handled similarly. We need to multiply the "optimal" Q by a function that is unity, except near the imaginary axis zeros. At these zeros the function is zero. A suboptimal solution is thus obtained.

NOTES AND REFERENCES

Example 9.8 is based on [20] and Example 9.19 is from [13]. The sections on solving the control problem as a minimum distance problem are based on [20], where this approach is covered in detail, including the matrix-valued case. State-space algorithms to solve this problem exist, and so in theory the foregoing technique can be applied to problems where T_1 and T_2 are matrices of high order. However, the degree of the system is increased at each step. Also, round-off errors often lead to $X \notin R\mathcal{H}_\infty$. Order reduction is generally needed at the end to remove spurious right-hand-plane poles. The sources [36] and [43] cover the interpolation approach.

EXERCISES

1.

$$T_1 = \frac{s}{s+2}, \quad T_2 = \frac{(s-2)(s-3)}{(s+1)(s+10)}.$$

(a) Using interpolation, find the smallest value of γ so that a $Q \in R\mathcal{H}_\infty$ exists satisfying

$$\|T_1 - T_2 Q\|_\infty < \gamma.$$

(b) Verify your solution by using the minimum-distance method.
(c) Calculate the corresponding Q using either method.

2. Repeat the previous question using

$$T_1 = \frac{s+1}{s+4}, \quad T_2 = \frac{(s-2)(s-4)}{(s+2)(s+9)}.$$

3. Consider the model of a flexible beam in Example 5.17.

(a) Calculate the transfer function for the given values of damping and frequencies. Use this and the Nevanlinna–Pick interpolation to calculate the optimal closed-loop sensitivity.

(b) Give two reasons, one theoretical and one practical, why this sensitivity is not achievable by a proper stabilizing controller.

4. Consider the model for an idle speed control system in Example 2.21.

(a) Show that with the given model, the optimal closed-loop sensitivity is zero. Is this optimal value achievable? Why or why not?

(b) In the derivation of the model, there is an induction to power delay of T s. The transfer function of this delay is e^{-Ts}. It was approximated by $1/(1+Ts)$. The exact transfer function is an inner function,

$$|e^{-jT\omega}| = 1,$$

for all frequencies ω. Approximate the delay by an inner function:

$$e^{-Ts} \approx \frac{1-\frac{T}{2}s}{1+\frac{T}{2}s}.$$

Plot the optimal sensitivity for various values of T.

5. A controller for a plant with transfer function

$$P(s) = \frac{5(s-4)}{(s+1)(s+2)}$$

is required so that $\|S\|_\infty$ is less than 2.

(a) Find a controller so that $\|S\|_\infty < 2$ and plot $|S(j\omega)|$ against ω.

(b) What is the minimum possible value of

$$\max_{0\le\omega\le5} |S(j\omega)|$$

if $\|S\|_\infty < 2$? (Hint: The Waterbed Effect)

(c) Calculate

$$\inf_{C\in S(P)} \|S\|_\infty.$$

APPENDIX A

Normed Linear Spaces

In order to "add" elements of a set, as well as to perform other operations an algebraic structure is needed.

Definition A.1: *A real linear space* X is a set $x, y, z, \ldots$ in which operations, called (1) addition and (2) scalar multiplication by real numbers α, β, etc., are defined and which is such that for all $x, y, z \in X$ and $\alpha, \beta \in R$,

1. $y + z \in X$
2. $\alpha y \in X$
3. $z + y = y + z$
4. $(x + y) + z = x + (y + z)$
5. $(\alpha\beta)y = \beta(\alpha y)$
6. $(\alpha + \beta)y = \alpha y + \beta y$
7. $\alpha(z + y) = \alpha z + \alpha y$
8. $1 \cdot y = y, 0y = \theta$

Furthermore, X contains an element θ, the zero element, and for all y in X,

9. $0y = \theta$.
10. There exists $(-y) \in X$ such that $y + (-y) = \theta$.

A familar real linear space is R^n. Another is $C[0, 1]$, the set of continuous real-valued functions on $[0, 1]$. This linear space of functions is given the familiar definitions of addition and scalar multiplication of functions.

The definition is actually valid for any field of scalars. For instance, if the real numbers are replaced by the complex numbers, we have a complex linear space.

A simple example of a complex linear space is C^n. Another example is $L_2(0, 1)$, the space of functions for which

$$\int_0^1 |f(t)|^2 dt$$

is well-defined and finite.

Definition A.2: A *normed linear space* $(X, \| \|)$(or simply X) is a linear space with a functional called the norm,

$$\|\cdot\| : X \to R^+$$

The norm satisfies the following axioms:

N1 $\|y\| \geq 0$
N2 $\|y\| = 0 \Leftrightarrow y = \Theta$.
N3 $\|\alpha y\| = |\alpha| \|y\|$
N4 $\|x + y\| \leq \|x\| + \|y\|$.

One example of a normed linear space is R^n with familar Euclidean norm:

$$\|x\|_2 = \sqrt{\sum_{\lambda=1}^{n} x_\lambda^2}.$$

Another normed linear space is R^n with the maximum norm,

$$\|x\|_\infty = \max_{1 \leq i \leq n} |x_i|,$$

where $x_i, i = 1 \ldots n$ are the components of the vector x.

Example A.3: Consider the real linear space $X = C[0, 1]$. By defining three different norms on X, we obtain three different normed spaces.

$$(X, \| \|_2): \quad \|f\|_2 = \left(\int_0^1 f(t)^2 dt\right)^{\frac{1}{2}}$$

$$(X, \| \|_w): \quad \|f\|_w = \left(\int_0^1 \omega(t) f(t)^2 dt\right)^{\frac{1}{2}} \quad \text{where}$$

$$\omega(t) \in C[0, 1], \omega(t) > 0 \forall t.$$

$$(X, \| \|_\infty): \quad \|f\|_\infty = \max_{0 \le t \le 1} |f(t)|.$$

We say a sequence $\{x_n\} \in X$ converges to x in the norm $\| \|$ if

$$\lim_{n \to \infty} \|x_n - x\| = 0.$$

If the norm is understood, we simply write $\lim_{n \to \infty} x_n = x$. If $\| \|_a$, $\| \|_b$ are two norms on a linear space X and there exist positive constants m, M such that

$$m\|x\|_a \le \|x\|_b \le M\|x\|_a$$

for all $x \in X$, then we say that $\| \|_a$ and $\| \|_b$ are *equivalent* norms. If two norms are equivalent, then

$$\lim_{n \to \infty} \|x_n - x\|_a = 0$$

if and only if

$$\lim_{n \to \infty} \|x_n - x\|_b = 0.$$

EXAMPLE A.4: In Example A.3,

$$[\min|\omega(t)|]\|f\|_2 \le \|f\|_w \le [\max|\omega(t)|]\|f\|_2.$$

The two norms are equivalent. Other equivalent spaces are $(R^2, \| \|_2)$ and $(R^2, \| \|_\infty)$.

If $\|f\|_a \le M\|f\|_b$ we say that $\| \|_b$ is a *stronger* norm on X than $\| \|_a$. [Convergence in $(X, \| \|_b)$ implies convergence in $(X, \| \|_a)$.

EXAMPLE A.5: In Example A.3,

$$\|f\|_2 \le \|f\|_\infty$$

for all continuous functions f. However, there is no constant M so that

$$\|f\|_\infty \le M\|f\|_2$$

for all continuous functions f. The norm $\| \|_\infty$ is stronger than the norm $\| \|_2$. Convergence in the $\| \|_\infty$ norm implies convergence in the $\| \|_2$ norm, but not vice versa. (In other words, uniform convergence implies mean square convergence, but not vice versa.)

Definition A.6: Let X be a linear space formed by the set of all linear combinations of elements of a set $\{l_i\}_{i=1}^n$. We say that X is *finite-dimensional.*

Theorem A.7: Let X be a finite-dimensional space and $\| \|_a$, $\| \|_b$ be any two norms on X. Then $\| \|_a$, and $\| \|_b$ are equivalent.

Definition A.8: A sequence x_n in a normed linear space is *Cauchy* if for any $\epsilon > 0$, there is an integer N so that for all $n, m > N$, $\|x_n - x_m\| < \epsilon$.

Definition A.9: A normed linear space in which every Cauchy sequence converges (to an element of the space) is said to be *complete.*

Complete normed linear spaces are often referred to as *Banach spaces.*

Definition A.10: A *complex inner product space* is a complex linear space X on which is defined a function $\langle , \rangle : X \times X \to C$, the inner product. The inner product satisfies the following rules: For every $x_1, x_2, y \in X$ and every scalars $\alpha, \beta \in C$,

1. $\langle \alpha x_1 + \beta x_2, y\rangle = \alpha\langle x_1, y\rangle + \beta\langle x_2, y\rangle$
2. $\langle x, y\rangle = \overline{\langle y, x\rangle}$
3. $\langle x, x\rangle \geq 0$
4. $\langle x, x\rangle = 0$

A real inner product space is defined identically, except that the scalars are the real numbers R instead of the complex numbers C. The inner product then maps $X \times X$ to R. Property (2) simplifies to $\langle x, y\rangle = \langle y, x\rangle$.

Every inner product space is a normed linear space through definition of the norm

$$\|x\| = \langle x, x\rangle^{1/2}.$$

A complete inner product space is generally referred to as a *Hilbert space.*

Also, any norm generated by an inner product satisfies the *parallelogram identity*:

$$\|x + y\|^2 + \|x - y\|^2 = 2\|x\|^2 + 2\|y\|^2. \tag{A.1}$$

A familiar Hilbert space is C^n with the usual dot product:

$$\langle x, y\rangle = \sum_{i=1}^{n} \bar{x}_i y_j.$$

Another Hilbert space is $L_2(0, \infty)$; consisting of functions for which

$$\int_0^\infty f(t)\overline{f(t)}dt$$

is well defined and finite with inner product

$$\langle f, g \rangle = \int_0^\infty f(t)\bar{g}(t)dt.$$

Theorem A.11: Let $\{x_n\}$ be a bounded sequence in a Hilbert space H,

$$\|x_n\| < M,$$

for some $M \geq 0$. Then $\{x_n\}$ has a subsequence $\{x_m\}$ with the property that there is $x \in H$ such that for each $y \in H$,

$$\lim_{m \to \infty} \langle x_m, y \rangle = \langle x, y \rangle. \tag{A.2}$$

Corollary A.12: If the Hilbert space is finite-dimensional, then the statement (A.2) can be strengthened to

$$\lim_{m \to \infty} x_m = x.$$

APPENDIX B

Algebra

B.1 LINEAR ALGEBRA

Theorem B.1 (*Cayley–Hamilton*): Any matrix A satisfies its characteristic equation. That is, defining

$$\kappa(s) = \det(sI - A),$$

then $\kappa(A) = 0$.

An important consequence of this is the following corollary.

Corollary B.2: Let A be an $n \times n$ matrix. For any $m \geq n$, A^m is a linear combination of A^i, $i = 0, \ldots, n-1$.

□ *Proof:* Since $\kappa(A) = 0$, A^n is a linear combination of $A^i, i = 0, \ldots, n-1$. ■

Theorem B.3 (*Cramer's Rule*): For an invertible $n \times n$ matrix A, let M_{ij} be the $(n-1) \times (n-1)$ matrix formed by deleting the ith row and jth column of A. Define also the adjoint matrix of A, with (i, j) entry:

$$[\text{adj}\, A]_{ij} = (-1)^{i+j} \det M_{ji}.$$

Then

$$A^{-1} = \frac{\operatorname{adj} A}{\det A}.$$

A matrix $A \in R^{m \times p}$ represents a map from R^m to R^p,

$$b = Ax,$$

where $x \in R^m$ and $b \in R^p$. If we indicate the Euclidean norm of a vector by $\| \ \|$,

$$\|b\|^2 = b^* b = x^* A^* A x,$$

where b^* indicates the transpose of b (or complex-conjugate transpose if b has complex-valued entries). Now

$$\sup_{\|x\|=1} x^* A^* A x = \lambda_{\max}(A^* A).$$

Thus,

$$\|Ax\|^2 \leq \lambda_{\max}(A^* A)\|x\|^2.$$

For the vector v with $\|v\| = 1$ corresponding to the largest eigenvalue of $A^* A$,

$$\|Av\|^2 = \lambda_{\max}(A^* A).$$

Thus, the norm of A, as a map from R^m to R^p (with the Euclidean norm) is

$$\|A\| = [\lambda_{\max}(A^* A)]^{1/2}.$$

We define the *singular values* σ_i of A to be the square roots of the eigenvalues of $A^* A$:

$$\sigma_i(A) = \lambda_i^{1/2}(A^* A).$$

With this notation,

$$\|A\| = \sigma_{\max}(A).$$

Lemma B.4: The spectrum of the Hamiltonian matrix

$$Z = \begin{bmatrix} A & R \\ -Q & -A^* \end{bmatrix}$$

is symmetric with respect to the imaginary axis.

□ *Proof:* Defining

$$P = \begin{bmatrix} 0 & -I \\ I & 0 \end{bmatrix},$$

$$P^{-1}ZP = \begin{bmatrix} -A^* & Q \\ -R & A \end{bmatrix} = -Z^*.$$

Thus, Z and $-Z^*$ are similar. Thus, if λ is an eigenvalue of Z, then so is $-\lambda$. The entries of Z are real, and so the spectrum is symmetric with respect to the real axis. That is, if $\lambda = x + jy$ is an eigenvalue, then so is $\bar{\lambda} = x - jy$. From above $-\bar{\lambda} = -x + jy$ is an eigenvalue. The result follows. ■

THEOREM B.5: Let A and B be $n \times n$ real matrices. The Sylvester equation

$$BX + XA = 0$$

has only the solution $X = 0$ if $\lambda_i(A) + \lambda_j(B) \neq 0$ for all $i, j = 1 \dots n$.

DEFINITION B.6: A symmetric matrix $A \in R^{n \times n}$ is *positive definite* $(A > 0)$ if

$$x^* A x > 0$$

for all nonzero $x \in R^n$. It is *positive semidefinite* or *nonnegative* $(A \geq 0)$ if

$$x^* A x \geq 0$$

for all $x \in R^n$.

The terms *negative definite* and *negative semidefinite* (or *nonpositive*) are defined similarly.

THEOREM B.7: Consider the Lyapunov equation

$$A^* X + XA + Q = 0$$

for real matrix A and symmetric positive semidefinite real matrix Q. Assume that (A, Q) is detectable. This equation has a symmetric positive semidefinite solution X, if and only if $\max \operatorname{Re}(\lambda(A)) < 0$.

□ *Proof:* If $\max \operatorname{Re}(\lambda(A)) < 0$, then

$$X = \int_0^\infty \exp(A^* t) Q \exp(At) dt$$

is well defined. This matrix is clearly symmetric and positive semidefinite. It can also be verified that X solves the Lyapunov equation.

Assume now the equation has a symmetric positive semidefinite solution X. Let λ be any eigenvalue of A and let v be a corresponding nonzero eigenvector. Multiply the Lyapunov equation on the left by v^* and on the right by v:

$$2\text{Re}(\lambda(v^* X v)) + v^* Q v = 0.$$

Suppose that $\text{Re}(\lambda) \geq 0$. Then $Qv = 0$. This implies that

$$[\lambda I - A \quad Q]v = [0 \quad 0]$$

and so (A, Q) is not detectable (See Corollary 3.25.) Thus, $\text{Re}(\lambda) < 0$. ■

Lemma B.8: Let $A \geq 0$, $B \geq 0$ be symmetric square matrices of the same size.

1. The nonzero number λ is an eigenvalue of AB if and only if it is an eigenvalue of BA.
2. All the eigenvalues of AB are real and nonnegative.
3. If 1 is not an eigenvalue of AB, then

$$A(I - BA)^{-1} = (I - AB)^{-1}A.$$

If the matrix $(I - AB)^{-1}A$ is positive semidefinite, then $r(AB) < 1$ (where $r(AB) = \max_{1 \leq i \leq n} |\lambda_i(AB)|$.)

□ *Proof:* Suppose that $\lambda \neq 0$ is an eigenvalue of AB with associated eigenvector $x \neq 0$. Then $y = Bx \neq 0$ since $ABx = \lambda x \neq 0$. Thus,

$$\begin{aligned} BAy &= BA(Bx) \\ &= B\lambda x \\ &= \lambda y. \end{aligned}$$

We now show that λ is real. We have $\lambda Bx = BABx$ and so

$$\lambda x^* Bx = x^* BABx \geq 0,$$

since $A \geq 0$. Since $Bx \neq 0$, and $B \geq 0$, $x^* Bx$ is real and positive. It follows that

$$\lambda = \frac{(Bx)^* A(Bx)}{x^* Bx} \geq 0.$$

Now assume that 1 is not an eigenvalue of AB so that $(I - AB)^{-1}$ is well-defined.

$$A - ABA = (I - AB)A = A(I - BA)$$

and so $A(I - BA)^{-1} = (I - AB)^{-1}A$.

We now prove the final statement by proving the contrapositive. Define

$$C = A(I - BA)^{-1} = (I - AB)^{-1}A.$$

Suppose that $r(AB) \geq 1$. Thus, since $(I - AB)^{-1}$ is defined, AB has a (real) eigenvalue λ, $\lambda > 1$. Let $x \neq 0$ be the eigenvector $ABx = \lambda x$. Let $y = Bx$ so $BAy = \lambda y$ as before:

$$\begin{aligned}(I - BA)y &= (1 - \lambda)y \\ \frac{1}{1-\lambda}y &= (I - BA)^{-1}y \\ y^* \frac{1}{1-\lambda} &= y^*(I - AB)^{-1} \\ \frac{y^* Ay}{1-\lambda} &= y^*(I - AB)^{-1}Ay.\end{aligned}$$

Since $1 - \lambda < 0$,

$$y^*(I - AB)^{-1}Ay < 0$$

and C is not positive semidefinite. Thus, if $C \geq 0$, then $r(AB) < 1$. ■

B.2 IDEALS

Let $\mathcal{P}[s]$ indicate the set of polynomials in the variable s over the complex numbers. We will add and multiply elements of $\mathcal{P}[s]$ in the usual manner. For $f(s) = a_m s^m + a_{m-1}s^{m-1} \ldots a_o$, $a_m \neq 0$, define

$$\deg(f) = m.$$

Indicate the zero polynomial by Θ.

Definition B.9: An *ideal* I in $\mathcal{P}[s]$ is a subset of $\mathcal{P}[s]$ such that rf, $fr \in \mathcal{I}$ for all $r \in \mathcal{P}[s]$, $f \in \mathcal{I}$.

Definition B.10: A *principal ideal* $\mathcal{I}$ is the set of all multiples rf, $r \in \mathcal{P}[s]$ of a fixed element $f \in \mathcal{P}[s]$. The element f is said to *generate* the ideal.

Theorem B.11:

1. If $f, g \in \mathcal{P}[s]$ with $f \neq 0$, then there are unique $q, r \in \mathcal{P}[s]$ such that

$$g = qf + r$$

where $r = \Theta$ or $\deg(r) < \deg(f)$.

2. Every ideal in $\mathcal{P}[s]$ is principal.
3. If f and g are coprime, then there are $a, b \in \mathcal{P}[s]$ such that

$$af + bg = 1.$$

4. Let $f, g \in \mathcal{P}[s]$ be coprime. For any $h \in \mathcal{P}[s]$ there are $a, b \in \mathcal{P}[s]$ with $\deg(a) < \deg(g)$ such that

$$af + bg = h. \tag{B.1}$$

□ *Proof:* (1) This is the result of "long division" of f into g.

(2) Let $\mathcal{I}$ be any ideal. $\mathcal{I} = \{\Theta\}$ is obviously principal, so assume $\mathcal{I}$ contains nonzero elements. Choose $f \in \mathcal{I}$ of lowest possible degree. The principal ideal generated by f is contained in $\mathcal{I}$. For any $g \in \mathcal{I}$ use long division to obtain $r = g - qf$. The remainder $r = \Theta$ since nonzero r would contradict the choice of f as an element of lowest degree. Thus, $g = qf$ and $\mathcal{I}$ is a principal ideal generated by f.

(3) Define

$$\mathcal{I} = \{af + bg \mid a, b \in \mathcal{P}[s]\}.$$

This is an ideal, and so by part (b) $\mathcal{I}$ is generated by some $d \in \mathcal{P}[s]$. Since $f, g \in \mathcal{I}$, $f = xd$, $g = yd$ for some $x, y \in \mathcal{P}[s]$. Since f, g are coprime, this implies that d is a constant polynomial. Without loss of generality, $d = 1$. Since $d \in \mathcal{I}$,

$$af + bg = 1$$

for some $a, b \in \mathcal{P}[s]$.

(4) From part (3) there is $a, b \in \mathcal{P}[s]$, so

$$af + bg = 1.$$

Letting $\tilde{a} = ah, \tilde{b} = bh$,

$$\tilde{a}f + \tilde{b}g = h. \tag{B.2}$$

If $\deg(\tilde{a}) < \deg(g)$, then (B.1) is satisfied. Otherwise, divide g into $\tilde{a}$ to obtain

$$\tilde{a} = qg + r$$

where $\deg(r) < \deg(g)$. Substituting this into (B.2) and rearranging, we obtain

$$rf + (qf + \tilde{b})g = h.$$

This proves (B.1). ■

APPENDIX C

System Manipulations

The following result can be easily verified by routine manipulations.

Theorem C.1:

1. Consider some system with state-space realization (A, B, C, E) and transfer function G. If E is invertible, then G^{-1}has state-space realization $(A - BE^{-1}C, BE^{-1}, -E^{-1}C, E^{-1})$.
2. Suppose G_1 has state-space realization (A_1, B_1, C_1, E_1) and G_2 has state-space realization (A_2, B_2, C_2, E_2). Then G_1G_2 has state-space realization

$$\left(\begin{bmatrix} A_1 & B_1C_2 \\ 0 & A_2 \end{bmatrix}, \begin{bmatrix} B_1E_2 \\ B_2 \end{bmatrix}, [C_1 \quad E_1C_2], E_1E_2\right),$$

or equivalently,

$$\left(\begin{bmatrix} A_2 & 0 \\ B_1C_2 & A_1 \end{bmatrix}, \begin{bmatrix} B_2 \\ B_1E_2 \end{bmatrix}, [E_1C_2 \quad C_1], E_1E_2\right).$$

BIBLIOGRAPHY

[1] B. D. O. Anderson and J. B. Moore, *Optimal Control Control—Linear Quadratic Methods*, Prentice-Hall, Englewood Cliffs, NJ, 1990.

[2] N. A. Bruisma and M. Steinbuch, "A fast algorithm to compute the $\mathcal{H}_\infty$ norm," *Systems and Control Letters*, Vol. 14, pp. 287–293, 1990.

[3] S. Boyd, V. Balakrishnan, and P. Kabamba, "A bisection method for computing the $\mathcal{H}_\infty$ norm of a transfer matrix," *Mathematics of Control, Signals and Systems*, Vol. 2, pp. 207–219, 1989.

[4] Frank M. Callier and Charles A. Desoer, *Linear System Theory*, Springer-Verlag, New York, 1991.

[5] Ewart R. Carson and Tibor Deutsch, "A spectrum of approaches for controlling diabetes," *Control Systems Magazine*, Vol. 12, No. 6, pp. 25–31, 1992.

[6] R. M. Chalasani, "Ride performance potential of active suspension systems part I: Simplified analysis based on quarter-car model," in *Symposium on Simulation and Control of Ground Vehicles and Transportation Systems*, AMD Vol. 80, pp. 187–204, AMSE, 1986.

[7] P. C. Chandrasekharan, *Robust Control of Linear Dynamical Systems*, Academic Press, New York, 1996.

[8] Chi-Tsong Chen, *Linear System Theory and Design*, Holt, Rinehart, and Winston, New York, 1984.

[9] Patrizio Colaneri, Jose C. Geomel, and Arturo Locatelli, *Control Theory and Design*, Academic Press, New York, 1997.

[10] Ruth F. Curtain and A. J. Pritchard, *Functional Analysis in Modern Applied Mathematics*, Academic Press, New York, 1977.

[11] Ruth F. Curtain and Hans Zwart, *An Introduction to Infinite-Dimensional Linear Systems Theory*, Springer-Verlag, New York, 1995.

[12] C. A. Desoer and M. Vidyasagar, *Feedback Systems: Input–Output Properties*, Academic Press, New York, 1975.

[13] John C. Doyle, Bruce A. Francis, and Allen R. Tannenbaum, *Feedback Control Theory*, Macmillan Publishing Co., New York, 1992.

[14] John C. Doyle, K. Glover, P. Khargonekar, and Bruce A. Francis, "State-space solutions to standard $\mathcal{H}_2$ and $\mathcal{H}_\infty$ control problems," *IEEE Trans. on Automatic Control*, AC-48, No. 8, pp. 831–847, 1989.

[15] John C. Doyle and Gunter Stein, 'Multivariable feedback design: Concepts for a classical/modern synthesis," *IEEE Trans. on Automatic Control*, AC-26, No. 1, pp. 4–16, 1981.

[16] John C. Doyle, "Guaranteed margins for LQG regulators," *IEEE Transactions on Automatic Control*, Vol. AC-23, No. 4, pp. 756–757, 1978.

[17] Richard C. Dorf, *Modern Control Systems*, Addison-Wesley Publishing Co., Reading, MA, 1986.

[18] Larry L. Dornhoff and Franz E. Hohn, *Applied Modern Algebra*, Macmillan Publishing Co., New York, 1977.

[19] Peter L. Duren, *Theory of H^P Spaces*, Academic Press, New York, 1970.

[20] B. A. Francis, *A Course in $\mathcal{H}_\infty$ Control Theory*, Vol. 88 in *Lecture Notes in Control and Information Sciences*, Springer-Verlag, New York, 1987.

[21] M. Green and D. J. N. Limebeer, *Linear Robust Control*, Prentice-Hall, Englewood Cliffs, NJ, 1995.

[22] Wolfgang Hahn, *Stability of Motion*, Springer, Berlin, 1967.

[23] Thomas Kailath, *Linear Systems*, Prentice-Hall, Englewood Cliffs, NJ, 1980.

[24] P. P. Khargonekar, I. R. Petersen, and M. A. Rotea, "$\mathcal{H}_\infty$-optimal control with state-feedback," *IEEE Trans. on Automatic Control*, AC-33, No. 8, pp. 786–788, 1989.

[25] Pramod P. Khargonekar and Eduardo D. Sontag, "On the relation between stable matrix fraction factorizations and regulable realizations of linear systems over rings," *IEEE Trans. on Automatic Control*, Vol. 27, pp. 627–638, 1982.

[26] Pramod P. Khargonekar and Allen Tannenbaum, "Noneuclidean metrics and the robust stabilization of systems with parameter

uncertainty," *IEEE Trans. on Automatic Control*, Vol. 30, pp. 1005–1013, 1985.

[27] Hans. W. Knobloch, Alberto Isidori, and Dietrich Flockerzi, *Topics in Control Theory*, Birkhauser Verlag, Switzerland, 1993.

[28] Stephen A. Lane and Robert L. Clark, "Improving loudspeaker performance for active noise control applications," *J. Audio Eng. Soc.*, Vol. 46, pp. 508–519, 1998.

[29] J. M. Maciejowski, *Multivariable Feedback Design*, Addison Wesley, Reading, MA, 1989.

[30] O. Mayr, *The Origins of Feedback Control*, M.I.T. Press, Cambridge, MA, 1970.

[31] K. A. Morris, "State feedback and estimation of well-posed systems," *Mathematics of Control, Signals and Systems*, Vol. 7, pp. 351–388, 1994.

[32] K. A. Morris, "Noise reduction achievable by point control," *ASME Journal on Dynamic Systems, Measurement and Control*, Vol. 120, No. 2, pp. 216–223, 1998.

[33] K. A. Morris and K. J. Taylor, "A variational calculus approach to the modelling of flexible manipulators," *SIAM Review*, Vol. 38, No. 2, pp. 294–305, 1996.

[34] K. A. Morris, "Justification of input/output methods for systems with unbounded control and observation," *IEEE Transactions on Automatic Control*, Vol. 44, No. 1, pp. 81–85, 1999.

[35] C. N. Nett, C. A. Jacobson, and M. J. Balas, "A connection between state-space and doubly coprime fractional representations," *IEEE Trans. on Automatic Control*, Vol. 29, pp. 831–832, 1984.

[36] Hitay Ozbay, *Introduction to Feedback Control*, CRC Press, Boca Raton, FL, 1998.

[37] William F. Powers, "Customers and controls," *IEEE Control Systems Magazine*, pp. 10–14, 1993.

[38] Dietmar Salamon, "Realization theory in Hilbert space," *Mathematical System Theory*, Vol. 21, pp. 147–164, 1989.

[39] Adel S. Sedra and Kenneth C. Smith, *Microelectronic Circuits*, Holt, Rinehart and Winston, New York, 1982.

[40] Stanley M. Shinners, *Modern Control System Theory and Design*, John Wiley and Sons, New York, 1992.

[41] A. Stoorvogel, *The $\mathcal{H}_\infty$ Control Problem: A State-Space Approach*, Prentice-Hall, Englewood Cliffs, NJ, 1992.

[42] Gilbert Strang, *Linear Algebra and Its Applications*, Academic Press, New York, 1976.

[43] A. Tannenbaum, "Frequency domain methods for the $\mathcal{H}_\infty$-optimization of distributed systems," in *State- and Frequency-Domain Approaches for Infinite-Dimensional Systems*, ed. R. F. Curtain, Springer-Verlag, New York, pp. 242–278, 1992.

[44] M. Vidyasagar, *Control System Synthesis: A Factorization Approach*, MIT Press, Cambridge, MA, 1985.

[45] S. J. Williams, D. Hrovat, C. Davey, D. Maclay, J. W. van Crevel, and L. F. Chen, "Idle speed control design using an $\mathcal{H}_\infty$ approach," *Proceedings of the ACC*, pp. 1950–1956, 1989.

[46] Jan C. Willems, "Least squares stationary optimal control and the algebraic Riccati equation," *IEEE Transactions on Automatic Control*, Vol. AC-16, No. 6, pp. 621–634, 1971.

[47] William A. Wolovich, *Automatic Control Systems: Basic Analysis and Design*, Holt, Rinehart and Winston, New York, 1994.

[48] Kosaku Yoshida, *Functional Analysis*, Springer, Berlin, 1968.

[49] D. C. Youla, J. J. Bongiorno, Jr., and C. N. Lu, "Single-loop feedback stabilization of linear multivariable dynamical plants," *Automatica*, Vol. 10, pp. 159–173, 1974.

[50] D. C. Youla, H. A. Jabr, and J. J. Bongiorno, Jr., "Modern Wiener–Hopf design of optimal controllers, part II: the multivariable case," *IEEE Trans. Auto. Control*, Vol. AC-21, pp. 319–338, 1976.

[51] G. Zames, "Feedback and optimal sensitivity: Model reference transformations, multiplicative seminorms and approximate inverses," *IEEE Trans. Automatic Control*, Vol. AC-26, No. 2, pp. 301–320, 1981.

[52] G. Zames and J. G. Owen, "Duality Theory of Robust Disturbance Attenuation," *Automatica*, Vol. 29, No. 3, pp. 695–705, 1993.

[53] Kemin Zhou, John C. Doyle, and Keith Glover, *Robust and Optimal Control*, Prentice-Hall, Englewood Cliffs, NJ, 1996.

[54] A. H. Zemanian, *Distribution Theory and Transform Analysis*, Dover Publications, New York, 1965.

INDEX